The Steel Square

The Steel Square

H.H. Siegele

Sterling Publishing Co., Inc. New York

Fifth Printing, 1982

Published in 1979 by
Sterling Publishing Co., Inc.
Two Park Avenue
New York, N.Y. 10016

ISBN: 0-8069-8854-1.
Previously
ISBN: 0-87749-0019-8.

Printed in the United States of America.

PREFACE

The steel square is so important in every branch of the building trades, that no mechanic can long follow his trade without having it among his collection of tools. But having a steel square does not imply that the holder knows all there is to know about its uses. There are so many things that can be done with the steel square that few tradesmen, if any, know them all. What higher mathematics means to the mathematician, the steel square means to the building tradesman—that is, if the relative principles are equally understood in both fields. And just as the mathematician can benefit by higher mathematics only so far as he understands it and uses it, just so the mechanic can benefit by the steel square only to the extent that he understands it and uses it. As in all of his other books, in this volume the author confines himself, as much as possible, to practical craft problems—especially problems that can be solved with the steel square. A great many of the problems treated in this work, the author has solved on the job with the steel square. And those problems in this book that he has not used in practice, he can solve, if called upon to do so.

The author is grateful for the kind receptions that were given to his other books, and especially for the many complimentary letters that were written about them by appreciative readers.

THE AUTHOR

TABLE OF CONTENTS

TABLE OF CONTENTS

Chapter I
ROOF FRAMING TABLE

The Steel Square.—Everybody knows what a steel square is; that is, he knows that it has a large arm and a smaller one branching off from it at a right angle. The large arm is 2 inches wide and 24 inches long, while

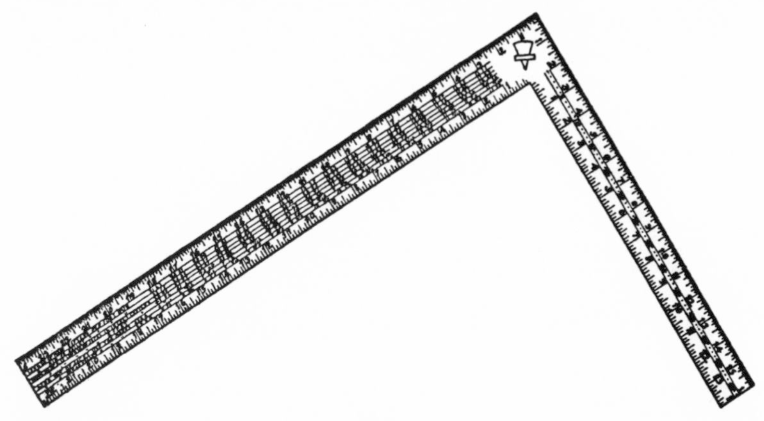

Fig. 1

the smaller arm is only 1½ inches wide and 16 inches long. The large arm is called the body or blade, and the small one is called the tongue. The point where the outside edges of the two arms meet is called the heel. The face side of a steel square is the side on which the manufacturer's name is stamped. The side opposite to the face side is called the back. The square shown by Fig. 1 has the face side up. In other words, if the body of the square is held in the left hand and the tongue in the right, keeping the square approximately on a level and the heel from you, the face side of the square will be up.

Roof Framing Table.—This writer has never used the roof framing table on the square for any practical purpose, but he will explain it. The student should obtain a square that has the table on it, as shown by Fig. 2, and lay it before him. He will find to the left on the body, the top line, these words: "Length of main rafters per foot run," just as shown by the drawing. Right under the figure 18 he will find 21.63 (21 and 63/100ths) which means that the main, or common rafter, per foot run and 18 inches rise, is 21.63 inches long. To find the length of the rafter in feet, multiply 21.63 by the number of feet in the run of the roof and divide by 12.

23 22 21 20 19	18	9	8	7	6
LENGTH OF MAIN RAFTERS PER FOOT RUN	21 63	15 00	14 42	13 83	13 42
" HIP OR VALLEY " " " "	24 74	19 21	18 76	18 36	18 00
DIFFERENCE IN LENGTH OF JACKS 16 INCHES CENTERS	28 84	20 00	19 23	18 52	17 87
" " " " " 2 FEET "	43 27	30 00	28 84	27 78	26 83
SIDE CUT OF JACKS USE THE MARKS	6 11/16	9 7/8	10 00	10 3/8	10 3/4
" " HIP OR VALLEY " " "	8 1/4	10 5/8	10 7/8	11 1/16	11 3/16
22 21 20 19 18 17		8	7	6	5

Fig. 2

Diagonal Scale.—Fig. 3 is a drawing of the diagonal scale, which is used to measure hundredths of an inch with a compass. The compass shown has the points set at *a* and *a*. Each space in the diagonal scale from left to right counts 10 one-hundredths of an inch. The distance then between the points of the compass, as shown, would be one whole inch plus four spaces of the diagonal

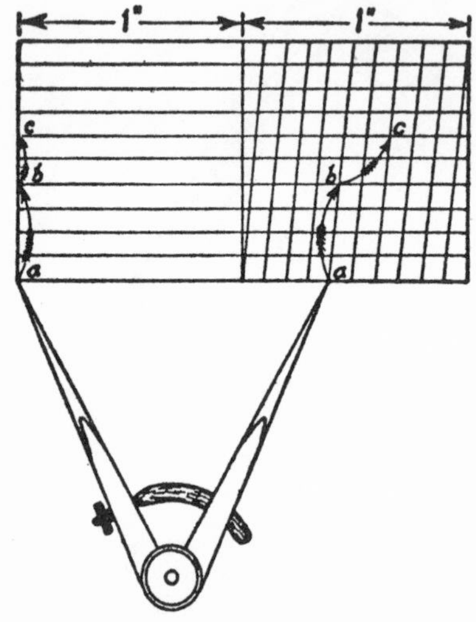

Fig. 3

scale, or 1.40 inches. If the compass were set four spaces up, as at points *b* and *b*, the distance would be 1.44 inches, since each space up adds one one-hundredth of an inch to the distance. In other words, the distance between *a* and *a* is 1.40 inches, but the four spaces up add .04 inch to it, making the distance between *b* and *b* 1.44 inches. If the compass were set at points *c* and *c*, the distance would be 1.66 inches. The student should use a compass and practice with it on measuring different distances to the hundredth part of an inch.

Application of the Square.—Fig. 4 shows the square applied to a timber, using 12 on the tongue and 18 on the body, to obtain the length of the rafter per foot run, which is 21.63 inches, just as the table gives it.

Hips and Valleys.—The second line of the table gives the lengths of hips and valleys per foot run of the common rafter. (The run for hips and valleys per foot run of the common rafter is

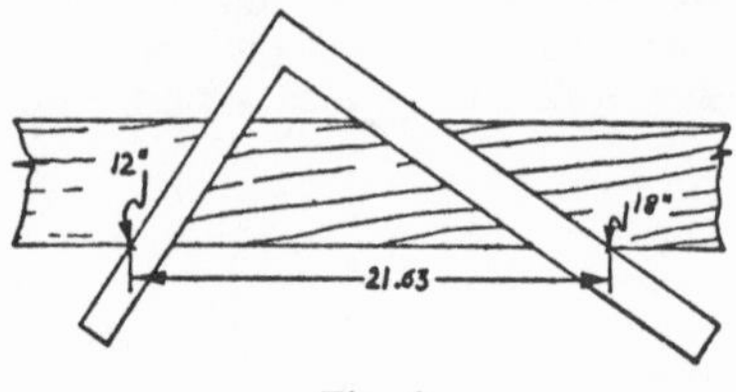

Fig. 4

17 inches minus—the exact figure being 16.97 inches.) Under the figure 18 on the body, the second line, will be found 24.74, or the length in inches of a hip or valley rafter per foot run of the common rafter. Fig. 5 shows the

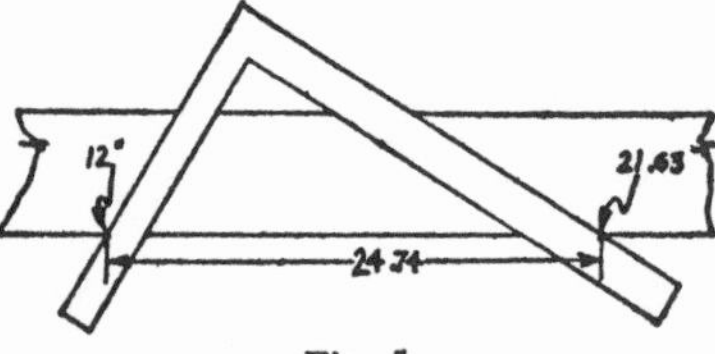

Fig. 5

square applied to a timber using 12 on the tongue and the length of the common rafter per foot run (21.63 inches) on the body of the square. The diagonal distance, as shown, is 24.74 inches, or the length of hips or valleys per foot run of the common rafter, the same as shown by the table.

Difference in Lengths of Jacks.—The third line of the table gives the difference in the lengths of jacks spaced 16 inches on center, or 28.84 inches. Fig. 6 shows the square applied to a

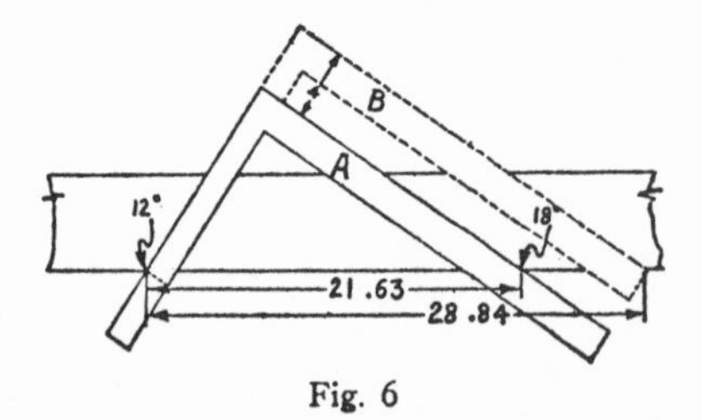

Fig. 6

timber showing how that figure is obtained. If the rafters were spaced 12 inches on center, the difference in the lengths of the jacks would be 21.63 inches, or the length of the common rafter per foot run, as shown by the square marked *A*. But the space is 16 inches, which is 4 inches more than 12, as shown by the square marked *B*. Four inches is one-third of 12, so by adding one-third of 21.63 inches to itself, the result will be the difference in the lengths of the jacks, or 28.84 inches, the same as the diagonal distance shown by the square in position *B*.

Jacks Spaced Two Feet.—The fourth line of the table gives the difference in the lengths of jacks, if spaced 2 feet on center, or 43.27 inches. Fig. 7 shows how this figure was obtained. Each of the two squares, *A* and *B*, gives the difference in the lengths of jacks if spaced 12 inches, or 21.63 inches. Then if they are spaced 2 feet, the difference in the lengths of jacks would be just twice 21.63 inches, plus, or 43.27 inches, as shown by the diagram.

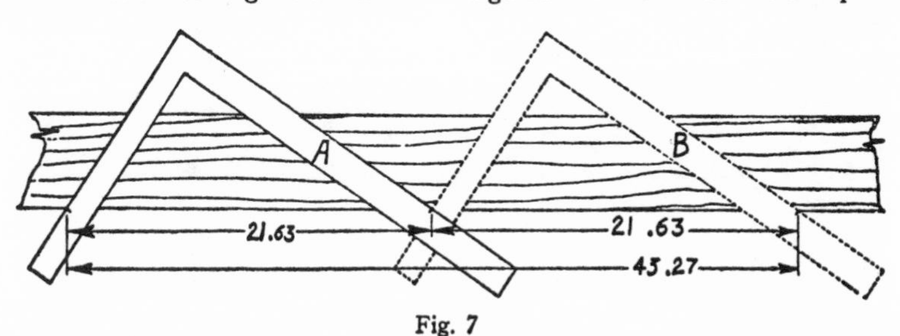

Fig. 7

Edge Bevel for Jacks.—The fifth line of the table gives the figure to be used with 12, in order to get the edge bevel for the side cut of jacks, or 6 11/16. How the square is applied is shown by Fig. 8 (square marked *A*). To the left, one of the two-way arrows points to the V-shaped mark that indicates the point to be used with 12. The enlarged part is a reproduction from Fig. 2, where to the right the figure 6 is shown. The square marked *B* shows how one foot run and the length of the rafter per foot run will give the same bevel.

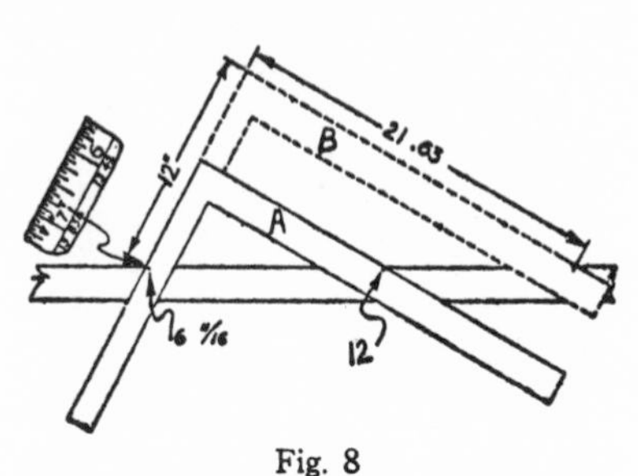

Fig. 8

Edge Bevel for Hips and Valleys.—The sixth line of the table gives the figure to be used with 12 to obtain the edge bevel for the side cut of hips and valleys, or 8¼. One of the two-way arrows (Fig. 9) shows the *X* that indicates the point to be taken. The enlarged part with the figures 7, 8, 9 on it is reproduced from the part to the right of Fig. 2—compare the two. The square shown by dotted lines, shows how 17 (the diagonal distance of 12 and 12) and the length of the hip or valley per foot run of the common rafter, will give the same bevel.

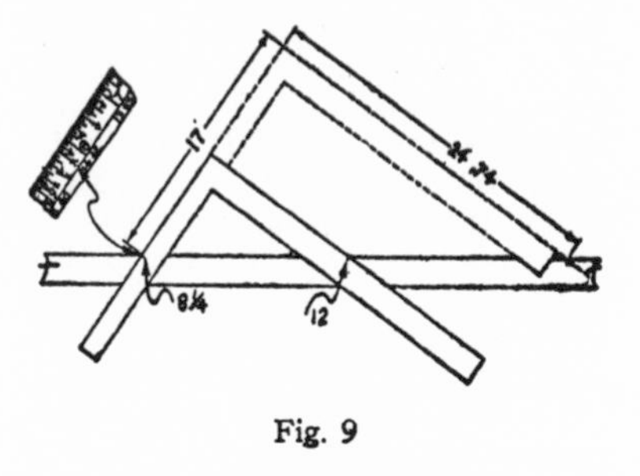

Fig. 9

Chapter 2

LUMBER AND OTHER TABLES

Lumber Table.—The lumber table is found on the back of the blade of the square. If possible, the student should obtain a square that has this table on it and lay it before him, so that he will have the whole rule to study. Fig. 10 gives two parts of the square, showing enough of the lumber table to serve the purpose here. The figure 12 is the base figure of the lumber rule. Now turn to the illustration and on the first line directly under the base figure you will find the figure 8. This means that a board 12 feet long and 8 inches wide will have 8 board feet of lumber in it. If the board were 13 feet long, it would have 8 and 8-twelfths board feet of lumber in it, as shown under the figure 13 on the first line. If the board were 14 feet long, as shown under the figure 14, it would have 9 and 4-twelfths board feet of lumber in it. A board 12 feet long and 9 inches wide, as shown on the second line under the base figure, would have 9 board feet of lumber in it, a board 10 inches wide would have 10 board feet of lumber, and so on down to the seventh line.

Now turn to the figure 3, shown toward the left of the illustration. Here you will find that a board 3 feet long and 8 inches wide has 2 board feet of lumber in it—if it were 6 feet long it would have 4 board feet of lumber in it. In the same way you can find the number of board feet in boards running from 2 to 24 feet long and from 8 to 15 inches wide, as shown directly under each edge figure on the back of the blade. The results will be the same by letting the edge figures represent the widths of the boards and the figures under the base figure 12 the lengths. For example, a board 8 feet long, 13 inches wide has, as we find under figure 13, 8 and 8-twelfths board feet of lumber in it. If it were 8 feet long and only 3 inches wide, it would have, as shown under the edge figure 3, 2 board feet of lumber in it.

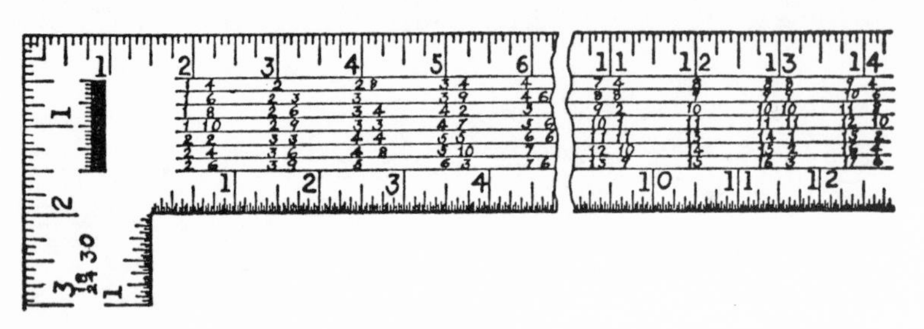

Fig. 10

The student should practice with different lengths and different widths of boards, until he thoroughly understands the table. For planks and heavy timbers, simply multiply the board feet of a board the length and width of the timber, by the thickness of the timber in inches, and you will have the number of board feet of lumber in the timber.

Figuring Board Feet with the Square.—Fig. 11, *A,* shows the square applied to a board, in such a way that 12 on the blade and 8 on the tongue of the square intersect the edge of the board. That means that a board 12 feet long, 8 inches wide has 8 board feet of lumber in it. The base figure 12 is always the starting point, and the figure used with it on the tongue always gives the width of the board. To find the amount of lumber in a board 6 feet long, and 8 inches wide, pull the square back until the blade is in position *B,* shown by dotted lines. This would make the edge figure 4 on the tongue intersect the edge of the board, indicating that there are 4 board feet of lumber in the board. Or if the board were only 3 feet long, then the blade would be brought to position *C,* shown by dotted lines, and the edge figure 2 would intersect the edge of tongue at point *b* will indicate that a board 14 feet long and 8 inches wide has 9 and 4-twelfths board feet of lumber in it. The examples that are used here correspond with some that were used in explaining Fig. 10. Compare and check the two methods. The same results can be obtained by moving the square to the right, as shown by Fig.

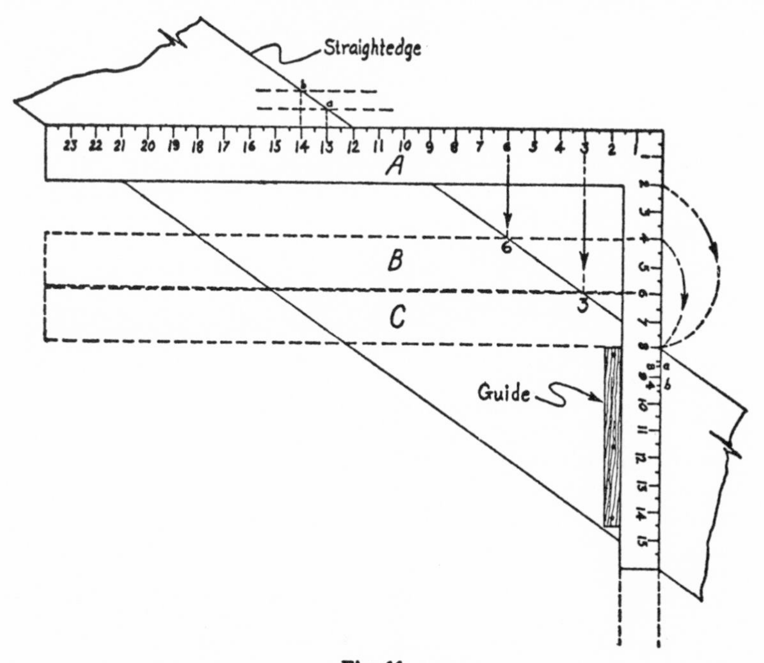

Fig. 11

the board, indicating that the board has 2 board feet of lumber in it. Directly above the edge figures 13 and 14 you will find dotted lines intersecting the edge of the board, marked *a* and *b* — also to the right of the tongue you will find two small figures, 8 and 4, at *a* and *b*. This means that if the square is slipped up so that the edge figure 13, on the blade, will intersect the edge of the board at *a,* the tongue of the square, at point *a,* will show that a board 13 feet long and 8 inches wide has 8 and 8-twelfths board feet of lumber in it. In the same way, if the edge of the blade is brought to point *b,* the 12. The square as applied to the board shows that a board 12 feet long and 8 inches wide has 8 board feet of lumber in it. But if the square is slipped from position *A* to position *B,* it will show that a board 14 feet long and 8 inches wide has 9 and 4-twelfths board feet of lumber in it. If the board were 13 feet long and 8 inches wide the tongue would show that it has 8 and 8-twelfths board feet of lumber in it. This is shown by short dotted lines, and pointed out with indicators to the right. These figures are the same as those found in both Fig. 10 and Fig. 11. Study and compare the methods used in the three illustrations. Problems that are easy to solve were purposely taken. But the results will be just as accurate when the problems are more difficult.

Brace Table.—Fig. 13, at *A,* shows one inch divided into one-hundred parts, which is necessary in getting the different lengths of braces. Directly under the edge figure 3 will be found

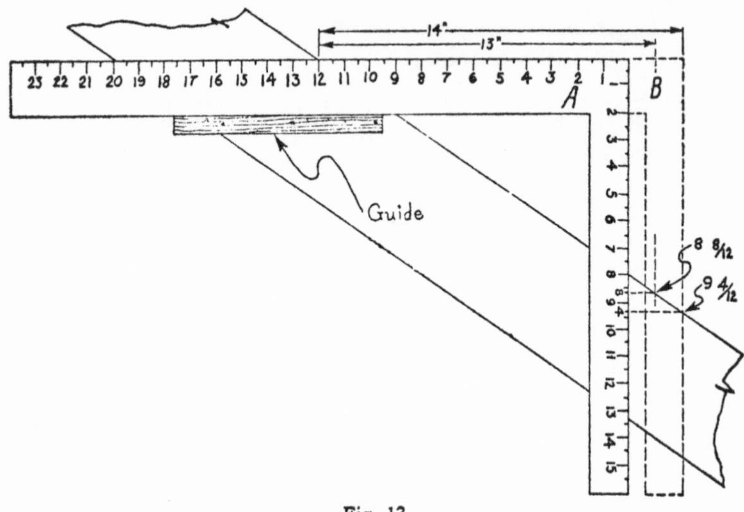

Fig. 12

these figures: 18 over 24, and to the right 30, which means that a brace joining the frame 18 inches one way from the angle, and 24 inches the other way, will have to be 30 inches long.

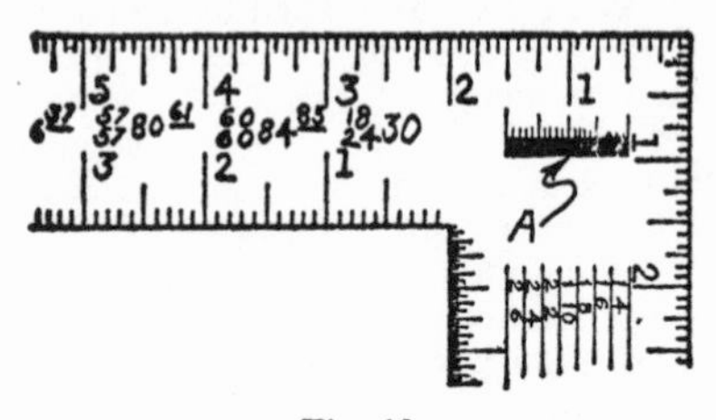

Fig. 13

This is illustrated by Fig. 14. Under the edge figure 4 (Fig. 13) you will find 60 over 60, and to the right 84.85, which means that a brace joining the frame 60 inches each way from an angle will have to be 84.85 inches long.

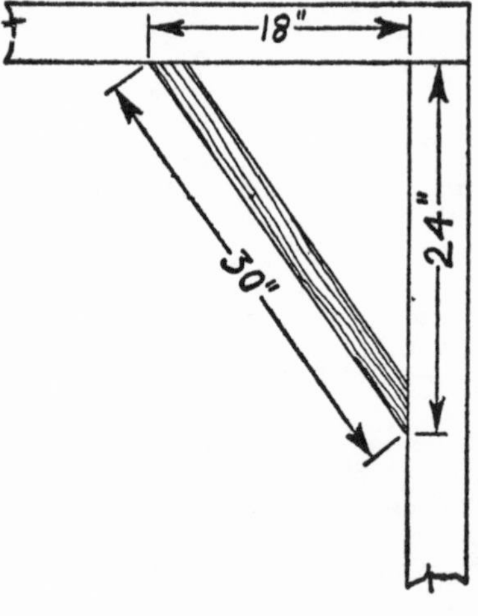

Fig. 14

Octagon Table. — Fig. 15 shows a part of the octagon table on the face side of the tongue. To describe an octagon, take as many spaces shown by dots in the table, as there are inches in one side of the square to be changed to an octagon. For example, if you want to make an octagon out of an 8x8, start at the dot pointed out at 0, count eight spaces to the right, and set the compass as shown, in part, by the drawing. Then set one leg of the com-

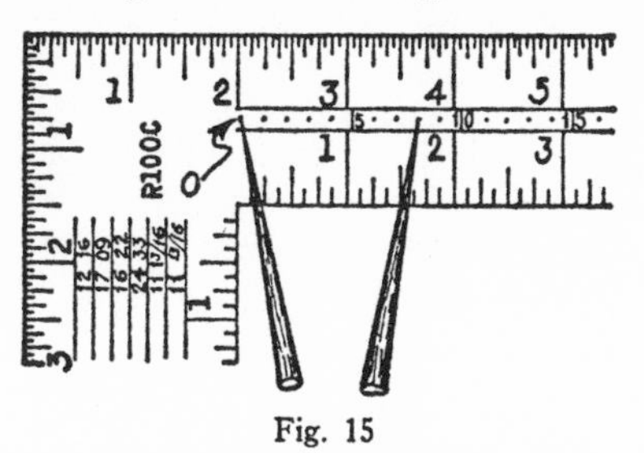

Fig. 15

pass at the center of one side, as at *C*, Fig. 16, and mark points 1 and 2, as indicated by the dotted half-circle. Now drop the two dotted lines from 1 to 3 and from 2 to 4, and from these four points mark off the four corners on a

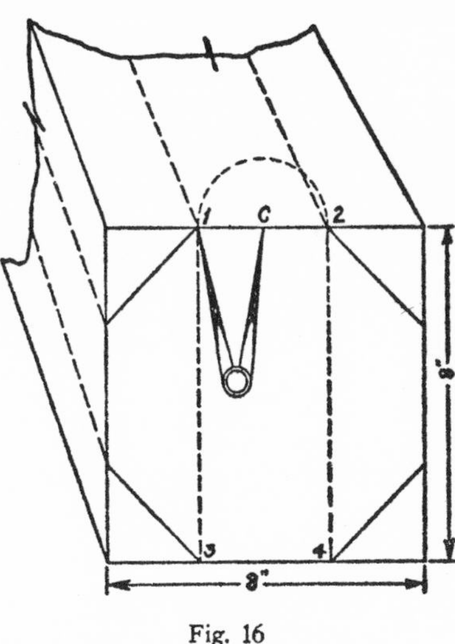

Fig. 16

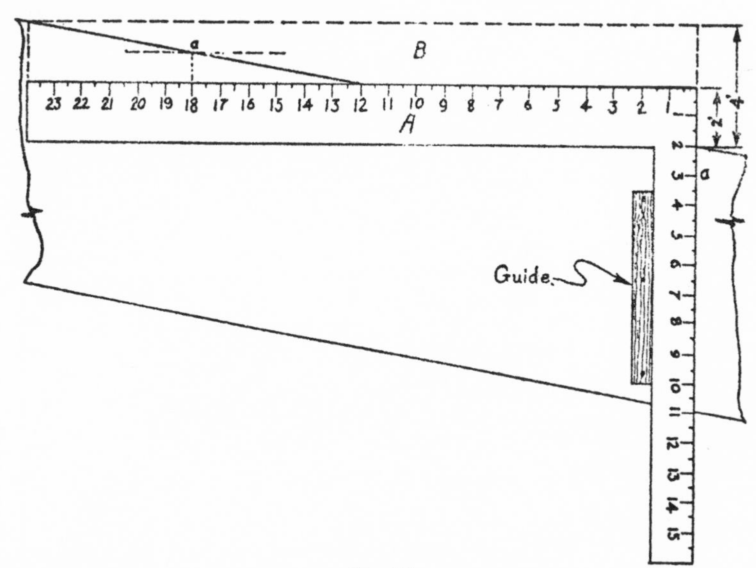

Fig. 17

45-degree angle, as shown by the drawing. That completes the laying out of the octagon.

Spacing Problem.—Fig. 17 shows a square placed on a board with 12 on the body and 2 on the tongue intersecting the edge of the board. The figures to the right between the arrows are always read as feet and the figures on the edge of the tongue are always read as inches, while the figures on the body of the square give the answer to the problem. The problem with the square in position *A*, would read: A distance of 2 feet divided into 2-inch spaces will have 12 spaces. That is easy, and the next problem is just as easy, which would read: If a distance of 2 feet will have 12 2-inch spaces, how many 2-inch spaces will a distance of 4 feet have? The problem is solved by moving the square from position *A* to position *B*. Where the edge of the body intersects the edge of the board, you will find the answer, which in this case is 24. One more problem: If there are 12 2-inch spaces in a distance of 2 feet, how many 2-inch spaces will there be in a distance of 3 feet? To solve the problem the square is moved up one inch, which will bring both the edge figure 18 on the body, and the edge figure 3 on the tongue, to the edge of the board. These points are marked *a* and *a*. The answer, which is 18, is found where the blade intersects the edge of the board.

Chapter 3

STEEL SQUARE PROBLEMS

In this chapter are taken up practical steel square problems that are related to circles, squares, and right angle triangles.

Finding Center of Circle.—Fig. 18, upper part, shows three heavy dots, numbered *1, 2, 3.* Now if you had to

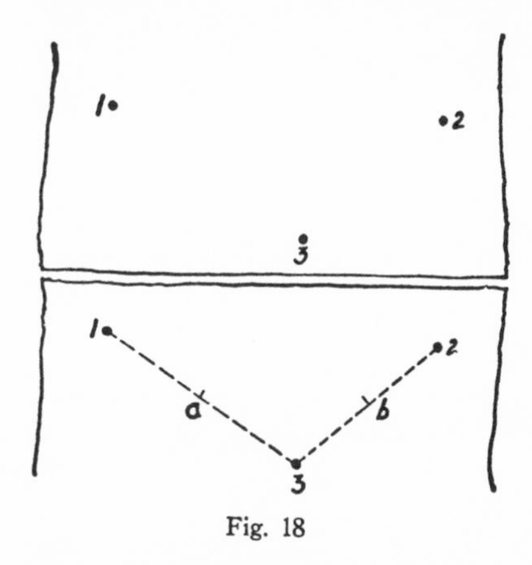

Fig. 18

build a round tank, silo, or some other circular structure, and the owner would give you three points that the structure would have to contact, somewhat as shown by the illustrations, how would you solve the problem? The solution can be found with the steel square. If the structure is large or small, draw a line from both points *1* and *2* to point *3*, as shown by dotted lines in the bottom part of Fig. 18. Mark each of these lines at the center, as shown at *a* and *b*. Then set a square at each of those points, as shown by Fig. 19. The point where the outside edges of the blades cross is the center of a circle that will cross the three points. What is shown in Fig. 19, is really a diagram drawn to a reduced scale. In cases of silos or tanks, a large square of wood is practical, which can be made by employing the 6-8-10 method of squaring. Otherwise a diagram should be made, using a reduced scale.

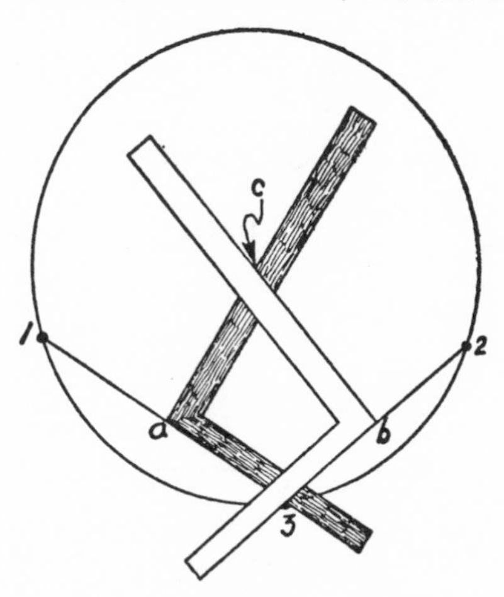

Fig. 19

The problem just explained is the same as finding the center of a circle, in which case you would mark off two segments anywhere on the circumference of a circle, and mark a perpendicular line from the center of each of the two chords. Where these two lines cross is the center of the circle. Fig. 19 shows how to apply the steel square to mark the perpendicular lines.

Another Way of Finding Center.—Fig. 20 gives another way of finding the center of a circle. Place the square on the circle in such a manner that the heel will contact the circumference, as at *a*. Then strike a line as from *b* to *b*. With a compass mark the points *d, d* from *b, b* and strike a line from *d* to *d*.

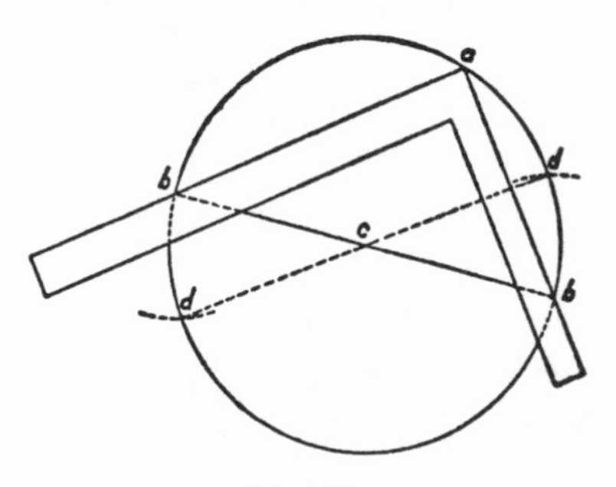

Fig. 20

Where this line crosses the other line (point *c*) is the center of the circle. The second line can be made with the steel square by shifting the square around to another position, with the heel contacting the circumference, and striking the line as before.

Describing a Circle.—Fig. 21 shows how to mark a circle with a steel square. Set two nails as far apart as the diameter of the circle you want, as

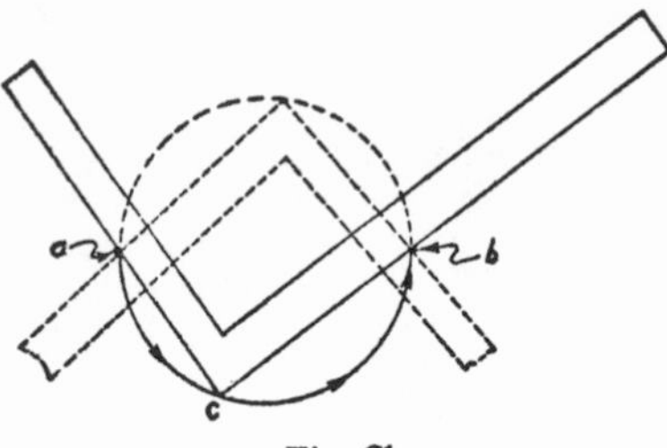

Fig. 21

at *a* and *b*. Then with a pencil held at the heel (point *c*) of the square, move the heel from *a* to *b*, as shown by the arrows, keeping the edges of the tongue and the blade against the nails constantly. This will describe a perfect half circle, if it is carefully done. The other half is described in the same way, as indicated by the dotted lines.

Finding Diameter.—Fig. 22 shows two circles. One has a diameter of 3 inches and the other has a diameter of

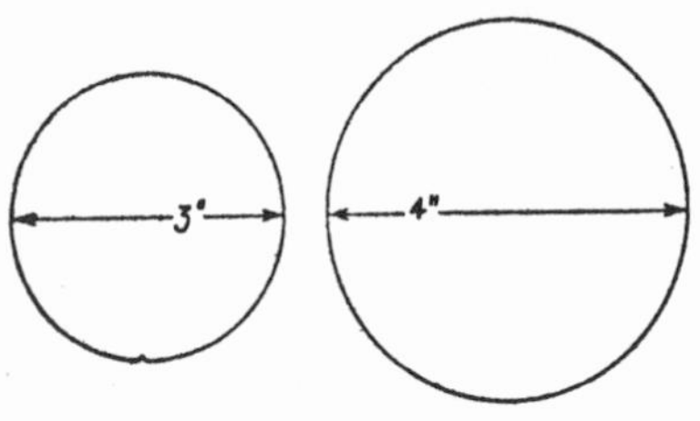

Fig. 22

4 inches. Inches are used for convenience—in practice it could be any units of measurement, inches, feet, yards, rods, or even miles. The reader can make his own scale with the units shown on the diagrams. With this in mind, if you were asked to build a tank that would have as much floor space as two smaller tanks with diameters like those shown in Fig. 22, how would you find the diameter of such a circular tank? Fig. 23 shows a simple way to do it with the steel square. Take the diameter of one tank on one arm of the square and the diameter of the

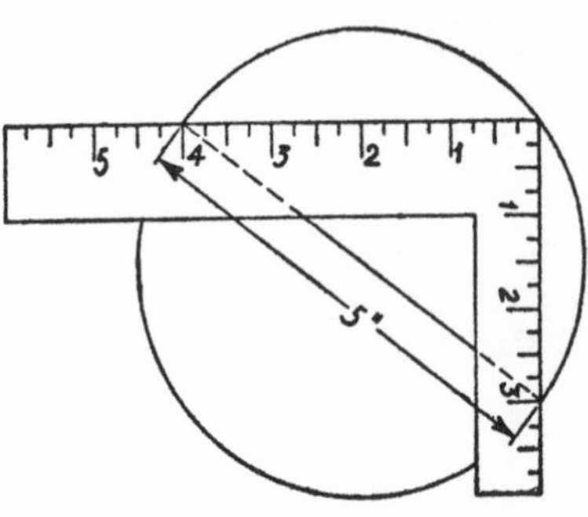

Fig. 23

other tank on the other arm—the diagonal distance between these two points will give the exact diameter of a tank with as much floor space as the two other tanks have. A machinist's square is used in these illustrations to simplify the matter. Whole numbers are used in this and the next two problems, that will make the diagonal distances in the problems come out in whole numbers, so that it will be easy for the student to prove the problems. But the results will be just as accurate in cases where all figures involve fractions.

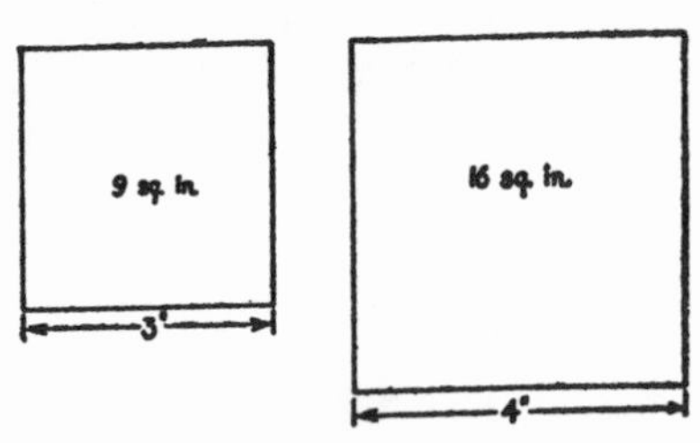

Fig. 24

Finding Area.—A similar problem is shown by Figs. 24 and 25. If you were asked to build one square bin that would have as much floor space as the two square bins have that are shown in Fig. 24—how would you do it? A simple way to solve the problem with the square is shown by Fig. 25. Take the distance of a side of one of the bins on one arm of the steel square, and the

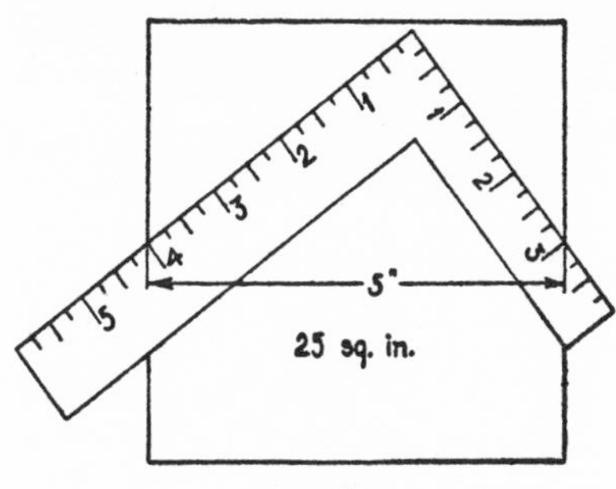

Fig. 25

distance of a side of the other bin, on the other arm of the square. The diagonal distance between these two points will give the distance of one side of a square bin that will have as much floor space as the two other bins have. The

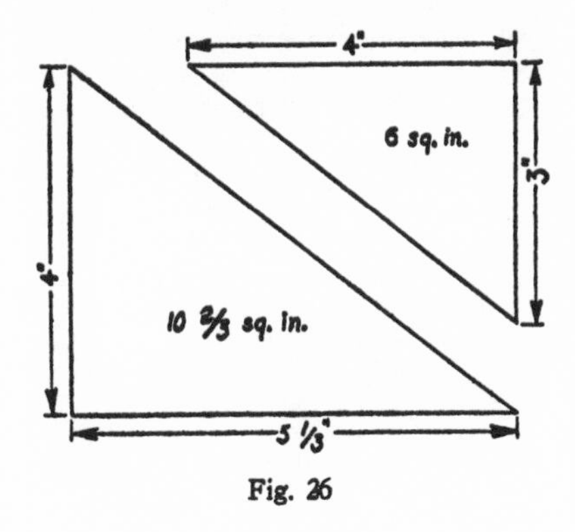

Fig. 26

area of the different bins is shown with figures.

Concerning Triangles.—If you had to describe a right angle triangle that would have as much surface as two triangles proportionately the same; how would you do it? The problem is again solved by taking, say, the shortest side of one of the triangles (Fig. 26) on one arm of the square and the shortest side of the other triangle on the other arm of the square—the diagonal distance between these two points will be the shortest side of a triangle that will have as much surface as the two other triangles have. This is shown, at a

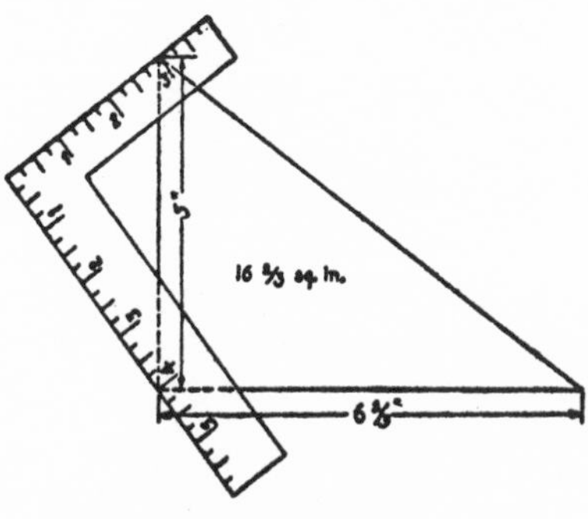

Fig. 27

smaller scale, by Fig. 27, where the area of the triangle is given in square inches, which is equal to the area of the two triangles shown in Fig. 26.

It should be remembered that in the last three problems the shapes of the surfaces must be proportionately the same, however, the areas of the two first surfaces may or may not be the same.

Chapter 4

SQUARE AND CIRCLE

In this chapter a number of problems will be treated in which both the square and the circle play important parts. Because this book deals with the steel square, the parts that belong to the square will be emphasized, while the parts concerning the circle will be treated as a secondary matter.

Obtaining Circumference.—A simple way to obtain the circumference of a circular layout, as a tank, a building, or a plot of ground, is illustrated by Fig. 28. If the diameter or radius is known, then the circumference can be obtained with the steel square. Using

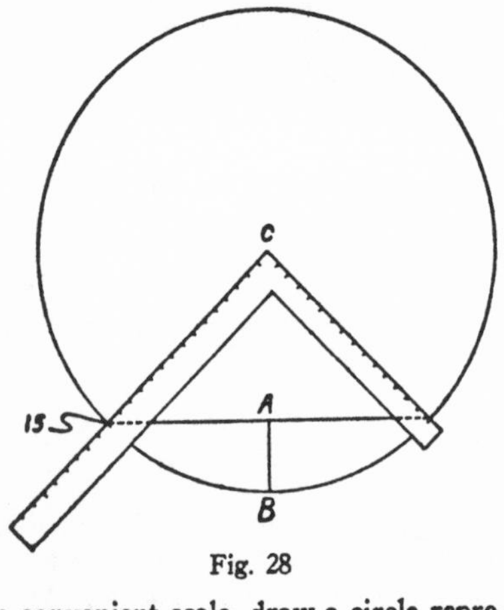

Fig. 28

a convenient scale, draw a circle representing the layout. Then place the heel of the square at the center of the circle, point *C*. From the point where the outside edge of the blade crosses the circumference draw a line to the point where the outside edge of the tongue crosses it, as shown. At the center of this line and at a right angle to it, draw *A-B*. Now multiply the radius by 6 and add the distance *A-B*, which will give you the circumference. If instead of multiplying the radius by 6, you multiply the diameter by 3 and add *A-B*, the answer will be the same. The same results can be obtained by mathematics, in which case you multiply the diameter by pi, or 3.1416.

Bell-Shaped Roof.—Fig. 29 shows a cross section of a sort of bell-shaped roof. The pitch of such a roof is de-

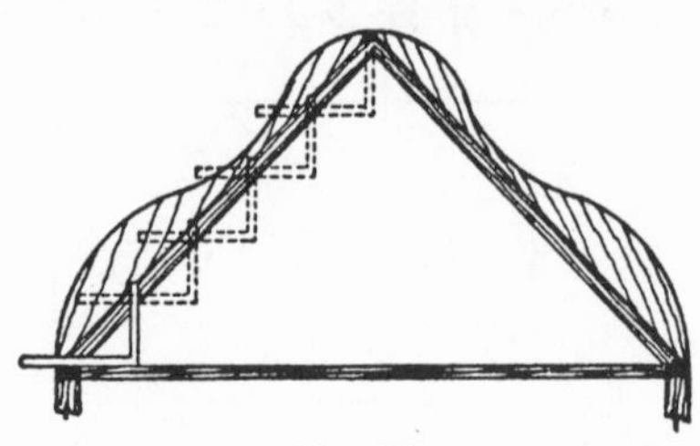

Fig. 29

termined by what could be called regular rafters, onto which the bell-shaped forms are fastened, or if the material will permit it, the shape of the roof can be cut on the upper edge of the rafters. In cases where the shape of such roofs are cut on the rafters, a measuring line should be used for marking the cuts and for stepping off the length. This measuring line could correspond with the upper edge of the rafter shown on the drawing. These rafters are distinguished from the curved forms by the shading. The pitch of the roof as shown by the drawing is a half pitch,

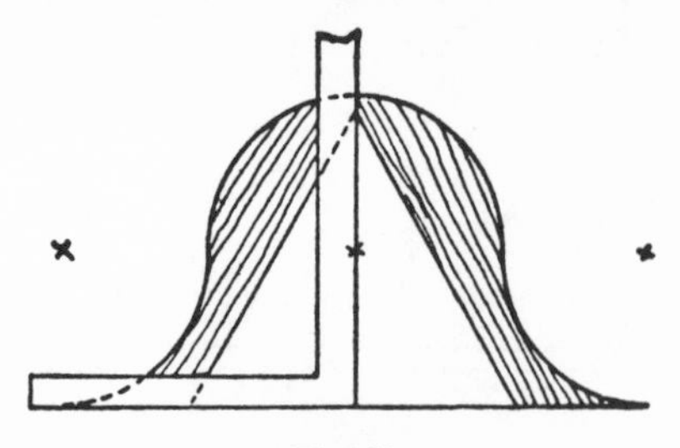

Fig. 30

and in this case the bottom edge of the rafter is taken as the measuring line. The five applications of the square indicate that the run is 5 feet, plus the length of the foot cut. Other roofs in which the rafters have curved upper edges are framed in the same way. A good example of this is shown by Fig. 30, where the bottom edges of the rafters are set at full pitch. The curves of these rafters are drawn with a radius pole. The upper half circle is made by pivoting the radius pole on the *X* shown at the center, while the quarter circles to the right and left are made by pivoting the radius pole on the *X*'s to the left and to the right.

Marking Curve with Trammel. — Fig. 31 shows two squares in part, in

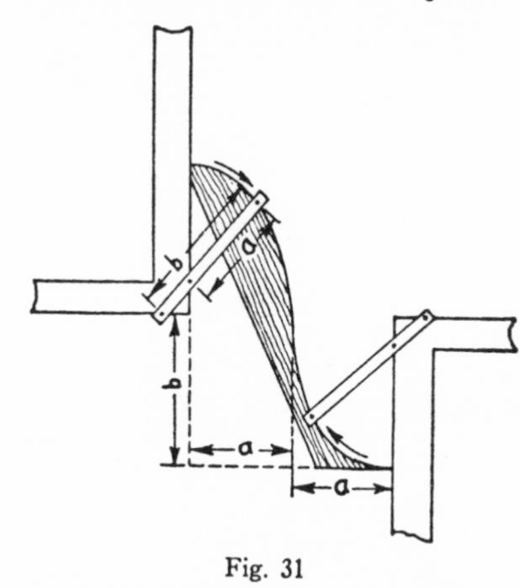

Fig. 31

positions for marking the curvatures of a rafter pattern with a trammel. Here the rise is longer than the run, but each is divided into two parts in

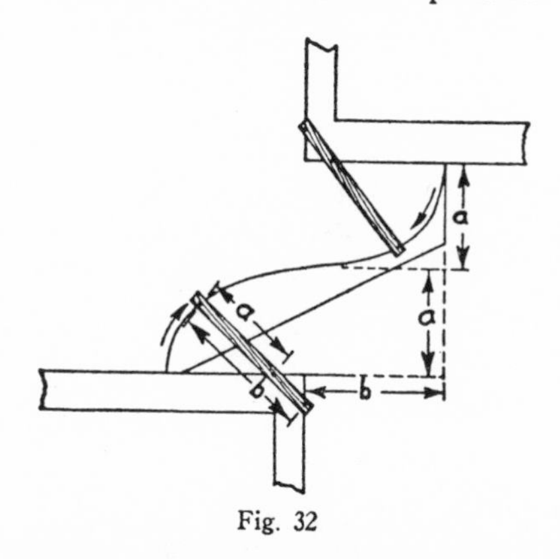

Fig. 32

order to obtain the ogee curvature. The distances *a* and *b*, shown by the upper trammel, are the same as the respective distances marked *a* and *b* in the run and rise. The marking is done by moving the trammel in the direction of the arrow, keeping the two points constantly in contact with the square. The pencil is fastened to the trammel by means of a small hole, into which it is wedged.

Fig. 32 shows the same kind of rafter, but its position in the roof will be in reverse order. The run here is divided into two equal parts, just as the rise is divided into two equal parts. The only difference between Fig. 31 and Fig. 32 is the reversal of the order. The reference letters are the same, and the explanation of Fig. 31 will apply to Fig. 32, but in reverse.

Another Trammel Problem. — Fig. 33 shows how to determine the different points of a trammel that are to be used for describing an oval-shaped hole on a pitch roof for a ventilating duct to

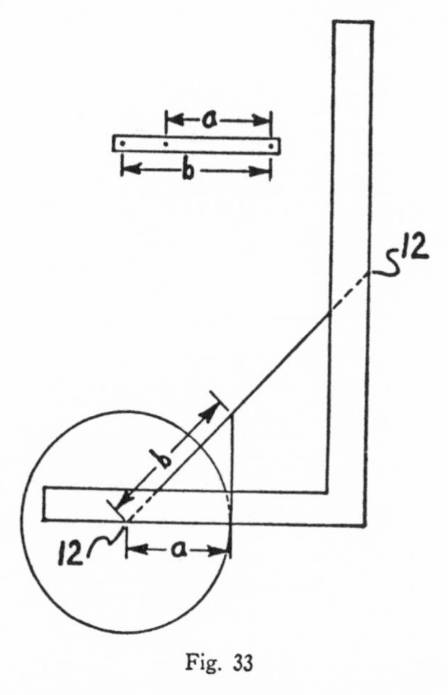

Fig. 33

pass through in a vertical direction. The square in the position shown has a diagonal line drawn from 12 to 12, which means that the roof is half pitch. Now set the compass at point 12 on the tongue, and with a radius one half the diameter of the duct, describe a circle, as shown. Where the circle crosses the outside edge of the tongue, raise a line parallel with the blade of the square. The distance *a* gives the short part of the trammel, as shown by the detail to the upper left, while the distance *b* gives the long part, also shown by the detail.

Fig. 34 shows the same square with the trammel in position for marking one quarter of the oval at *A*. When this quarter is marked on stiff building

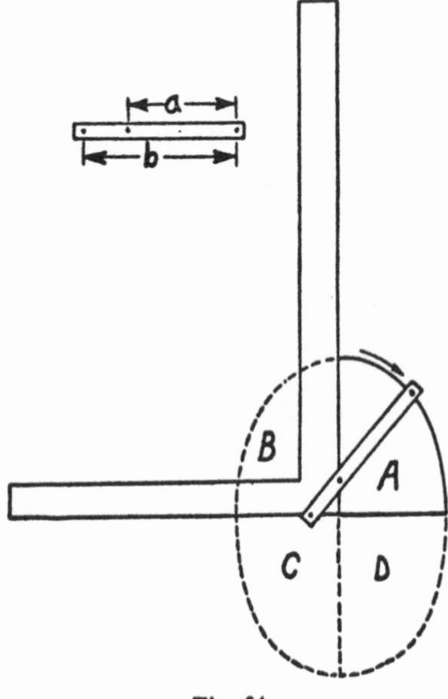

Fig. 34

paper, cut it out for a pattern with which the whole oval can be marked, as indicated by dotted lines at *B*, *C*, and *D*. The points of the trammel that contact the square have two short metal pegs, while the marking is done with a pencil fastened to the trammel by means of a hole, into which the pencil is wedged.

Describing an Oval.—Another method for marking a hole in a pitch roof for a duct that is to pass through it in a vertical direction is shown by Figs. 35 and 36. Fig. 35 shows the square in such a position that the edge of the blade holds the same pitch as the roof, in this case a half pitch. At a convenient point on the tongue, set the compass, and with a radius one half the diameter of the duct, describe a

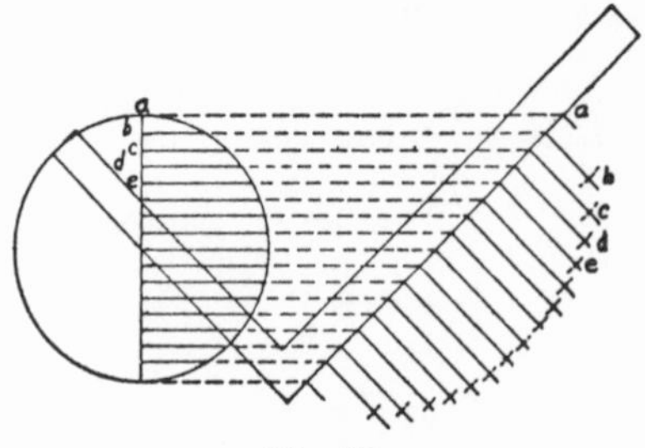

Fig. 35

circle, as shown. Now draw a vertical line through the center of the circle, and divide it into any number of spaces, somewhat as shown. From these points draw horizontal lines to the outside edge of the blade, as shown by the

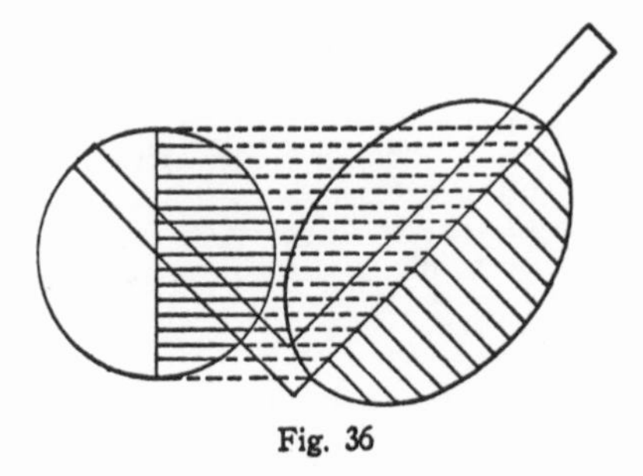

Fig. 36

straight-and-dotted lines. Where these lines contact the edge of the blade, draw indefinite lines at a right angle to it, as shown. This done, transfer the lengths of the lines, *b*, *c*, *d*, *e*, etc., of the circle to the respective right-angle lines, *b*, *c*, *d*, *e*, etc., of the blade. When these distances are all transferred, starting at point *a*, draw a curved line through these points, which will give you one half of the oval, as shown to the right of the blade in Fig. 36. This marking should be done on stiff building paper, and when one half is marked, use it as a pattern to mark the hole in the roof for the duct.

Chapter 5
POLYGONS

Polygons.—Almost every carpenter in his experience as a tradesman will be called on to lay out polygons. This is especially true when it comes to regular polygons, which is to say, polygons whose sides are all equal in length. Polygons in which the sides are not equal in length, are rarely called polygons on the job, but they are indirectly present in almost every steel square problem. For instance, the right-angle triangle is present in every cut that is laid out with the square, excepting the square-across cut.

Triangle.—Fig. 37 shows a simple way of laying out an equilateral triangle. The figures to use on the square

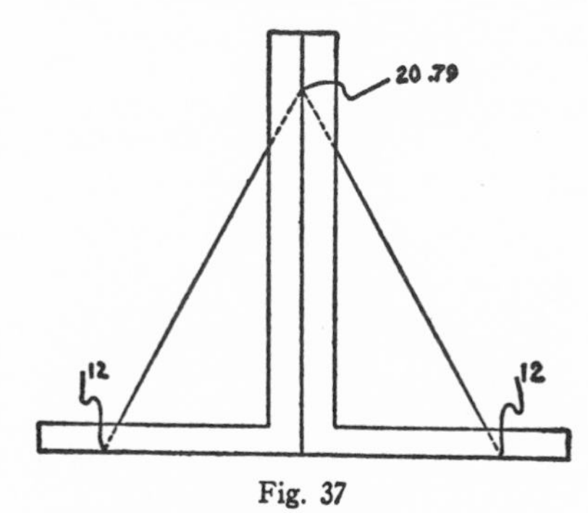

Fig. 37

are 12 and 20.79. In this instance two applications of the square are necessary to locate the three points of the triangle, as shown by the diagram.

Fig. 38 shows the same triangle, which was laid out by using 12 on the tongue and 20.79 on the body of the square. Besides these figures there are shown two other couples of figures that will give the butt and miter cuts for triangles. They are 8 and 13.86, and 4 and 6.93. These figures are used on the tongue and body of the square, respectively. In practice 4 and 7 are often used.

A simple way to obtain the figures to use for marking the butt and miter joints of a triangle is as follows: Mark off the length of one side of any regular triangle on the edge of a board (if the sides are longer than the square will reach, use a convenient scale) and take half of the length of one side on the tongue of the square, holding it to one of the marks, and bring the body

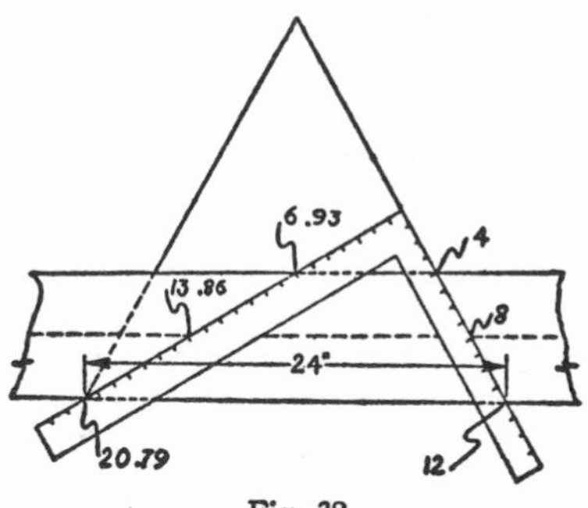

Fig. 38

of the square to the other mark. Then the body of the square will give the miter cut, while the tongue will give the cut for the butt joint. This is shown in both Fig. 38 and Fig. 39. Fig. 39 also shows the 4 and 7 applica-

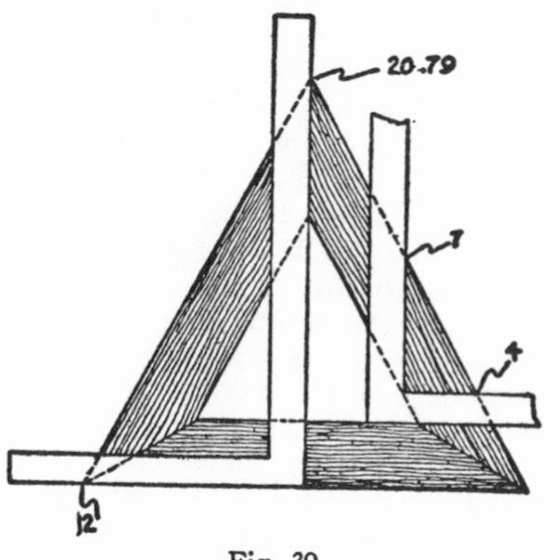

Fig. 39

tion of the square to the right, using the inside edges of the square. These figures, as mentioned before, will answer for all practical purposes, especially where only rough joints are involved. The inside edges of the square are used in this case for convenience. The student should learn to use the inside edges of the square whenever these edges are more convenient than the outside edges.

Rectangular Square.—Fig. 40 shows how a rectangular square 16 by 16 can be laid out with just two applications of the square. It also shows how a 24

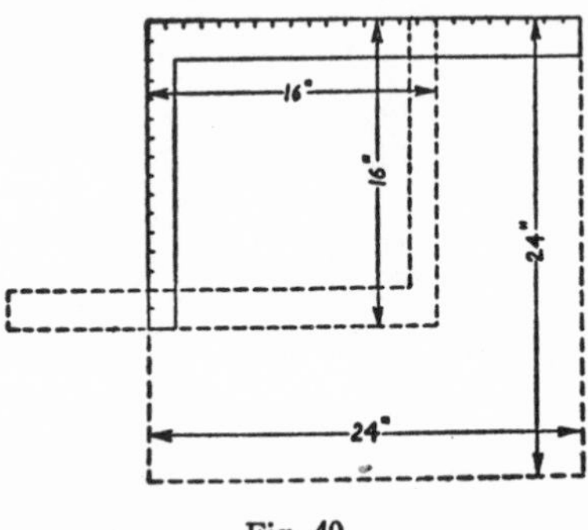

Fig. 40

by 24 square can be laid out with one application of the square by extending the line of the tongue, as shown by dotted lines, and locating the other two sides by measurements — these two sides are also shown by dotted lines. From what has just been said, it is evident that any size square can be laid out by using one or the other method just explained, but it should be remembered that the larger the square to be laid out is, the less practical these methods become.

Fig. 41 shows a butt joint at *A*. At *B* the square is shown applied for marking a butt joint. At *C* the square is shown in part, applied for marking a miter cut. The figures used are 12 and 12, but as shown, 6 and 6 can also be used. In fact, any other equal figures used on the body and tongue of the square will give the true miter cut. At *D* is shown a completed miter joint.

Pentagon. — Fig. 42 shows how to lay out a pentagon with a square. Draw line *X-X* and apply the square, using 8 on the tongue and 11 on the body, as shown by the squares numbered *1* and *2*. Then make each, *a-b* and *a-c*, as long as one side of the pentagon. This done, find the centers of *a-b* and *a-c* and apply the square, bringing the heel

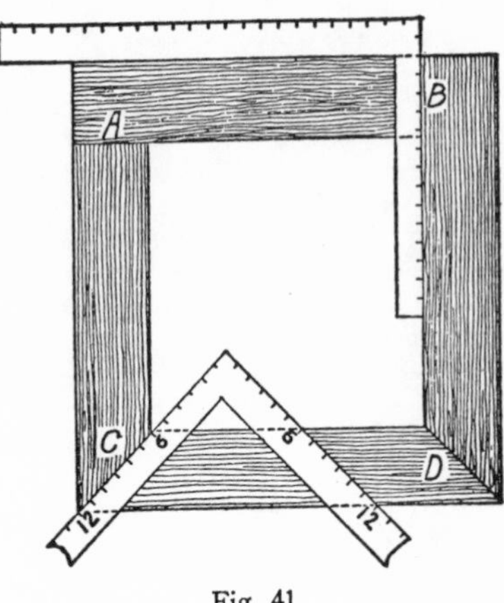

Fig. 41

to the center points, as shown by squares numbered *3* and *4*. Where the outside edges of the bodies of these squares cross, point *O*, is the center of a circle within which the laying out of the pentagon can be finished. From

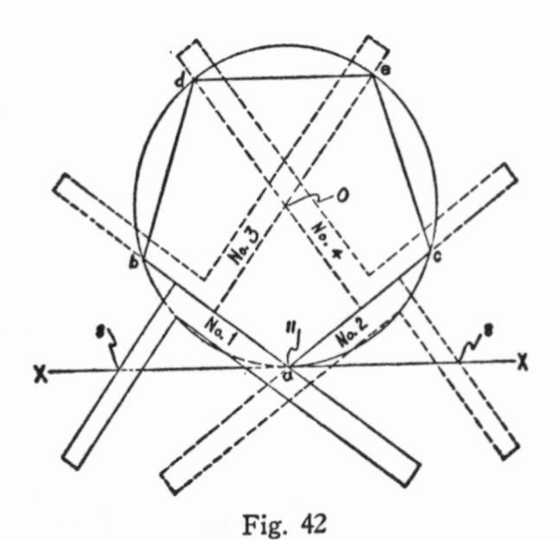

Fig. 42

point *O* describe the circle. Where this circle crosses the outside edges of the bodies of squares *3* and *4*, are the two remaining points of the pentagon. Now mark *b-d*, *d-e* and *e-c*, which completes the layout.

Fig. 43 shows a pentagon with three miter joints and two butt joints. Square *A* is applied for marking the miter joints, using 11 on the body and 8 on the tongue—the tongue giving the

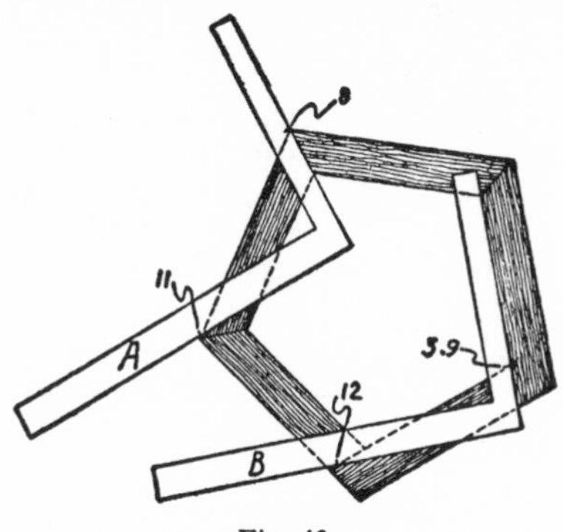

Fig. 43

cut. Square *B* is applied for marking the butt joints, using 12 on the body and 3.9 on the tongue—the tongue giving the cut.

Hexagon.—To lay out the hexagon shown in Fig. 44, draw line *X-X,* and perpendicular to it strike the dotted line *a-e.* Make *a-O* equal to the length

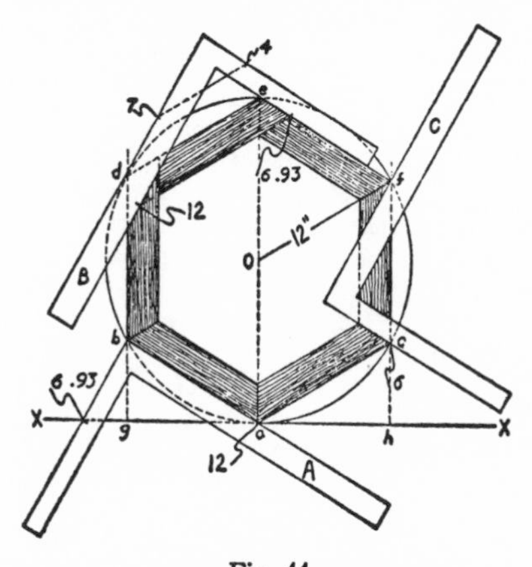

Fig. 44

of one side of the hexagon, which gives the radius for describing the circle. Then apply the square marked *A* as shown, using 12 on the body and 6.93 on the tongue. Now mark *a-b,* and in the same way mark *a-c.* Perpendicular to line *X-X,* strike a dotted line from *h* through *c* to *f,* and another from *g* through *b* to *d.* Where the three perpendicular dotted lines cross the circle, at *d, e,* and *f,* are the other points for marking the remaining sides of the hexagon, *b-d, d-e, e-f,* and *f-c.*

The square marked *B* is applied for marking the butt joints. Here it will be noticed that the inside edges of the tongue and body are again used for convenience. The figures 12 and 6.93 are used, as shown on the drawing, which are the same as those used in the application shown at *A.* These figures are also used for the miter joints. Ordinarily, 7 and 4 are accurate enough for all practical purposes.

The hexagon shown in Fig. 44 has sides that are 12 inches long, as the square marked *A* will show. If you cannot remember the figures that are given by the diagram, you can get the cuts for any hexagon by this simple method: Mark off the length of a side of an assumed hexagon on the edge of a board. Then take one half of it on the tongue of the square, hold this point to one of the marks, and bring the edge of the body to the other mark. The point where the body contacts the mark will give the figure to use on the body of the square. The square marked *C* illustrates this. Here we have taken one half of one side, or 6, on the tongue of the square, and brought the body to point *f,* which gives the point on the body to be used with 6 on the tongue. The square as shown is applied for marking the miter joint, but the same figures will make the butt joint.

Heptagon.— Fig. 45 is a diagram showing how a heptagon can be laid out with a square and the aid of a circle. Draw *X-X,* and perpendicular to it strike the line *a-h* from *a* through *o* to *h.* With *a-o* as the radius, strike the circle within which the heptagon is to be laid out. Now apply the square marked *A,* as shown, using 12 on the body and 5.78 on the tongue and mark *a-b.* Mark *a-c* in the same way. With half the length of one side of the heptagon, say half of *a-b,* apply the square marked *B* to *a-h,* as shown, and mark half of the upper side from the heel of

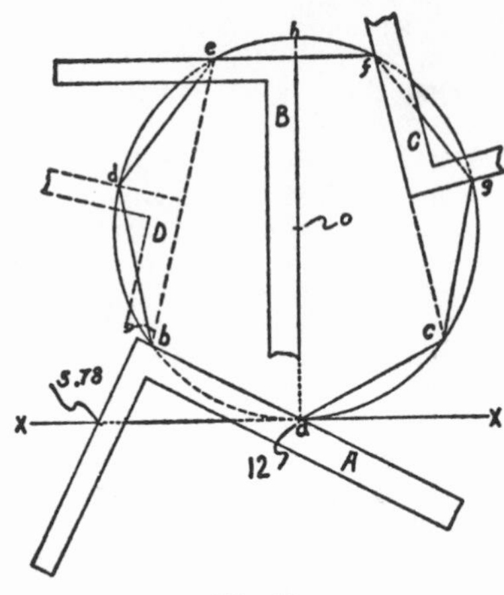

Fig. 45

the square to *e*. Mark the other half in the same way, completing *e-f*. To get the two remaining points strike *b-e* and *c-f*. With the heel of the square centering *b-e* and *c-f,* locate points *g*

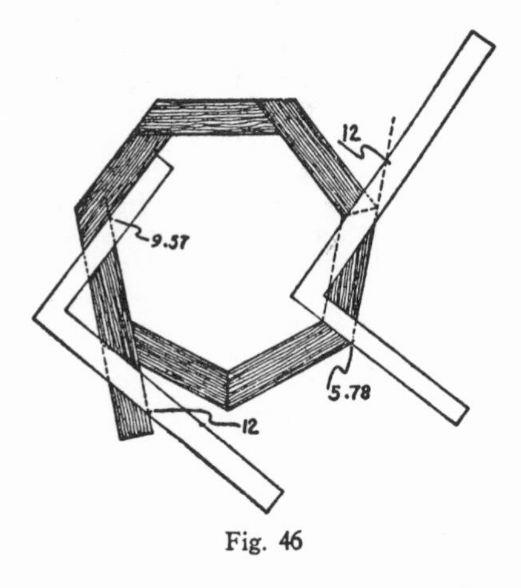

Fig. 46

and *d,* as shown by the squares marked *C* and *D*. Having these points, mark *b-d, d-e, f-g,* and *g-c,* which completes the heptagon.

Fig. 46 shows the plates in place for the heptagon just laid out, showing to the right the square applied for the miter joint, and to the left for the butt joint. The figures used for the miter joint are the same as those used in Fig. 45 for the first application of the square. The lug shown to the left was extended for convenience in applying the square.

Octagon.—Fig. 47 shows how to lay out an octagon with the square and the aid of a circle. First strike a circle with a radius as long as half the distance across the octagon. Then mark

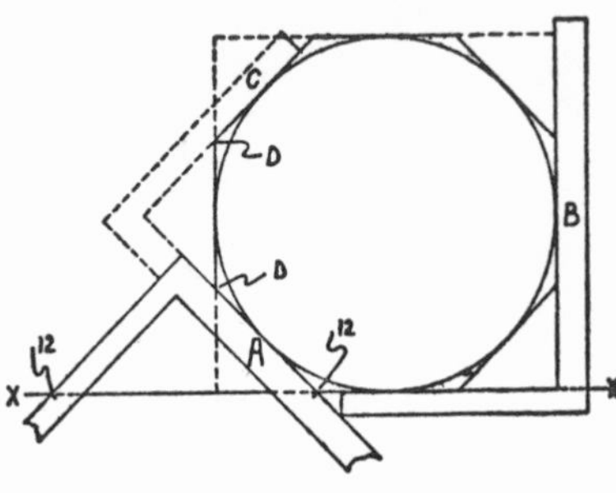

Fig. 47

line *X-X* so it will contact this circle. With the aid of line *X-X,* mark a square around the circle, as shown in part by dotted lines. Square *B* is applied for marking the right side of the square. Mark the left side in the same way. Make the upper side parallel to line *X-X*. Using 12 and 12 apply the square marked *A,* as shown, and mark off the two corners at the bottom. The other two corners can be marked off by using 12 and 12 on the square, or by using equal numbers at points *D, D* on the inside edges of the square marked *C*. Laying out an octagon in this way is so simple that the student can find several other ways to apply the square for marking the sides.

Fig. 48 shows the plates in place for the octagon that was laid out in Fig. 47, and two applications of the square for making the joints. The square to the right shows how to mark the butt joint by using 6 and 6, or as shown by dotted line, 12 and 12. To the left the square is in place for marking the miter joint, using 12 and 4.97. The lugs shown at *A* and *B* were left for convenience. At *A* the lug is used to aid

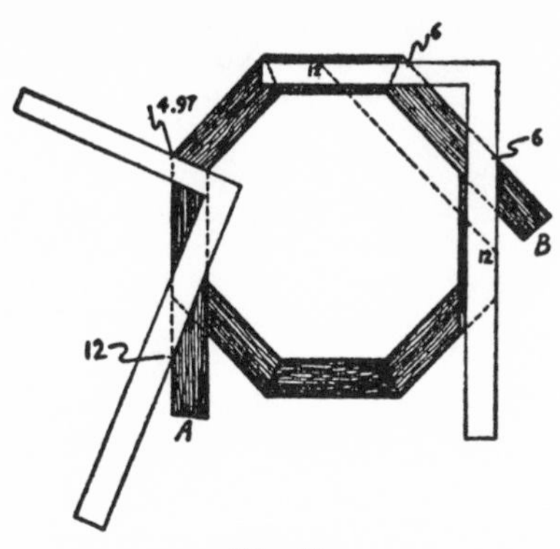

Fig. 48

in the application of the square, while at *B* it shows the part that is to be cut off.

Sparmaker's Rule.—Fig. 49 shows two applications of the square for marking timbers that are to be made into octagons, using the Sparmaker's

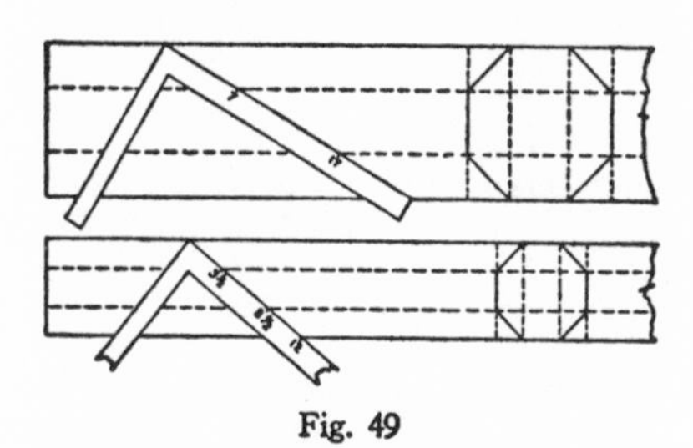

Fig. 49

rule. The upper application shows that by using the full length of the blade of the square, 7 and 17 will give the points for marking a square timber that is to be made into an octagon. To the right the diagram shows an end view of such a timber. This method of marking timbers that are to be made into octagons is so nearly correct that it will serve for all practical purposes. The application of the square shown at the bottom is the same, excepting that only half the length of the blade is used—the four points being the heel, 3½, 8½, and 12. The smaller figures are used when the timbers are rather small.

Octagon-Top Opening.—Fig. 50 shows how to lay out an octagon-top opening. On a 45-degree angle strike a line from each corner, which will cross at point *C*. From point *C*, with

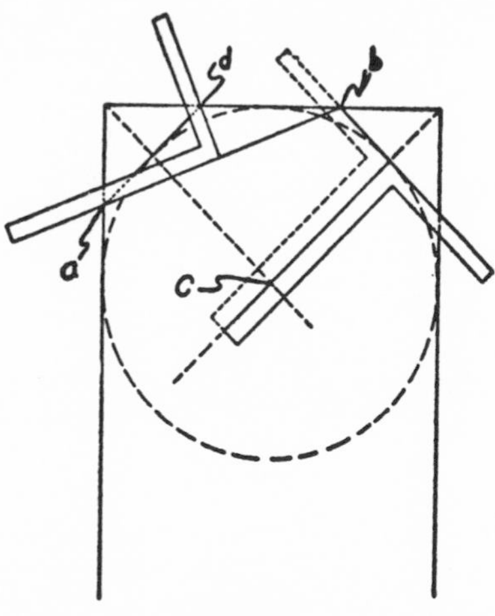

Fig. 50

a radius half the width of the opening, describe the circle shown by dotted line. Now mark off the two corners so that the marks will contact the circle, as shown by the two squares to the right. This will make a true octagon top. To get the miter cut, draw line *a-b*, and apply the square as shown. The figures on the square at points *a* and *d* will make the miter cut. This cut can also be obtained by using 12 and 4.97 on the square. The octagon-top opening framed is shown by Fig. 51. The dotted half-circle shows what a simple matter it is to change an octagon-top opening to a circle-top opening, by merely filling in the angles as the dotted half-circle shows.

The octagon is the most practical of the regular polygons, excepting the triangle and the square, which are so commonly used that they are in a class by themselves. There are two other polygons of minor importance, which should be named, the nonagon (nine sides) and the decagon (ten sides). To get the miter cut for the nonagon, use 12 and 4.37 on the square — the latter gives the cut. For the butt joint use 12 and 14.30 on the square—cut on the latter. For the miter cut of a decagon use 12 and 3.90—the latter gives the cut. For the butt joint use

Fig. 51

12 and 16.52. The number of polygons, of course, are unlimited, but when you get above the ten-sided regular polygon, they are of little practical value.

Polygons Unlimited.—Fig. 52 is a diagram giving eight different polygons, but the number of polygons are unlimited. Those shown, from center out, are a triangle numbered *3*, because it has three sides; a square, numbered *4;* a pentagon, *5;* hexagon, *6;* heptagon, *7;* octagon, *8;* nonagon, *9;* and decagon, *10*. The methods used in this diagram for laying out polygons will work on any polygon. With the square in the position shown, take 12 on the tongue for the basic center, from which describe a circle within which the polygon you want can be laid out, making all angles contact the circle. It will be observed that all of the polygons shown are laid out within circles shown by dashed lines. The upper square shows eight straight lines radiating from the basic center, 12. Where these lines contact the outside edge of the blade of the square will be found the figures to be used with 12, in order to mark the

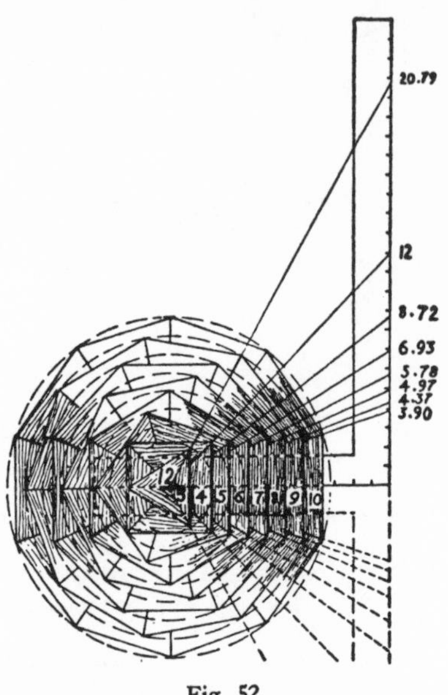

Fig. 52

miter cuts for the joints of the different polygons. Reading from the top down, 20.79 for the triangle, 12 for the square, 8.72 for the pentagon, 6.93 for the hexagon, 5.78 for the heptagon, 4.97 for the octagon, 4.37 for the nonagon, and 3.90 for the decagon. At the bottom of the diagram is shown in reverse order, a part of a square by dotted lines. Joining this square will be found dotted lines radiating from the basic center, corresponding to the lines joining the upper square. Where the respective upper and lower lines cross the circle around the polygon they are used with, are the points that mark the ends of one side of the polygon. The distance between these points is the length of the sides of the particular polygon, as indicated by the heavy lines to the right side of the drawing.

Methods of Procedure.—Fig. 53 shows two methods of procedure in laying out polygons, one for polygons with an even number of sides, and the other for polygons with an odd number of sides. The larger of the two polygons shown in this diagram is a decagon; that is, it has an even number of

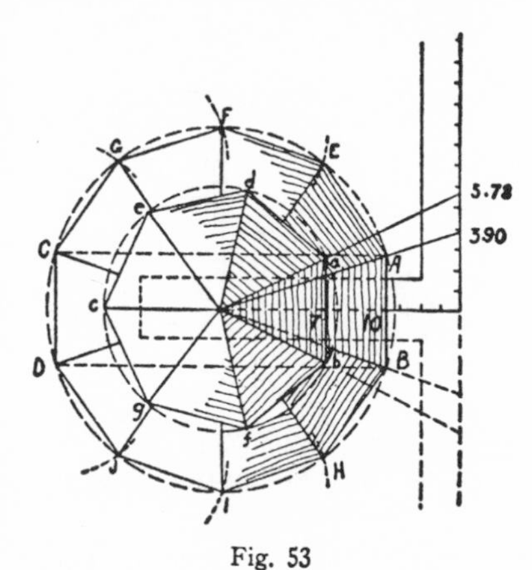

Fig. 53

10 sides. The respective radiating lines to be used in laying out this polygon cross the circle around it at points *A* and *B*. The distance between these two points is the length of the sides. From points *A* and *B* draw two horizontal lines, establishing points *C* and *D*, as shown. With the compass set at point *C*, locate *G*, from *G* locate *F*, and from *F* locate *E*. In the same way, starting at *D*, locate *J*, *I*, *H*.

The smaller of the two polygons shown is a heptagon, which means that it has seven sides, or an odd number of sides. The two respective radiating lines to be used cross the circle around this heptagon at points *a* and *b*. The distance between these two points is the length of the sides. Establish point *c* by extending the horizontal center line until it contacts the circle, as shown. Having these three points, with the compass set at *a*, locate point *d*, from *d* locate *e*. Then set the compass at *b* and locate *f*, from *f* locate *g*. Now join these points with lines, as shown, and the heptagon is complete.

To find the number of degrees for one side of any polygon within a circle is simple—divide 360 by the number of sides in the polygon in question. Try that on the polygons shown by Fig. 52. For the triangle—3 sides—360 divided by 3 gives 120, or 120 degrees. For the square—4 sides—360 divided by 4 gives 90, or 90 degrees. In the same way, for the pentagon, 72 degrees; the hexagon, 60; the heptagon, 51$\frac{3}{7}$; the octagon, 45; the nonagon, 40; the decagon, 36, and so on indefinitely.

It should be remembered that in laying out polygons, accuracy in measurements and in making the lines is of vital importance. Without accuracy the results are not dependable.

Chapter 6

CUTS FOR BRACES

Braces.—The most important part of a building is its foundation. But almost on a par with the foundation is the bracing of a building. The foundation might hold the building on a level and at the proper elevation, but if it does not have substantial bracing, it will soon twist or lean in some way, if not fall to the ground completely. A great deal of the bracing of light frame buildings is found in the boxing of the walls. In fact, what is known as balloon framing depends entirely on the boxing to hold the structure in its proper upright position. But the braces that are taken up in this chapter are timbers of some kind that serve the specific purpose of holding buildings in their proper positions.

Twelve-Step Method.—A simple way to get the length and cuts of a brace, such as is shown by Fig. 54, is to let inches on the square represent

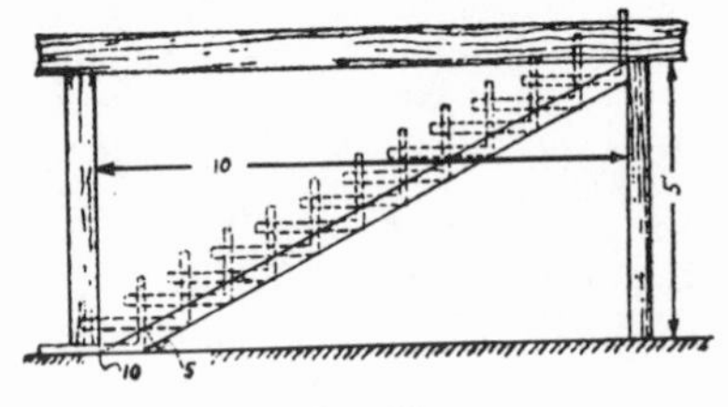

Fig. 54

feet in the run and rise of the brace, and take 12 steps. For instance, the run of the brace shown is 10 feet and the rise is 5 feet. Now take 10 on the blade of the square and 5 on the tongue, and take 12 steps, as shown by the dotted outlines of squares on the drawing. The first application of the square, shown at the bottom left, gives the foot cut, while the last application, to the upper right, gives the top or plumb cut.

Reducing Number of Steps.—Using 10 and 5 is a little tedious, because of the short steps. This can be avoided by doubling both figures, which gives 20 on the blade and 10 on the tongue, as shown by the first application of the square at the bottom left in Fig. 55. With these figures, only half the number of steps are taken, or 6. The blade again gives the horizontal cuts, while the tongue gives the plumb cuts.

Use of Measuring Line.—The brace in Fig. 55 is not framed like the one

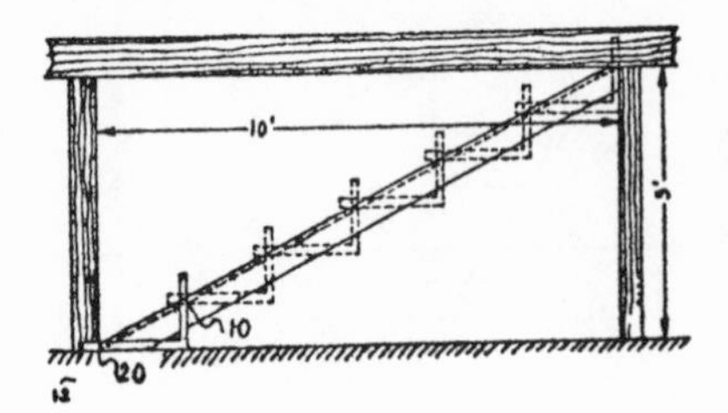

Fig. 55

shown in Fig. 54. Here the square is applied to the dotted measuring line. It will be noticed that both the top and the bottom cuts are partly horizontal and partly plumb, so that

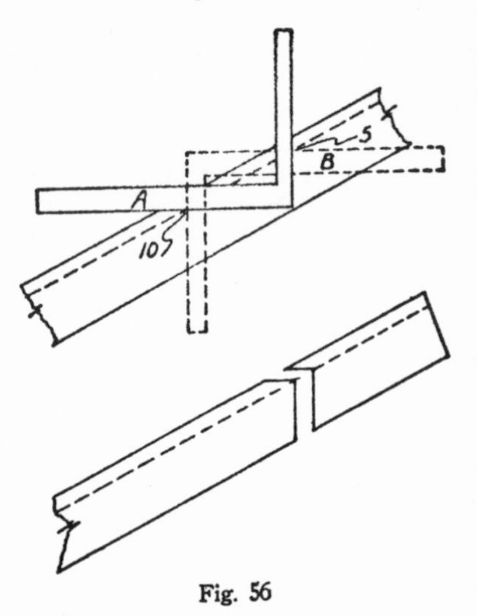

Fig. 56

the brace will fit into the angles. In Fig. 56 the upper drawing shows the square marked *A* in position for marking the plumb part of the upper cut, while the square marked *B* is in position for marking the horizontal part of the same cut. The same drawing gives the applications of the square for the bottom cut. The square marked *A* gives the horizontal cut, and the one marked *B* gives the plumb cut. The squares are applied to the dotted measuring line. Study the drawing. The bottom drawing of Fig. 56 shows the timber cut in two. Notice that the dotted measuring line joins the plumb and horizontal lines where they form the angle.

Finding Figures to Use.— Fig. 57 shows a brace with a rise of 48 inches and a run of 40 inches. To obtain the figures to use on the square for stepping off the length of the brace, divide both the run and the rise by the same figure, say 4, which gives 12 and 10. With these figures on the square, take four steps, as shown. The blade gives the plumb or bottom cut, while the tongue gives the horizontal or top cut. Figures that will divide in whole numbers are purposely used so as to make the point clear. In practice this cannot always be done, but it is just as accurate when fractions are involved as it is with whole numbers.

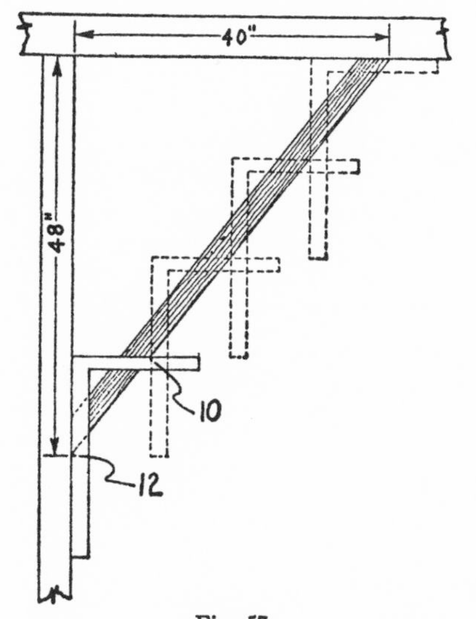

Fig. 57

Fraction of Step.—A little different approach is shown by Fig. 58. Here is shown a brace with a 48-inch rise and a 48-inch run. Dividing 48 by 12 gives a result of 4, or the number of steps to be taken. Then using 12 and 12 on the square take four steps, which will give the length of the brace. If instead of 48 inches both the run and the rise had only 40 inches, with 12 and 12 on the square three steps would be taken. This would bring the square up to position A, with 4 inches left over. To take care of the 4 inches the square is moved from position *A* to position *B*, or up to the dotted horizontal line. The tongue of the square in this case gives the top or horizontal cut. In case both the run and the rise of the brace were 44 inches, then after taking three full steps, as shown by the unshaded squares, take a fraction of a step, using 8 and 8 on the square, as shown by the black square in Fig. 58. If the square is moved from this point to position *C* the fourth full step is complete. Study this illustration.

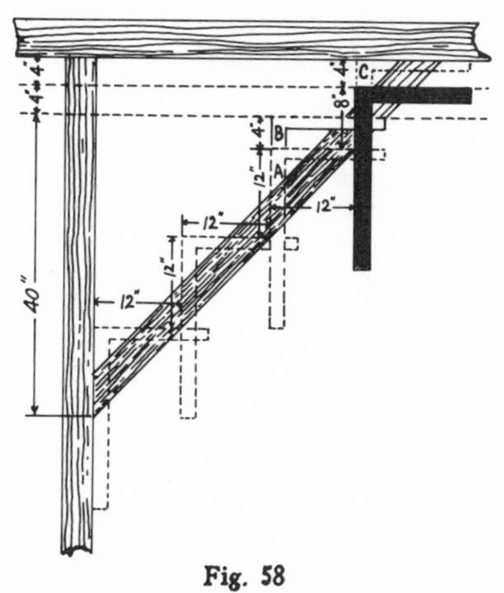

Fig. 58

Cuts for Joining Braces.— Fig. 59 shows part of a brace with a 12 and 6 cut. The tongue of the square to the upper right gives the plumb or top cut, while the blade will give the horizontal or foot cut. The square to the upper left shows how the same figures used on the square will give the cut for a timber joining such a brace horizontally. The blade, as shown, gives the cut. The square at the bottom center shows how the same figures will give

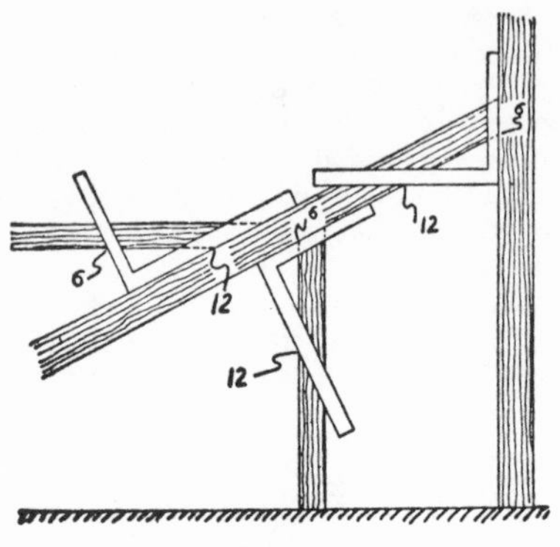

Fig. 59

the cut on a timber joining the brace in a plumb position. The tongue in this case gives the cut.

Bridging Cuts.—How to find what figures to use on the square for obtain-

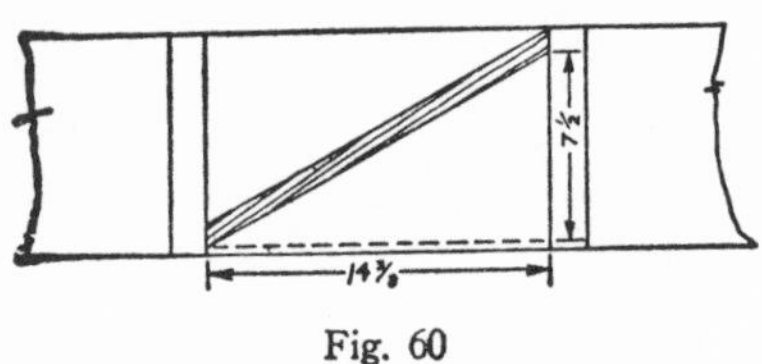

Fig. 60

ing the bevel of bridging is shown by Fig. 60. Here is shown a part of a 2x8 timber with two joists spaced 16 inches on center marked on it with a piece of bridging, shaded, in place. The bridging does not come quite to the bottom edge of the joist, as indicated by the dotted line. To the right is given the distance from the bottom of the top joint down to the bottom of the bottom joint, or 7½ inches. At the bottom is shown the distance between

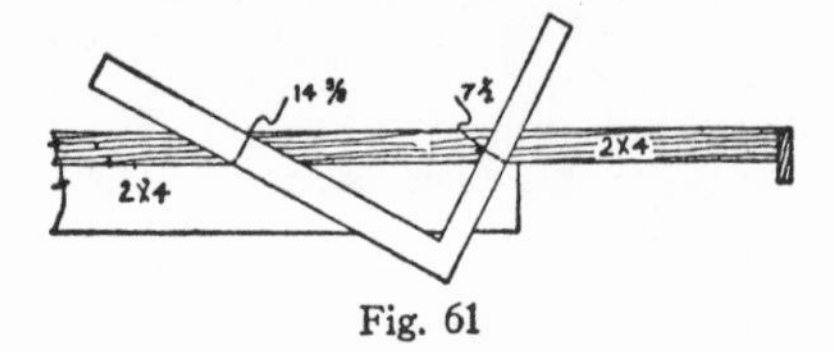

Fig. 61

the two joists, which is 14⅜ inches. With 14⅜ on the body of the square and 7½ on the tongue—the tongue will give the bevel of the bridging, as shown by the application of the square in Fig. 61. The square here is applied to an improvised miter box for cutting bridging. Fig. 62 shows a piece of bridging material in the box with the end cut to

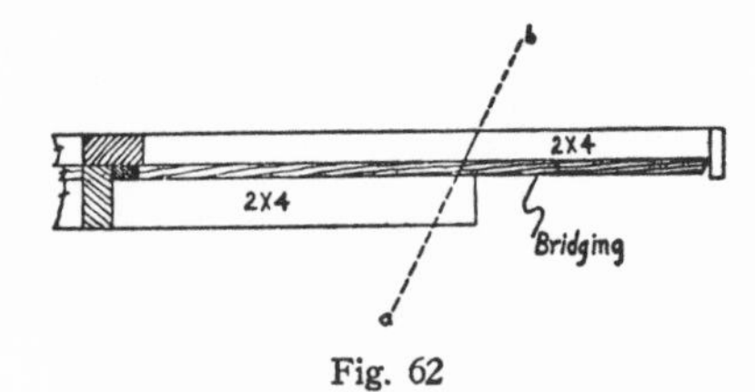

Fig. 62

a bevel, ready for making the second cut. The dotted line between *a* and *b*, gives the position of the saw for cutting the bridging. A cross section of the box is shown to the left, inset.

Chapter 7

HOPPER CUTS

Frequently one hears that making a hopper is the same as framing a hip roof, but that is only partly true. The basic principles, though, are the same. The various cuts for a hopper are the same as the cuts involved in cutting hip roof sheeting, that is, when the sheeting is cut accurately. As a rule, roof sheeting is cut by guess.

Hopper Cuts.—Fig. 63 shows a cross section of a hopper, made of square-edged material. To the left is shown a square applied for marking the bevel for the bottom edge, while to the right the square is applied for marking the bevel for the upper edge. It should be mentioned here that the squares shown are drawn to a scale in which 4 inches on the square equals 1 inch on the hopper material. By doing this, the applications of the square can be made clearer, because a much greater proportion of the square can be shown on the drawings. This proportion is carried out in all of the illustrations for this chapter, excepting the last one. The figures given in connection with the different applications of the square represent inches on the square. At the right of Fig. 63, by dotted lines is shown how the figures used with the squares were obtained. The figures show, using the small scale, a run of 8 inches and a rise of 14 inches, just as the two squares do.

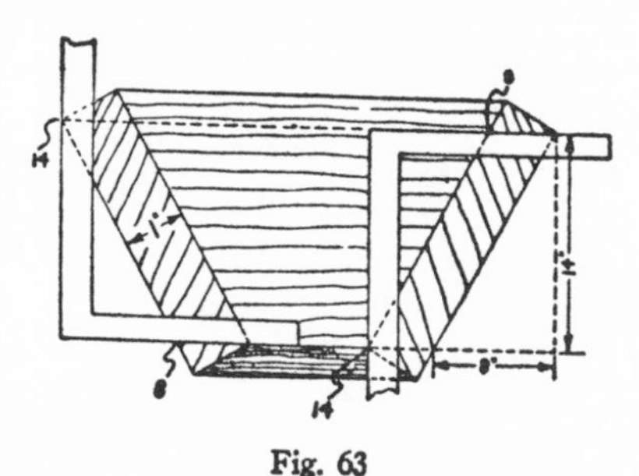

Fig. 63

Miter and Butt Joints.—A cross section of the hopper shown in Fig. 63, is shown by the upper drawing in Fig. 64, but in this drawing the edges are shown beveled. The bottom drawing shows a top view, where the square shown to the upper right is applied for marking the miter cut, using 12 and 12 on the square. To the left a miter joint is shown. The square shown at the bottom left is applied for marking a butt joint, while to the right a butt joint is shown.

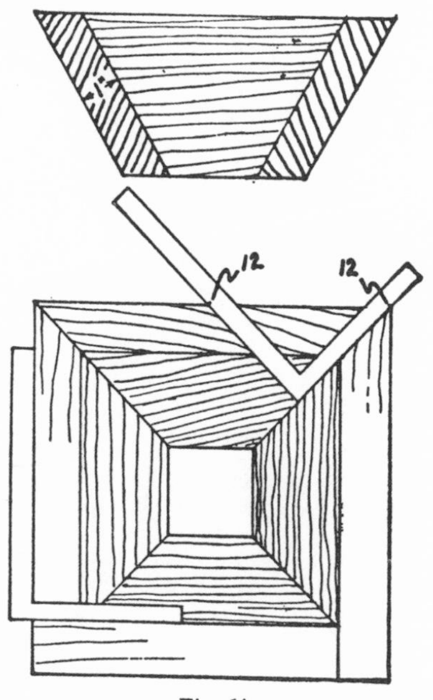

Fig. 64

Fig. 65 is a reproduction of the left part of Fig. 63, excepting the squares. The square shown by dotted lines is in position for marking the bevel on the hopper material, but what one wants to know here is how to get the figures for marking the miter cut on square-edged hopper material. In roof framing terms, this cut is obtained by using the length of the rafter on the blade of the square, and the rise on the tongue—the tongue giving the cut. How the figures are obtained is shown by the full-line square to the left, where 16⅛ (length of rafter) is shown on the blade and 14 (rise) on the tongue. The square

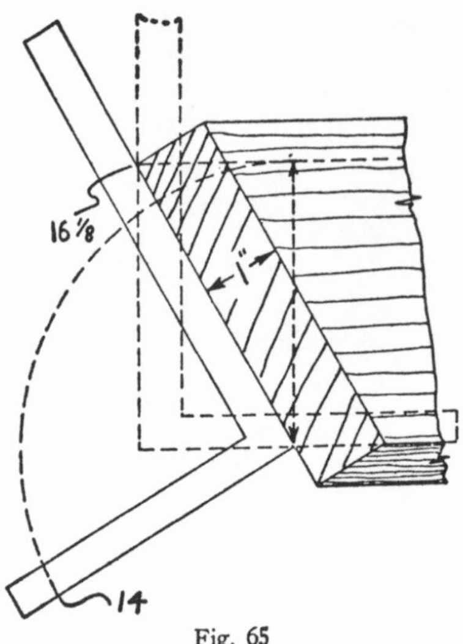

Fig. 65

applied to the edge of the material for marking the miter cut is shown by Fig. 66.

Fig. 67 shows how to obtain the figures to be used on the square for marking the cut for a butt joint on

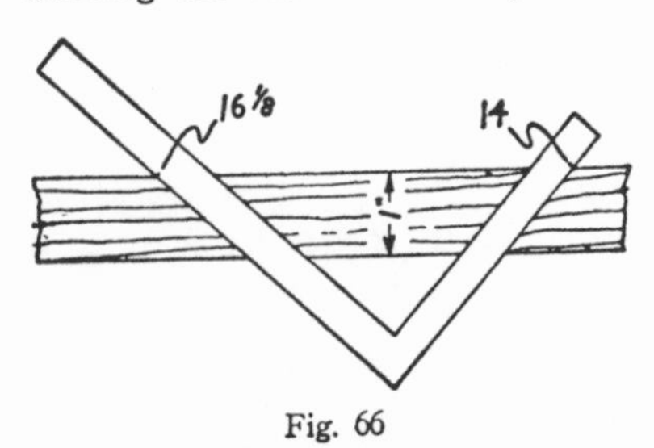

Fig. 66

square-edged hopper material. In roof framing language, the cut is obtained by taking the length of the rafter on the blade of the square, and the counter-rise on the tongue (study the rise and counter-rise on the drawing). This means, as shown by the drawing, that 16⅛ is taken on the blade of the square and 4½ plus on the tongue—the tongue giving the cut. How the square is applied to the edge of the material for marking the bevel is shown by Fig. 68.

In the same way Fig. 69 shows how the figures are obtained for marking

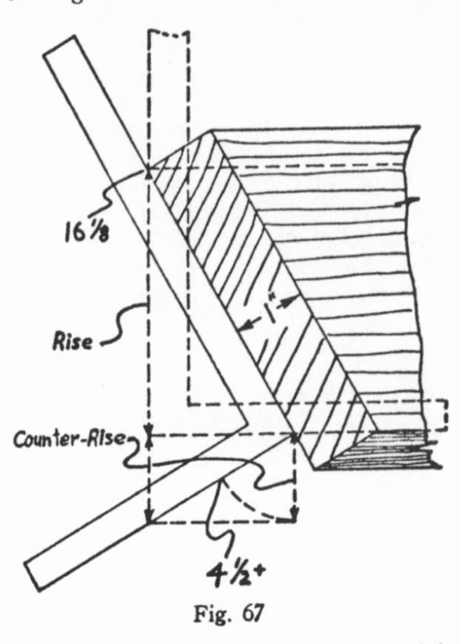

Fig. 67

the face cut on the hopper material. In other words, the length of the rafter and the run, will give the face cut, or 16⅛ on the blade of the square and 8 on the tongue—the tongue giving the cut. The application of the square to

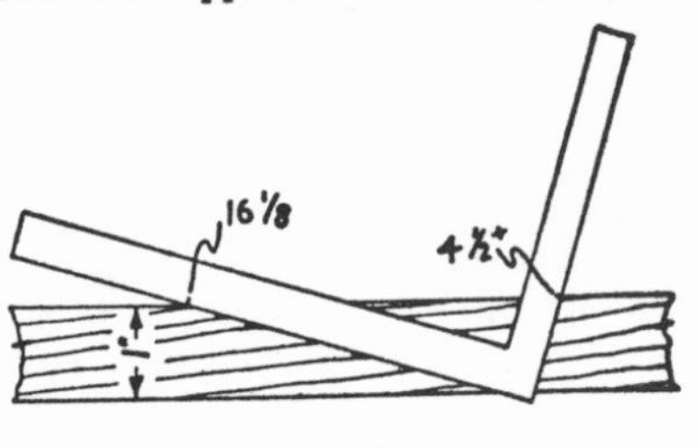

Fig. 68

the face of the material is shown by Fig. 70. Here it should be mentioned again, that 4 inches on the square equal 1 inch on the material. The dotted line continues the bevel across the face of the board.

Diagram for Hopper Cuts.—Fig. 71 is a diagram showing how to obtain the figures to be used on the square for the different cuts on square-edged hopper material. The shaded square gives the run, 12 inches, and the rise, 16 inches, while the unshaded square is in position for obtaining the length of the

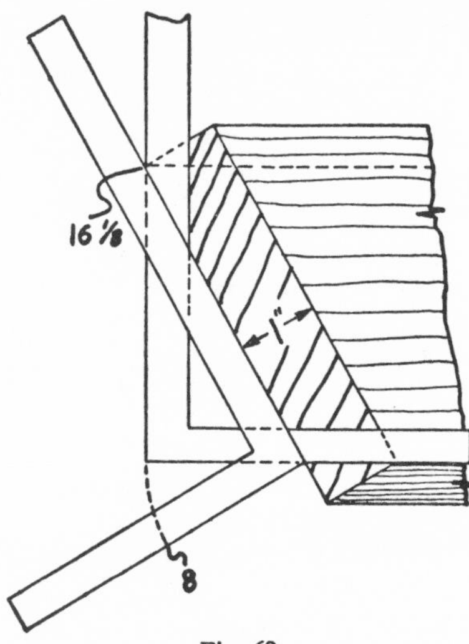

Fig. 69

rafter which, as shown, is 20 inches. With these figures and the squares in the positions shown, all of the points for the different cuts can be established.

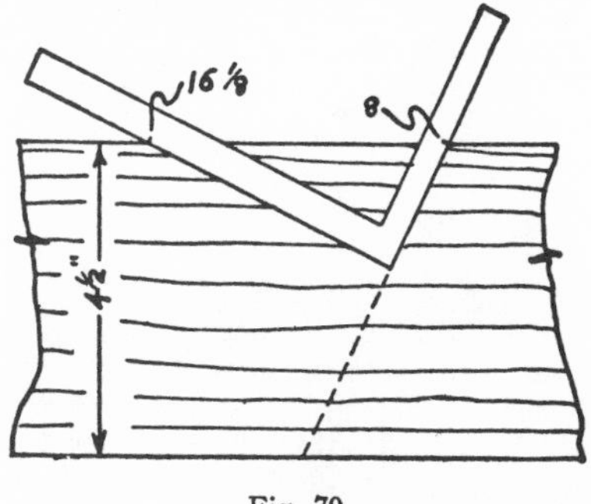

Fig. 70

A cross section of the square-edged hopper material is shown on the blade of the unshaded square. Because the rafter is 20 inches long per foot run, 20 is used on the blade of the square as the constant figure. To obtain the key points on the tongue of the square, project *M* to *M′*; *B* to *B′*; and *F* to *F′*. Now with 20 on the blade of the square and point *M′* on the tongue — the tongue will give the miter cut. For the butt joint, use 20 on the blade and point *B′* on the tongue—the tongue giving the cut. The face cut is obtained by using 20 on the blade and point *F′* on the tongue. Mark along the tongue for the cut.

Reference letters are used on the diagram to indicate the points to be taken on the tongue of the square, but in practice figures should be used on both

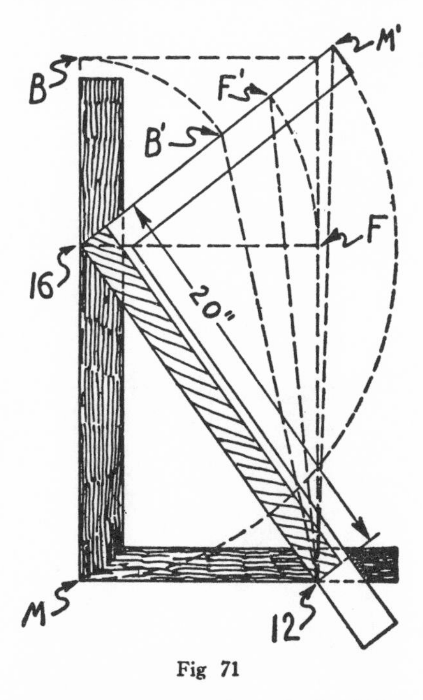

Fig 71

the blade and the tongue. It should be remembered that the diagram shows how to obtain the cuts for a hopper with a run of 12 inches and a rise of 16 inches. These figures were chosen in order to make the diagram as simple as possible. The principles involved in the various cuts shown by the diagram will work on any kind of square hopper, regardless of the size or pitch. In cases of small hoppers, the workman could use a full-size diagram, which would simplify the process of obtaining the figures to be used on the square.

Chapter 8

THE COUNTER PITCH

Hopper and Plancher Cuts. — The rule for getting the various cuts for hoppers is the same as that used for cutting the plancher board for hip cornices that follow the line of the rafter tail. If you can cut the plancher board for such a cornice so that it will fit snugly, then you can make the cuts for a square or oblong hopper, but you will have to work on the hopper in an upside-down position. In other words, you will have to imagine that the hopper (upside down) is a little hip roof, and make the cuts accordingly.

The Counter Pitch.—Fig. 72 shows a small scale cross section of a building with a roof that is in a counter-pitch position with the hopper boards.

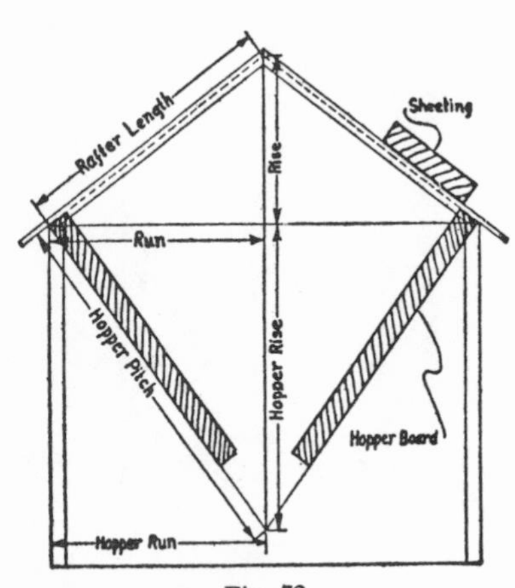

Fig. 72

That is, the rafters and the hopper boards are joined at a right angle. The run, the rise, and the length of the rafters are indicated on the drawing. To the right, in a much larger scale and shaded, is shown a cross section of a sheeting board. Below the roof are shown two cross sections of hopper boards in position. These, like the sheeting board, are shown shaded and are drawn to a much larger scale than the parts representing the building. The hopper run, the hopper rise, and the hopper pitch are indicated. If you were to frame a square square-edged hopper in a rightside-up position, the face sheeting cut for the counter-pitch hip roof would be the miter cut on the edge of the hopper boards, while the edge cut of the sheeting would be the face cut for the hopper boards. On the other hand, if you would turn the hopper over, then the cuts for the hopper boards would be just the same as the cuts for hip roof sheeting.

Cuts by Diagram. — Fig. 73 shows the principal parts of the left half of

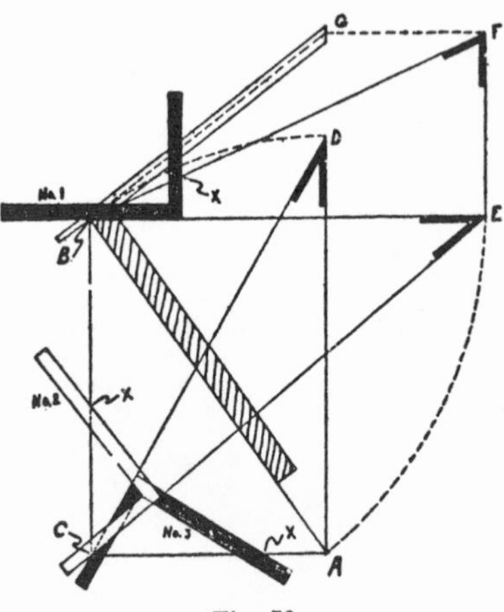

Fig. 73

the diagram shown in Fig. 72. The line *A-B* contacts the hopper board and gives the hopper pitch, while the dotted line *B-G* gives the counter pitch, which is the roof pitch. To develop the diagram, strike the arc *A-E,* with the compass set at *B*. Establish point *F* by drawing lines, *E-F* and *G-F*. Continue by drawing lines *B-E* and *F-B*. In the same way draw the other lines, including the arc, *B-D*. Now apply square No. *1* in such a manner that 12 on the blade will be on point *B*. Then the key point for marking the butt joint will be found at point *X*.

The tongue will give the cut. To obtain the miter cut for the hopper, apply square No. *2* in such a way that 12 on the tongue will come on point *C*. The key point will them be found at *X*. The tongue will give the miter cut. To obtain the face cut, apply square No. *3* in such a way that 12 on the tongue will come on point *C*. The key point will be found at *X*. The tongue will give the face cut. The shaded bev-

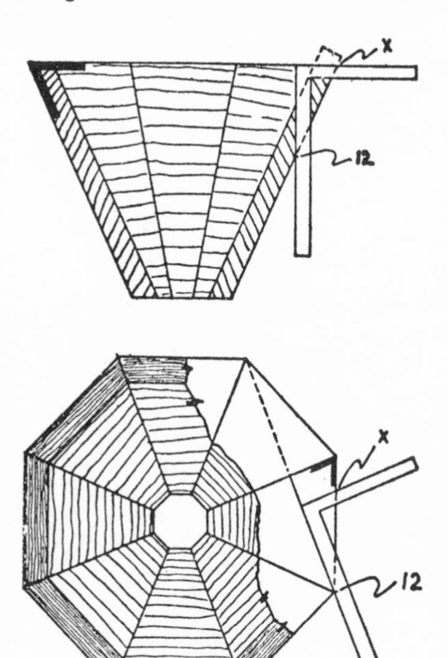

Fig. 74

els at *F*, *E*, and *D* are the respective bevels for the three cuts just explained.

Cuts for Octagon Hopper.—Fig. 74 shows at the top a cross section of an octagon hopper. The shaded bevel to the upper left is the same as the bevel on the edge of the hopper boards. To obtain the point to be used on the tongue of the square for marking this bevel, apply the square as shown to the upper right, using 12 on the body, which will give the key point, *X*, on the tongue. The tongue will give the bevel. The bottom drawing gives a plan of the hopper. The unshaded part

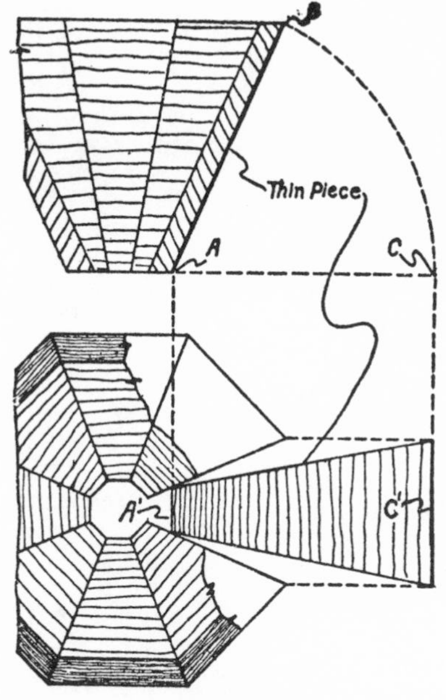

Fig. 75

to the right shows how to make a diagram for obtaining the key point on the square for the miter joint when the hopper boards are beveled. The square is applied to the diagram, using 12 on the body, as shown—then the key point will be found at *X*.

Face Cut.—Figs. 75, 76, and 77 are diagrams illustrating how to obtain the key point on the square for the face cut of an octagon hopper. In the upper drawing between *A* and *B* on the right a thin piece is used, as a makeshift pattern instead of the regular hopper board. This piece is dropped down, as shown by the dotted lines, bringing point *B* to point *C*. The bottom draw-

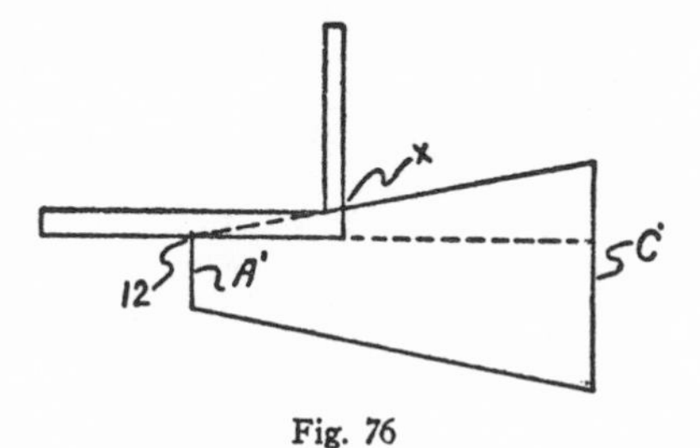

Fig. 76

ing shows a full view of this piece between A' and C'. Fig. 76 shows the pattern piece with the square applied, using 12 on the body in order to obtain the key point X. Fig. 77 shows the square applied to a board for marking the face cut of one of the octagon hopper boards. The dotted lines give the full outline. It should be remembered that the squares shown in these illustrations are drawn at a smaller scale than the drawings for the hoppers and hopper material. This prevents the

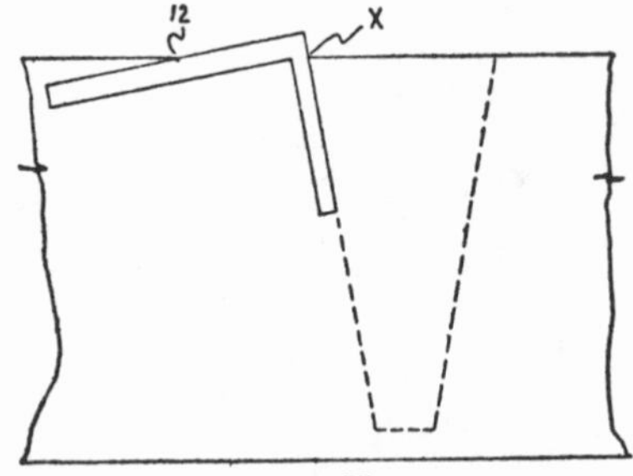

Fig. 77

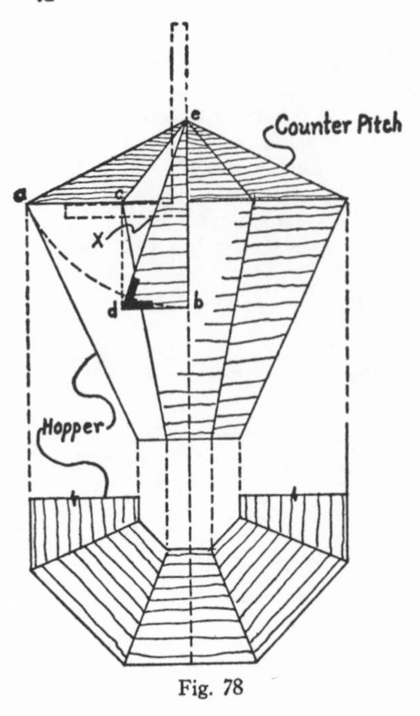

Fig. 78

drawings for the squares from taking up too much space.

Marking Miter Joint.—Fig 78 shows at the bottom a plan of one half of the octagon hopper that was shown in previous illustrations, and just above it, connected with dotted lines, is an elevation of the same hopper with a counter-pitch roof. The counter-pitch roof is the important feature of this illustration, which shows how to obtain the figures on the square for marking a miter joint on square-edged hopper material. To develop the part of the diagram that gives this, draw the arc *a-b,* as shown by the dotted line. Then draw the perpendicular dotted line *c-d*. This done, draw the lines *b-e, e-d,* and *d-b*. This triangle represents one half of one side of the counter-pitch roof, looking straight at it. The bevel that we want is shown, shaded, at *d*. Now apply the square in such a way that 12 on the body will come on point *e,* and the figures to be used on the tongue will be found at X . The tongue of the square will give the miter cut.

Fig. 79 is a reproduction of the counter-pitch roof shown in Fig. 78, giving a simple way of obtaining the same results. Make *a-b* equal in length to *c-d,* and join *b* with *e*. The bevel shown at *b* is the bevel that will make

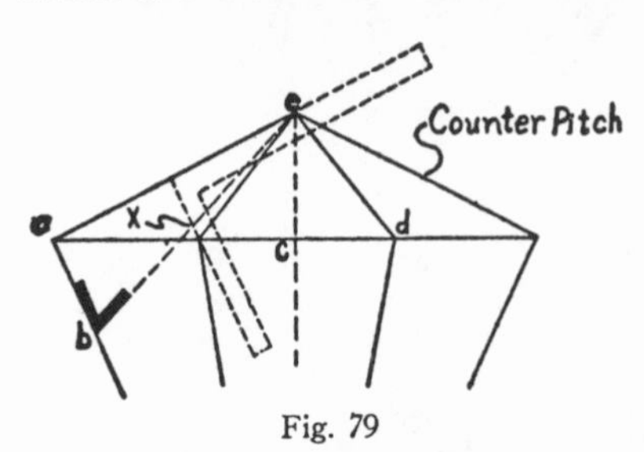

Fig. 79

the octagon miter cut for the hopper in question. Apply the square as shown, making 12 on the blade come on point *e*. Point X is the point to be used on the tongue.

The principles involved in the problems dealing with the octagon hopper in this chapter, will apply to any other polygon hopper.

Chapter 9

SIMPLE ROOF FRAMING CUTS

Steel Square Method.—The steel square, while it is used for a great many things, is primarily a roof framing tool. Roof framing tables and gadget squares which have everything thought out for the workman, are all based on the principles of the steel square. In fact, roof framing tables are made by means of the square. This is also true of gadget squares that have the tables stamped on them, showing the points to be used for the different cuts. Now these things are all right in their place, but if you really want to learn how to frame all kinds of roofs, then you should use the steel-square method of roof framing and stick by it.

Simple Cuts.—Fig. 80 shows a cross section of a small building. The shaded

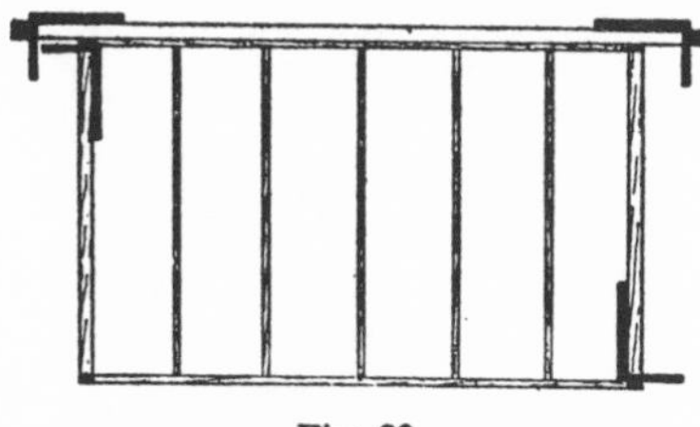

Fig. 80

squares show the square-across cut, which is the only cut that is needed in framing such a building. The roof, for example, is perfectly flat, and the ends of the roof joists, or let's call them rafters, are cut square. The shaded parts that stick out beyond the squares are to be cut off.

Low-Pitch Cuts.—Fig. 81 shows a sort of diagram cross section of a building that is 12 feet wide, which has a single pitch roof with a 4-foot rise. The pitch of this roof would be a 12 and 4 pitch, which is to say that the run is 12 feet and the rise is 4 feet, or every 12-inch run has a 4-inch rise. By letting feet be represented by inches on the square, or 12 on the blade and 4 on the tongue, you would have the points for stepping off the rafter. This is shown by the shaded square to the upper left, which is in position for the first step. The squares represented by dotted lines, show four additional steps,

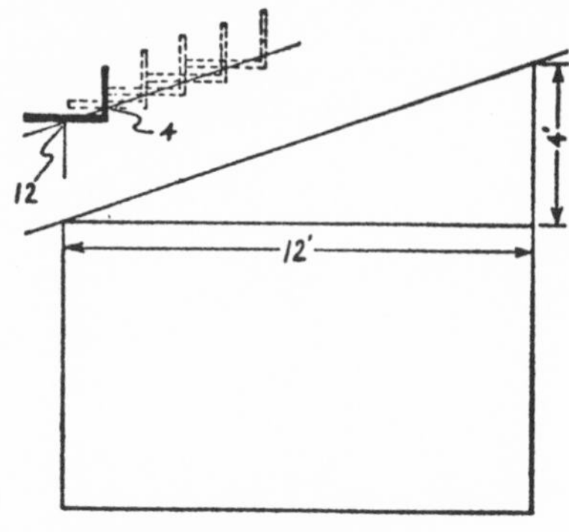

Fig. 81

and if these were continued up to 12 steps, or as many steps as there are feet in the width of the building, you would have the length of the rafter. To this length you would have to add the end projections, or overhangs, of the rafter to get the full length.

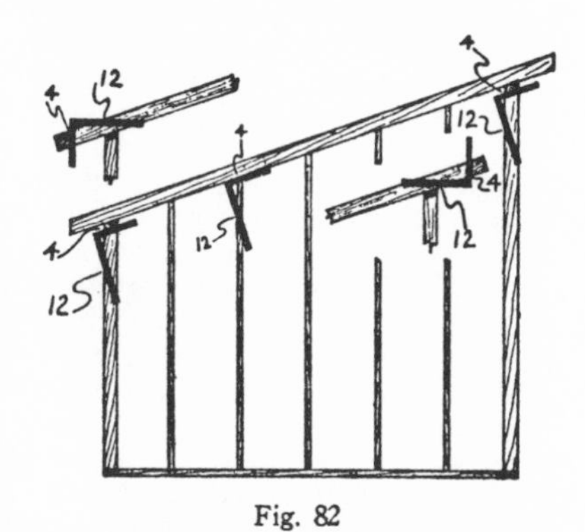

Fig. 82

Fig. 82 shows a skeleton cross section of the building shown by the diagram in Fig. 81. Here the squares are applied to the studding and to the rafter ends, showing that in all of these cuts, the same figures are used, or 12 and 4. This shows, after a little study, that roof framing is really simple. Practically all of the cuts are based on the principle of the run and the rise. In Fig. 82 the rafter has no seat cuts, for the plates slope with the slope of the roof. However, the same figures, 12 and 4, would be used in framing the seat cuts of the rafters, if the plates were in a level position. This is shown by Fig. 83, where the bottom drawing shows a rafter in part, in position. The upper drawing shows how the seat cut is laid out. Square No. *1* is in position for marking the plumb cut of the seat, using 12 and 4, while square No. *2* is

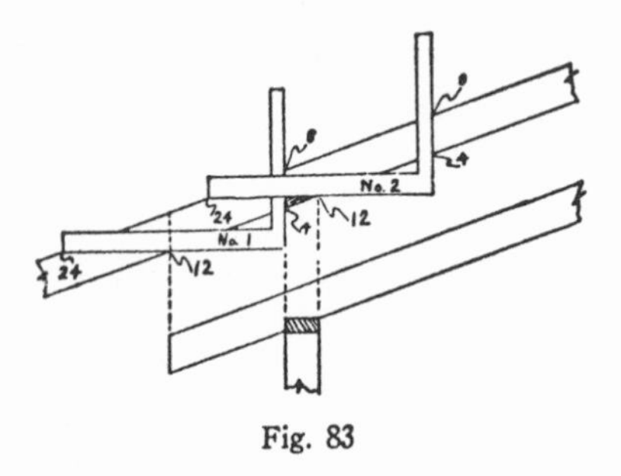

Fig. 83

applied for marking the level cut of the seat. The same cuts can be obtained by using 24 and 8, instead of 12 and 4, which is shown by the two squares where they intersect the upper edge of the timber.

Third Pitch Cuts.—Fig. 84, *A*, shows the application of the square for marking the seat cuts of a third pitch rafter. Square No. *1* is in position for marking the plumb cut, while square No. *2* gives the level cut of the seat. It will be noticed that these applications of the square are the same as the applications shown in Fig. 83, excepting that the figure used on the tongue, or key figure, is 8 instead of 4. In roof framing 12 is the base figure, which is usually used on the blade of the square, in pitches below the half pitch, but not always. In pitches above the half pitch, 12 is more frequently used on the tongue. No rule can be laid down, excepting that the workman should study the situation under which he is operating, and use the figure 12 on whichever arm of the square that will make the handling of the square the most convenient. At *B*, Fig. 84, the square is shown applied for marking the end of the rafter lookout, or the tail piece.

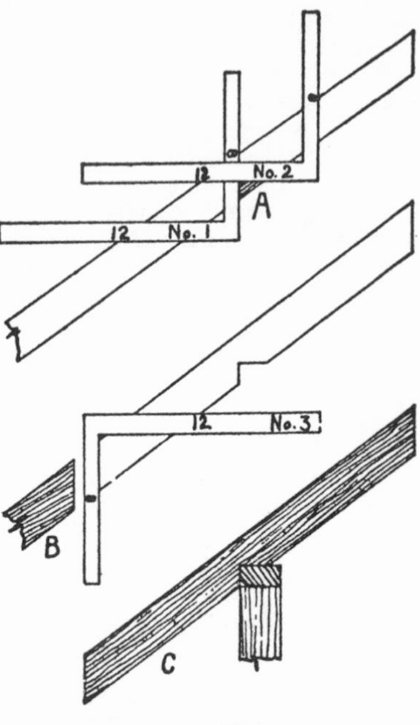

Fig. 84

At *C* the rafter is shown in position, giving both the plumb and the level cuts of a common rafter. The rafter is purposely shown rather short, in order to make the drawing more nearly complete, using a convenient scale.

More Third Pitch Cuts. — Fig. 85, *A*, shows by dotted lines the stepping off of a rafter having a 2-foot run and a 16-inch rise, but without a lookout. At *B* the rafter is shown in position. A different seat cut is shown at *C*, where the two squares by dotted lines are applied for marking the plumb and the level cuts of the seat. The figures used are 12 and 8, or the figures for a third pitch. At *D* the seat end of the rafter is shown joined to the plate.

Tie Beam. — Fig. 86 shows a side view of a pair of rafters for a third pitch roof with a tie beam at the bottom. The bottom drawing shows the

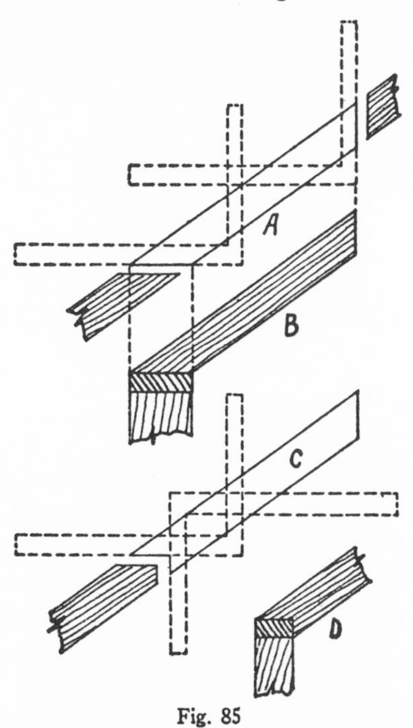

Fig. 85

square applied to a timber for marking an end of the tie beam. It will be observed that the figures on the square that are used are 12 and 8, which are

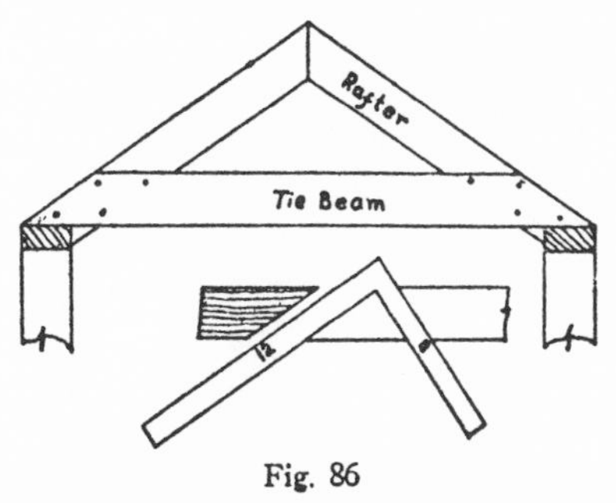

Fig. 86

the same as were used in laying out the rafter.

Determining Which Arm to Use.— The joint of any timber joining a roof, either horizontally or perpendicularly can be marked with the steel square, using the figures on the square that were used in framing the roof. The thing that the workman must determine is which of the arms of the square will give the cut.

Fig. 87 shows a cross section of a building that has a roof with a steep pitch on one side and a somewhat flat pitch on the other side. This layout was chosen purposely in order to bring out two things, the use of the base figure, 12, and how to tell which arm gives the cut.

The Base Figure.—The use of the base figure, sometimes puzzles the beginner, because in the ordinary pitches it is always used on the blade of the square. So long as the rise for a foot run is below 16 inches, the base figure, 12, can conveniently be taken on the blade of the square, but when the rise goes above 16 inches to the foot run, then the base figure should be used on the tongue of the square. Fig. 87, to the right shows the squares applied with the base figure taken on the blade of the square, while to the left the base figure is taken on the tongue of the square. Convenience in performing with the square should be the major factor in deciding which arm to use the base figure on. The key figure, which is the figure that represents the rise for a foot run, always determines

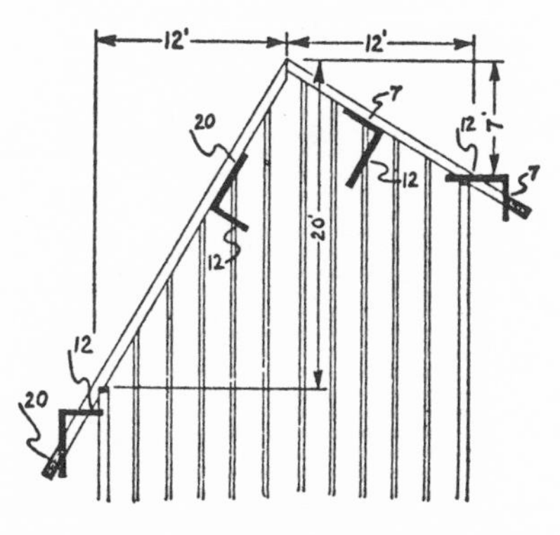

Fig. 87

the position of the square in laying out rafters.

Another thing that often puzzles the beginner is how to tell which arm of the square gives the bevel, once he has the figures to be used for the cut. This can also be shown by Fig. 87. When the point of the bevel has an angle that is below 45 degrees, which means a sharper than 45-degree angle, then the arm on which the larger of the figures is used will give the bevel for the cut. This is shown by the two squares to the left, where the key figure, 20, is taken on the blade of the square, which gives the cut. To the right the point of the bevel has an angle that is above 45 degrees, and, as can be seen, the key figure, or the smaller of the two figures, is used for marking the cut.

Chapter 10

STEPPING OFF RAFTER PATTERNS

Stepping Off Lengths of Rafters.— While there are a number of ways to obtain the lengths of rafters, this writer believes that the stepping off method, with the steel square, is the most practical. At any rate, the roof framer who uses this method and understands it, can easily prove whether or not the rafters will fit when put in place, by counting the number of steps taken. The way to do that is to mark each step on the timber where the steps join each other. If there are as many steps as there are feet in the run of the roof, the rafter will fit when put in place. Of course, it is understood that the stepping off must be done

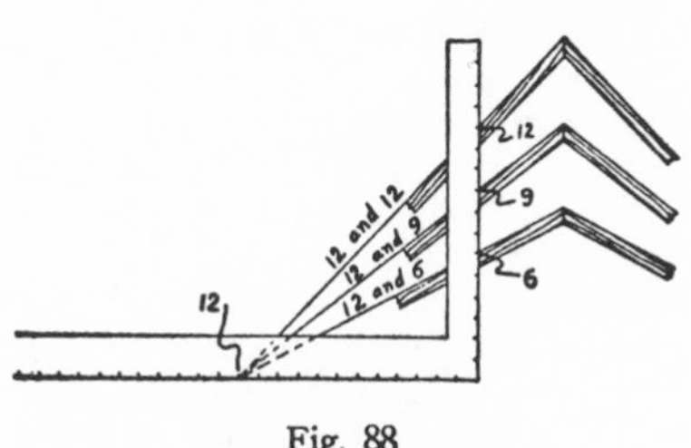

Fig. 88

carefully, and if there is a ridge board, the proper deduction must be made for it, and the cuts must be made in keeping with the pitch of the roof. What has just been said will apply in all cases of stepping off, where 12 is used as the base figure.

Base and Key Figures. — Fig. 88 shows a large square applied to three pitches, drawn to a much smaller scale. The steepest pitch shown is a 12 and 12 pitch, commonly called a half pitch, which means that for every 12-inch run there is a 12-inch rise. The next lower pitch shown is a 12 and 9 pitch, which is the simplest pitch to frame, because there are no fractions of an inch involved in the length of the rafter per foot run. The lowest pitch shown here is a 12 and 6 pitch, also called a fourth pitch. In roof framing 12 is the base figure, with some exceptions. The figure used with 12, that determines the pitch or bevel, is the key figure. For example, the run is represented by the base figure, 12, while the rise is represented by the key figure, which gives the rise in inches for a foot run.

Stepping Off Illustrated. — Fig. 89 shows how to step off a rafter having a run of 6 feet 8 inches, and a rise of 6 feet 8 inches, which means that the pitch is a 12 and 12 pitch. At the bottom are shown six runs of 12 inches, and an 8-inch fraction of a run. The

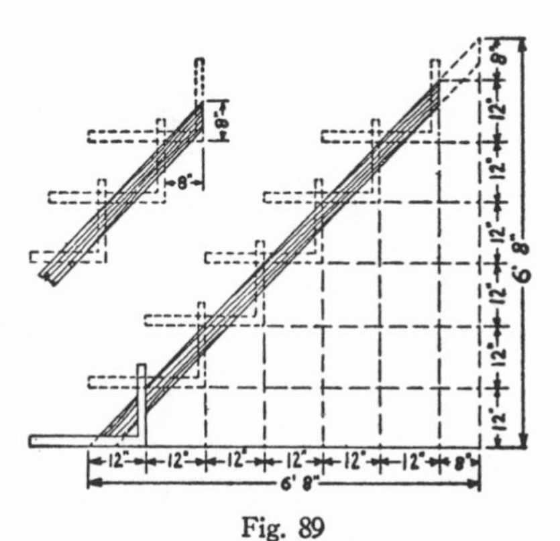

Fig. 89

rise to the right from the bottom up has six 12-inch rises and an 8-inch fraction of a rise. In stepping off the length of the rafter, each application of the square measures one 12-inch run and one 12-inch rise. The fraction of 8 inches in both the run and the rise is stepped off as shown to the upper left, where a run of 8 inches and a rise of 8 inches is taken on the square to obtain the fraction of a step. In effect this would be the same as taking an extra full step and then pulling the square back 4 inches. This will be explained more fully in another place.

Stepping Off a 12 and 6 Pitch. — Fig. 90 shows the same method of stepping off used in getting the length of the rafter, but on a 12 and 6 pitch. Six

12-inch runs and a 6-inch fraction of a run are shown at the bottom, while six 6-inch rises and a 3-inch fraction of a rise are shown to the right. Each step measures 12 inches of the run and 6 inches of the rise, as a little study of the drawing will reveal. The fraction of a step is obtained by pushing the square forward 6 inches, after the last full step has been taken. Square No. *1* is in position for the first step, square No. *2* is applied for the last full step, and square No. *3*, in a larger scale, shows how to obtain three different

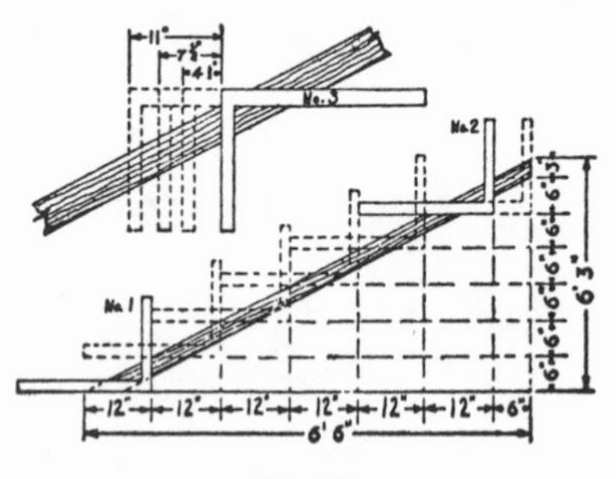

Fig. 90

fractions of a step. In this drawing, in order to economize in space, the square is in reverse or upside down order. After the square has been applied for the last full step, for a 4½-inch fraction of a step, push the square toward the heel, or to the left, 4½ inches, for a 7½-inch fraction, push it toward the heel 7½ inches, for 11 inches push it forward 11 inches, or if any other fraction of a step is to be obtained, push the square forward the required distance. Study the drawings.

Allowing for Ridge Board.—Fig. 91 shows how to deduct for a ridge board. Step off the full length of the rafter, and when the square has been applied for the last step, pull it back one half the thickness of the ridge board, which in this case would be 13/16 of an inch, as shown in figures on the drawing. To the upper left a cross section of a ridge board is shown, with two parts of rafters joining it. The dotted lines show how the thickness of the ridge board is divided into two parts, one belonging to one run and the other to the other run. This is also shown by dotted lines to the right of the drawing.

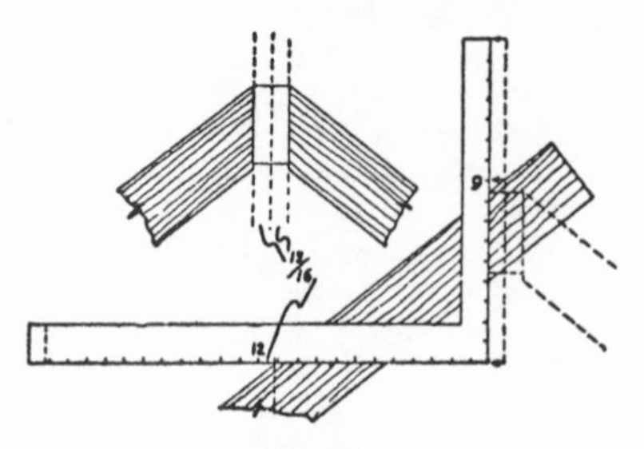

Fig. 91

The pitch in this case, as shown on the square, is a 12 and 9 pitch. The three small arrows indicate how the square is pulled from the dotted-line position to the full-line position of the square.

Three Pitches.—Fig. 92 shows a square applied to three pitches, drawn to a much smaller scale. The highest pitch, 12 and 8, is a commonly used pitch, or a third pitch. The next highest, 12 and 7, is also commonly used, especially when asphalt shingles are

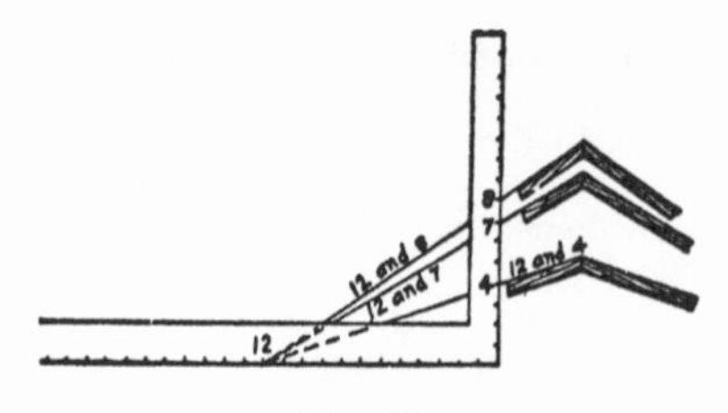

Fig. 92

used for roofing. The lowest pitch shown is a 12 and 4 pitch, which is suitable for roofs on which roll roofing is to be used.

Double Steps.—Fig. 93 shows how to take double steps with the square. This is especially suitable for stepping off rafter lengths for low pitches. Instead of taking the base figures, such as are shown in Fig. 92, and the key figure, you double each figure and use them, taking just one half as many double steps as there are feet in the run, as shown in the main drawing. To the upper left, in a little larger scale, is shown how to obtain fractions of steps. Apply the square for an extra double step, which would bring the square to position *1,* shown in part by dotted lines. If the fraction of the step is, say, 18 inches, pull the square back from position *1* to position *2,* but if the fraction is 9 inches, pull the square back to position *3*. In other words, make an extra application of the square for a double step, and then pull the square back to the point that will give the fraction of a step that is needed, and mark the cut.

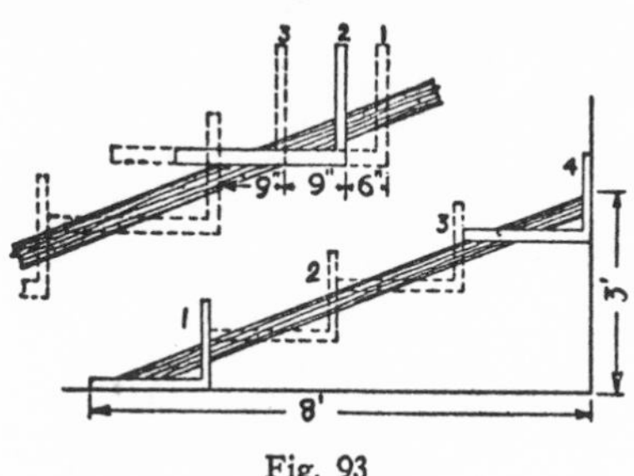

Fig. 93

Simple Stepping Off Method.—Fig 94 gives a very simple method of stepping off the lengths of rafters, especially when there are inches and fractions of inches involved, both in the run and in the rise. Let inches on the square equal feet for the run and for the rise, as shown by the large square on the drawing, and take twelve steps. The shaded small square at *1* is in position for the first step, while the shaded square at *12* is in position for the last step. The small squares, both shaded and dotted lines, show how the stepping off is done. The diagram to the upper left shows the run and the rise of the layout in feet and inches. The dotted lines show where the rafter is to be placed when cut.

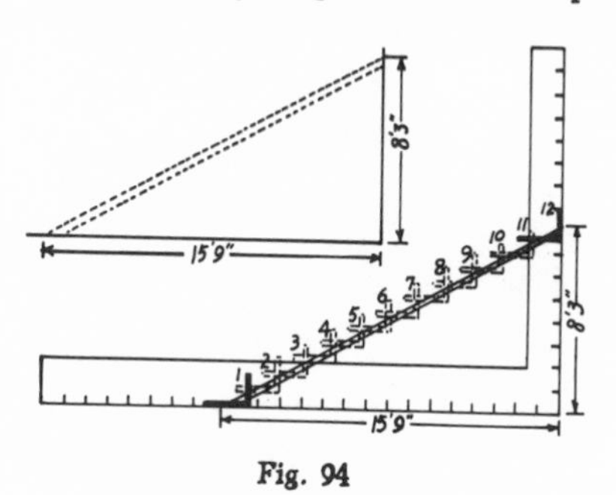

Fig. 94

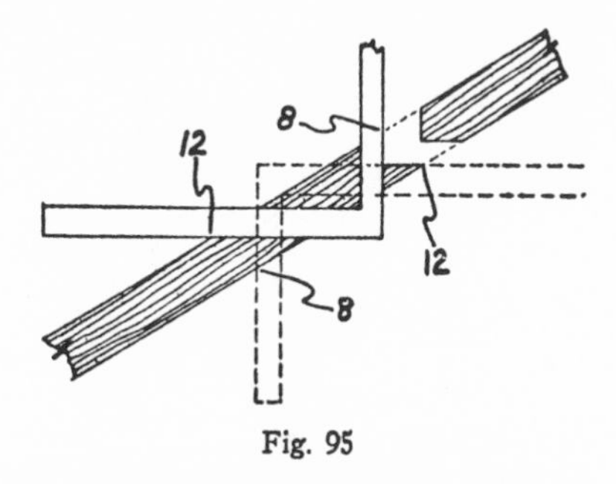

Fig. 95

Right-Angle Cuts. — Fig. 95 shows the applications of a square for a right-angle rafter cut that varies a great deal in different cases. Fig. 96, to the upper left, shows the cut in two forms used to join rafters or braces to a plate or sill, while to the bottom right, the same cuts are shown joining rafters or braces to a frame at the upper end.

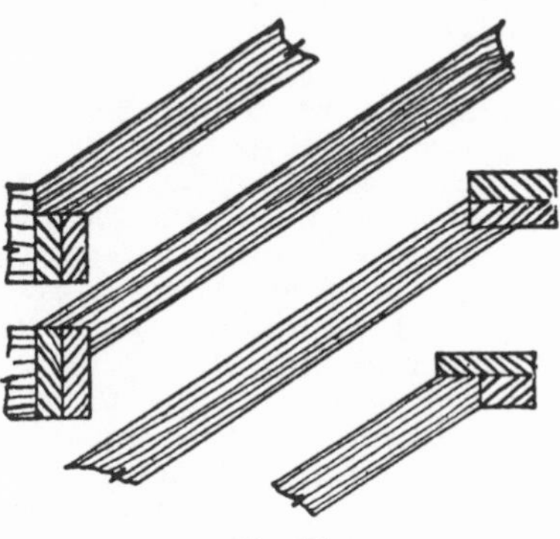

Fig. 96

Chapter 11

FRAMING HIPS AND VALLEYS

The run and the rise of the common rafter constitute the basis for getting the lengths and cuts of the hip and valley rafters. When this principle is thoroughly understood by the student, the framing of hips and valleys will be no more difficult than framing the common rafter. What makes this framing seem more difficult is the change in the base figure. Where 12 is used as the base figure in framing the common rafter, 17 is used in framing the hip and valleys rafters. To make this point clear it is necessary to show just how the 17 is arrived at.

How 17 Is Obtained.—Fig. 97 is a perspective view of a corner on a skeleton frame that, let's say, is to have a

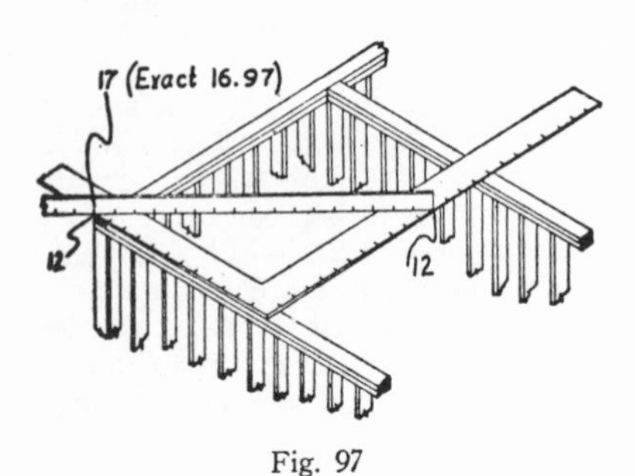

Fig. 97

hip roof. The run of this roof is to be 12 feet, and the rise 8 feet. This would mean that the figures to be used on the square for stepping off the common rafter and for marking the cuts would be 12 and 8. For the hips and valleys, 17 and 8 would be used. How the figure 17 is obtained is illustrated by the drawing, where a large square is shown laid on the plates in such a manner that the 12 on the tongue intersects the corner of the frame, while the 12 on the blade indicates the run of the common rafter. By placing a rule, as shown, from 12 to 12, it will be seen that the diagonal distance is practically 17 inches (in this case, 17 feet). The exact distance in inches is 16.97. For marking the horizontal and plumb cuts for the hips and valleys, 17 and 8 are used—for 17 is accurate enough for practical purposes.

Run and Rise.—In Fig. 98 the square shown in the previous illustration is re-

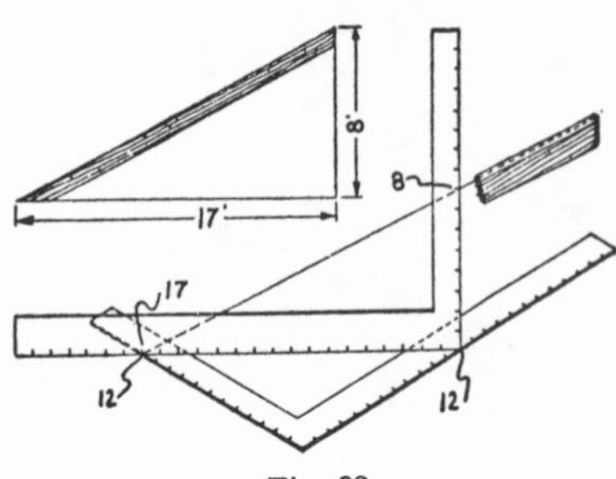

Fig. 98

produced, with the rule removed and a square in its place, giving the run, 17, of the hip and valley rafters, and the rise, 8, which is the same as the rise used for the common rafters. The diagonal line from 17 to 8 gives the incline of the rafter. To the right of this square is shown the upper end of a rafter cut to fit in between two common rafters that meet on a right angle at the comb of the roof. To the upper left is shown a side view of a rafter with a 17-foot run and an 8-foot rise,

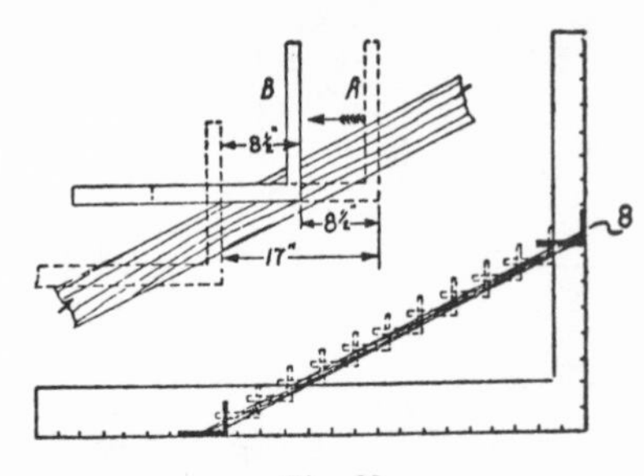

Fig. 99

which is drawn to the same scale as the squares, in the sense that inches represent feet.

Stepping Off and Fraction of Steps.—Fig. 99 shows a large square, inches representing feet, with a rafter running from 17 on the blade to 8 on the tongue. The shaded small square at 17 is in position for the first step, while the other shaded square at 8 is in position for the last step. The squares by dotted lines give the steps between the first and last steps. Each step measures 17 inches of run and 8 inches of rise. The steps should be marked at the intersections and numbered, *1, 2, 3,* and so forth. In this way the roof framer can tell when he makes the last step, the number of which should be the same as the number of feet in the run of the common rafter; in this case it would be 12. To the upper left in a larger scale is shown how to take a fraction of a step. The drawing shows a half step, or 8½ inches, which is one half of 17, or the diagonal distance of 6 and 6. To obtain this, take a full step with the square, as shown by dotted lines at *A,* and then pull the square back 8½ inches. In other words, pull the square back from position *A* to position *B* and mark the plumb cut along the edge of the tongue.

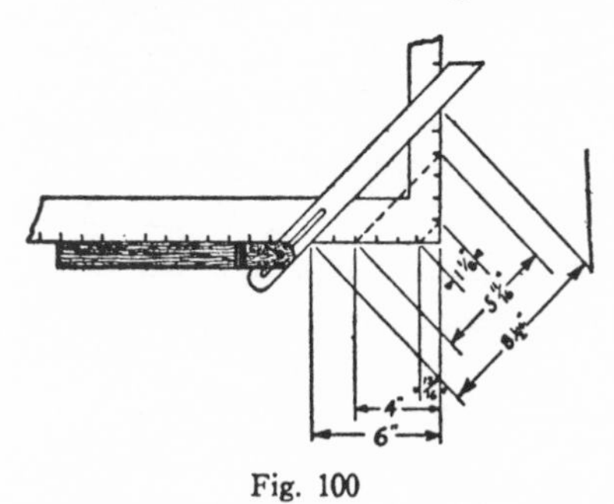

Fig. 100

Fractions of Runs.—How to get the diagonal distances of fractions of runs is shown by Fig. 100. At the bottom, right, are shown three fractions of runs, a 6-inch run, a 4-inch run, and a 1³⁄₁₆-inch run. To the right, angling, are shown the respective diagonal distances, or 8½ inches, plus; 5¹¹⁄₁₆ inches, and 1⅛ inches. How these fraction of runs are deducted with the square is shown by Fig. 101, where the square in position *A,* represents the last step. To deduct 1⅛ inches, pull the square back from position *A* to position *B,* shown by dotted lines; to deduct 5¹¹⁄₁₆ inches, pull the square back from the original position, to position *C,* and to get the 8½-inch deduction, pull it back to position *D.* The figures just

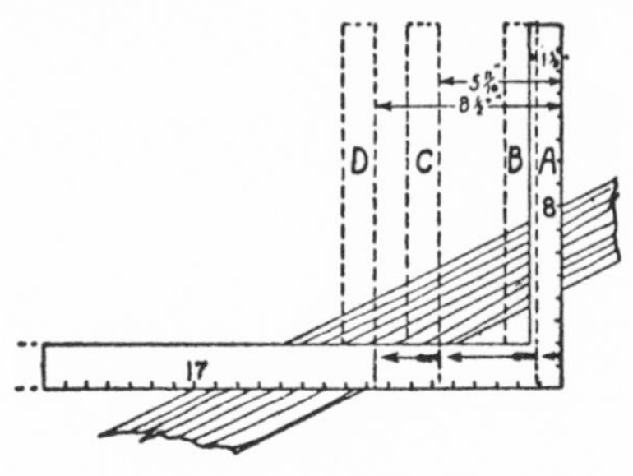

Fig. 101

used are shown at the top of the drawing in the order given. The arrows at the bottom indicate the direction the square is pulled. When the fraction of a step is to be added instead of deducted, then the square is pushed forward instead of pulled back.

Seat of Hip Rafters. — Fig. 102 shows a square in position for marking the plumb part of a seat cut for a hip rafter. The part of a square by dotted lines is in position for marking the horizontal part of the seat cut. It will be noticed that in both applications

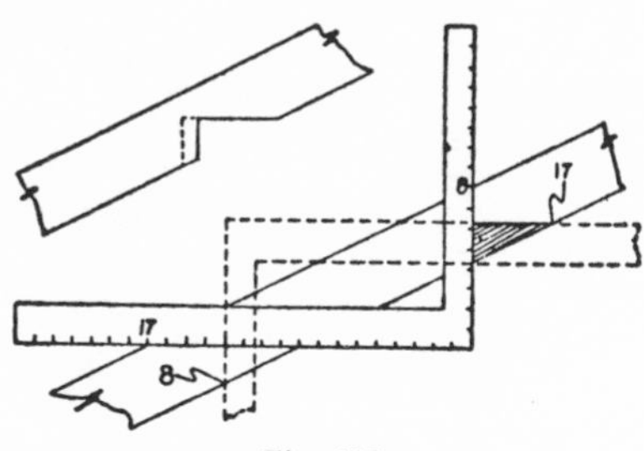

Fig. 102

17 and 8 are used respectively as base and key figures. The part to be cut out is heavily shaded. To the upper left is shown the seat cut completed. The dotted lines indicate the depth of the housing for the corner of the plate.

Cut for Tail. — Fig. 103 shows the square in position for marking the plumb cut of the tail for a hip rafter. Again the figures 17 and 8 are used. The upper drawing, which is connected with the main drawing by dotted lines,

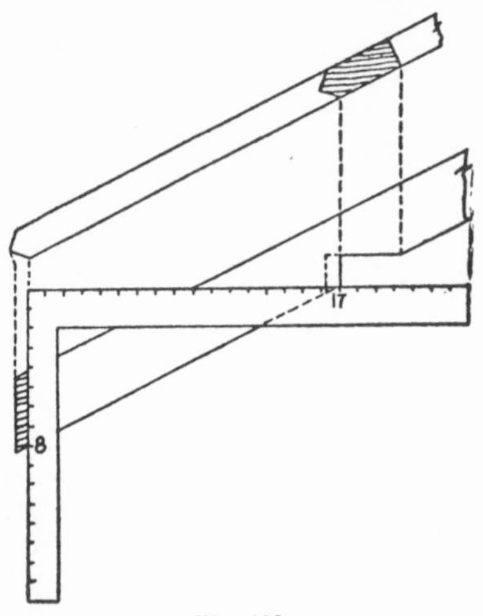

Fig. 103

shows the bottom of the rafter in part, with the seat shaded. Here the housing for the corner of the plate is clearly shown—it is also indicated on the main drawing by dotted lines.

Seat Cut for Valley Rafters.—Fig. 104 is a reproduction of Fig. 102, ex-

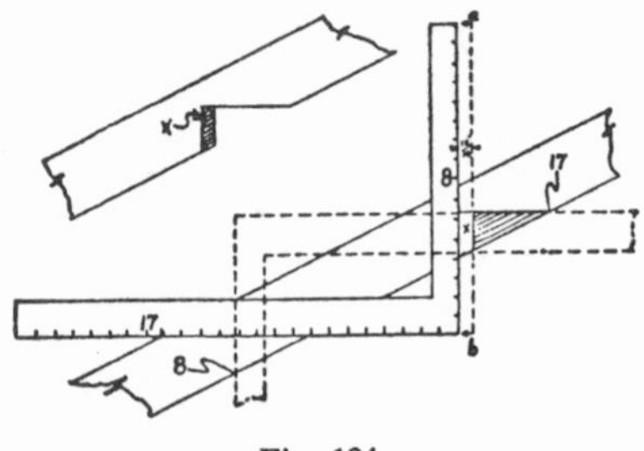

Fig. 104

cepting that the square is in position for marking the plumb part of a seat for a valley rafter, rather than for a hip rafter. The shaded part to the right of the dotted line, *a-b,* is the same as the seat for the hip rafter shown in Fig. 102. In marking the seat for a valley rafter, when you have the seat for a hip, simply pull the square back $13/16$ of an inch, or from the perpendicular dotted line, *a-b,* to the position of the square where it is now shown. The unshaded part marked *X,* when the seat is cut out, will be the V-shaped part that fits into the corner of the plate. This is pointed out on the drawing to the upper left at *X.* The part of a square shown by dotted lines in Fig. 104 is in position for marking the horizontal part of the valley seat.

Difference in Valley and Hip Seats. —In case the order is reversed, that is,

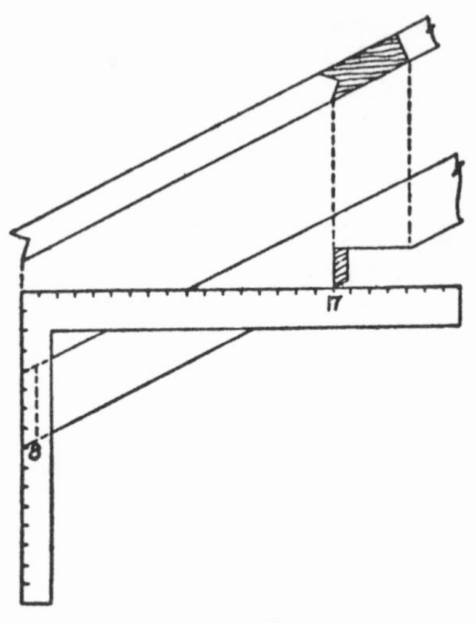

Fig. 105

you have the seat for the valley rafter, as shown in Fig. 104, to get the plumb part of the seat cut for the hip, you push the square forward $13/16$ of an inch, or from the position of the square shown, to the perpendicular dotted line, *a-b,* which will give you the plumb cut of the hip seat. The unshaded part marked *X,* will have to be housed out to receive the corner of the plate. In simple language, when you have the seat cut for the hip rafter, pull the square back one half the thickness of the rafter material, to get the plumb cut of the valley seat. But if you have the seat cut for the valley rafter, push the square forward one half the thickness of the rafter material, and you will have the plumb cut of the seat for the hip. What has been said here about the seat cut for valley rafters applies only to valley rafters that are not backed.

Marking Tail of Valley Rafters.—Fig. 105 shows the square applied for marking the end of the tail for a valley rafter. The same base and key figures are again used, or 17 and 8. The upper drawing shows the bottom of the rafter, which is connected with the main drawing by dotted lines. A good view of the seat, shaded, and the end of the tail, V-shaped, are given here. The depth of the V-shaped end is indicated on the main drawing by dotted line.

Chapter 12

TANGENT AND EDGE BEVELS

Simplicity of All Cuts.—The level and plumb cuts for hips and valleys, that were treated in the last chapter, are the cuts that are marked on the sides of the rafter material, and are as simple as the cuts for common rafters. But hips and valleys usually join decks, ridge boards or common rafters, and therefore must have bevels marked on the edges to make them fit where the joint is made. While those bevels, or edge bevels, as they are called, might seem difficult to obtain with the framing square, they are just as simple as any of the cuts that have been treated in previous chapters—that is, if you

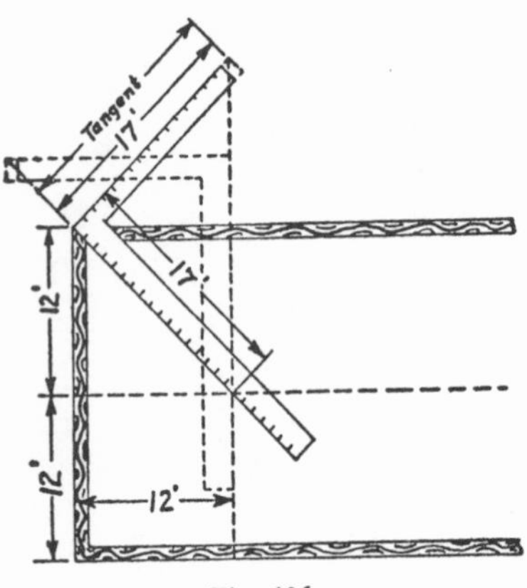

Fig. 106

know for sure why you do what you do in getting these bevels.

The Tangent. — Fig. 106 shows a plan of one end of a building that is to have a hip roof. The width of this building is 24 feet, or twice 12 feet, as shown by figures. The run of the hip rafter is on a 45-degree angle, as shown by the blade of the square. In this case inches represent feet. The diagonal distance of 12 and 12 on the square is 17 inches, or to be exact, 16.97 inches. On the drawing it is given as 17 feet, which is the run of the hip rafter. Now to get the edge bevel of a hip rafter that has no pitch, that is, if the rafter were on a level, you would take 17 on the blade of the square and what is called the tangent on the tongue, which, as shown on the drawing is also 17, and place the square on the rafter material, as shown by the dotted-line square, to get the edge bevel. The rafter material would run in the direction of the edge of the blade of the main square, and extend beyond the corner of the building, as shown by the dotted line. The blade of the square gives the bevel.

The student, perhaps, has already discovered that 17 and 17 on the square will give a 45-degree bevel, and therefore the cut can be marked by any equal figures — 12 and 12 are often used. Some carpenters use 16 and 16

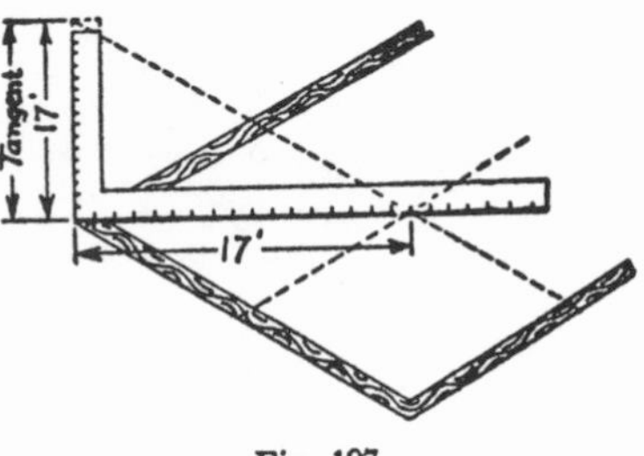

Fig. 107

on the square, to mark the 45-degree bevel, locating the tongue of the square with the feel of the fingers, and adjusting the blade by bringing it to the 16-inch mark.

Tangent in Perspective.—What has been shown in plan in Fig. 106, is shown in perspective in Fig. 107, where the run of the hip rafter is again shown as 17 feet, and the tangent on the tongue is also 17 feet. Now imagine that the tongue, which represents the tangent, forms a sort of hinge and you were to lift the blade of the square up enough to give it a rise of 8 feet, as shown by Fig. 108. This will show why the length of the hip rafter and the tangent will give the edge bevel. The distance on the blade of the square is 18¾ feet, or the length of the hip rafter. To get the edge bevel for a hip or valley rafter, take the length of the rafter and the tangent, or in this

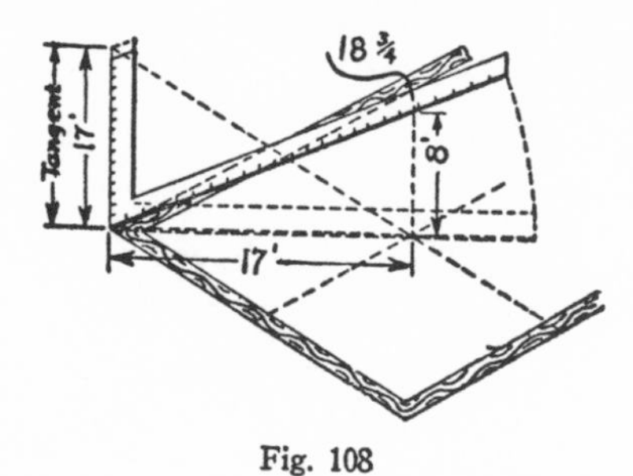

Fig. 108

case, 18¾ on the blade of the square, and 17 on the tongue, and apply the square to the rafter material as shown by the dotted-line square in Fig. 106. The blade of the square will give the bevel. Fig. 109 shows the square applied to the line *a-b*, as the square should be applied to the timber for obtaining the bevel. These figures, 18¾

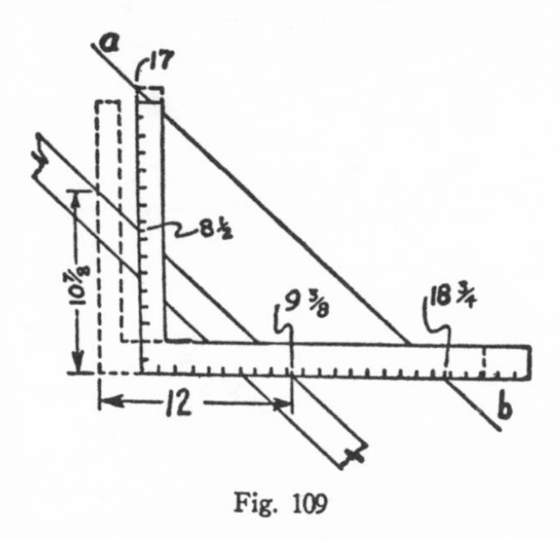

Fig. 109

and 17, are inconvenient and in practice they are used merely as starting points. The roof framer simply divides each of the two figures by 2, giving him 9⅜ and 8½, which he takes on the square and applies it as shown by the drawing toward the left. The blade gives the bevel. Some roof framers carry this process one step farther — they push the square forward until 12 on the blade intersects the edge of the timber, using it as the base figure, and taking the figure shown on the tongue as the key figure. This is shown by the dotted lines to the left. In other words, to get the edge bevel, the roof framer takes 12 on the blade of the square, and 10⅞ on the tongue —the blade gives the bevel.

Marking Edge Cuts.—Fig. 110 shows a main square and a dotted-line square applied to the timber for marking the edge cut of a hip or valley rafter, using 12 on the blade and 10⅞ on the tongue, as given in Fig. 109. It will be noticed that the edge cut takes two applications of the square in this case. The cut

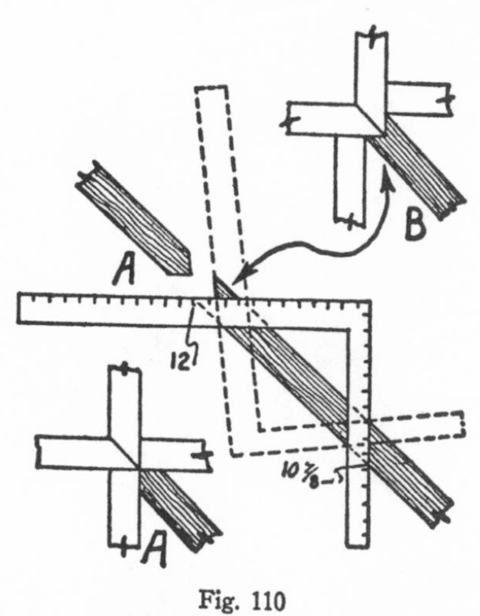

Fig. 110

shown to the upper left, at *A*, is the edge cut of a hip or valley rafter that will fit into an angle such as is shown at *A*, to the bottom left. At *B* the same edge cut in reverse is pointed out with the double indicators. This edge cut is necessary when the rafter joins a corner of a deck, as shown to the upper right.

Use of Bevel Square.—While it is good practice for the beginner to mark the edge bevels with the steel square, even though it takes a little more time, the experienced roof framer simply uses the square to obtain the bevel,

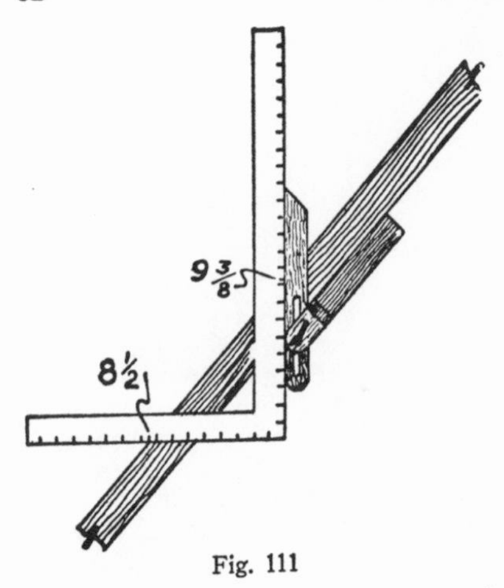

Fig. 111

and then he sets his bevel square and uses it to do the marking. Fig. 111 shows how the square is applied to the edge of the rafter material to get the bevel—it also shows how the bevel is transferred to the bevel square, which is then used for marking the edge cuts. Fig. 112 is a detail, showing two applications of a bevel square for mark-

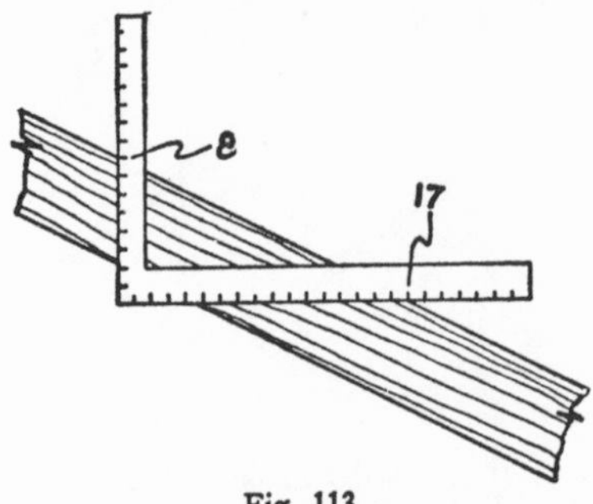

Fig. 113

ing the edge cuts of hip and valley rafters. The plumb and level cuts of hips and valleys are made by using 17 on the blade and the rise on the tongue, which in this case is 8, as shown by Fig. 113.

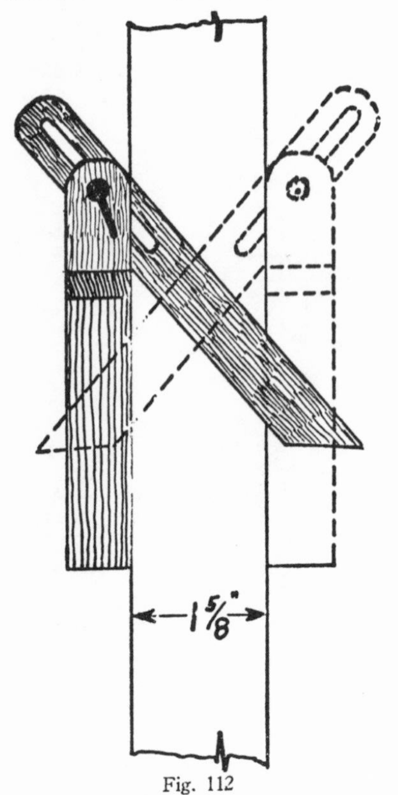

Fig. 112

Single Bevel and Square Edge Cuts for Hips and Valleys. — Fig. 114 is a plan of a ridge and two valley rafters in part. At *A* the square is applied for marking the edge bevel of a valley

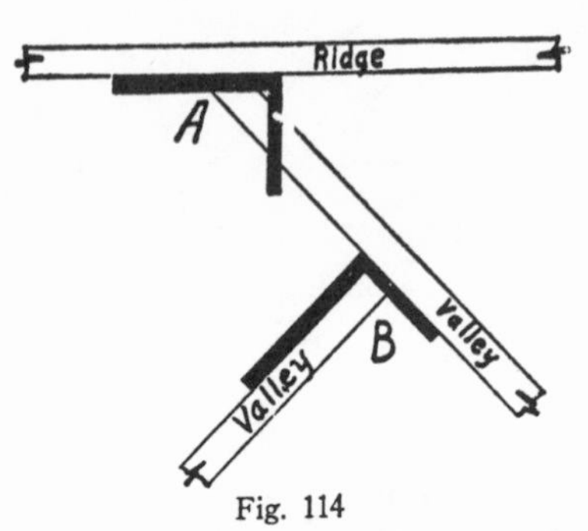

Fig. 114

rafter that joins a ridge board, while at *B* the square is shown applied for marking the edge for the top cut of a valley rafter that joins another valley rafter. It will be noticed that this cut is square across, due to the fact that the short valley joins the long valley at a right angle, as shown in the plan.

Chapter 13

USE OF TANGENT EMPHASIZED

Tangent Avoided.— In all of the previous treatments of roof framing or the steel square, this writer has purposely avoided the use of the term, "tangent," excepting perhaps two or three times. The average carpenter does not use the word when he frames hip and valley roofs. Perhaps because of the old rule for obtaining the edge bevel, which is:

Take the run of the rafter (17 for hips and valleys, and 12 for jacks) and the length of the rafter on the square —the length of the rafter gives the cut.

This rule will work in all cases involving a regular hip or valley roof. But when it comes to the irregular hip or valley roofs, it will not work. The reason is that it is a misstatement. It works in a regular hip or valley roof, because in such roofs the run of the rafter and the tangent are always the same. The rule that will work on both regular and irregular hip and valley roofs is this:

Take the tangent on one arm of the square, and the length of the rafter on the other — the latter gives the edge bevel.

The term tangent is a sort of scare word to many carpenters, although it is a legitimate term, and when properly understood it will simplify hip and valley roof framing. Since this is true, the term tangent will be used in all cases where it will simplify the problem.

Tangent Explained.—Fig. 115 shows a drawing of one end of a building that is to receive a hip roof. A large square, on which inches represent feet, is applied in such a way that the tangent comes on the tongue of the square, and the run of the rafter is shown on the blade. It will be noticed that in this case both the tangent and the run are shown as being 12 feet. This is true, because the roof is a regular hip roof. The diagonal distance between 12 and 12 is the run of the hip, and is given as 17 feet (16.97) which will serve all practical purposes. Now if a perfectly level roof, or pitchless roof, were to be framed like a hip roof, the edge bevel of the rafters joining the hip or diagonal timber, would be made by using

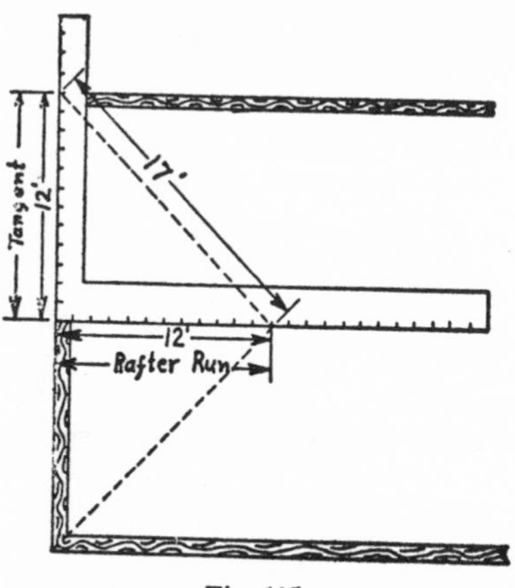

Fig. 115

12 and 12 on the square. In this instance either arm of the square will give the bevel, but just as soon as the framing involves some pitch, the bevel must be made with the arm on which the length of the rafter is used, which is usually the blade of the square.

Perspective Views of Layout.—Fig. 116 shows a perspective view of the

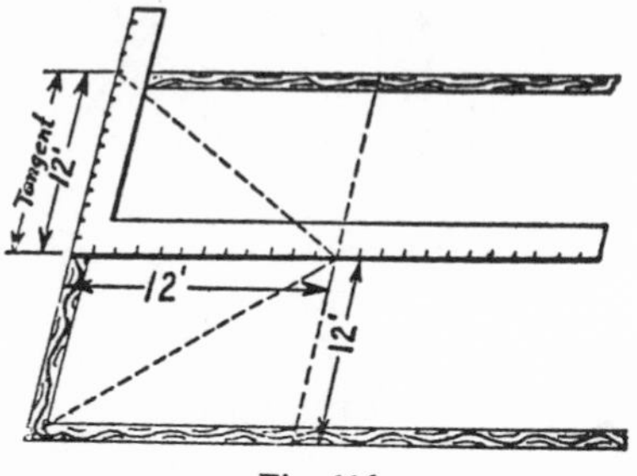

Fig. 116

same layout, in which the square is shown out of proportion, because of the perspective viewpoint. If the student will imagine the square hinged at the outer edge of the tongue, and the blade lifted enough to give it an 8-foot rise, he will see why the length of the rafter is used in getting the edge bevel of jack rafters. This is shown by Fig.

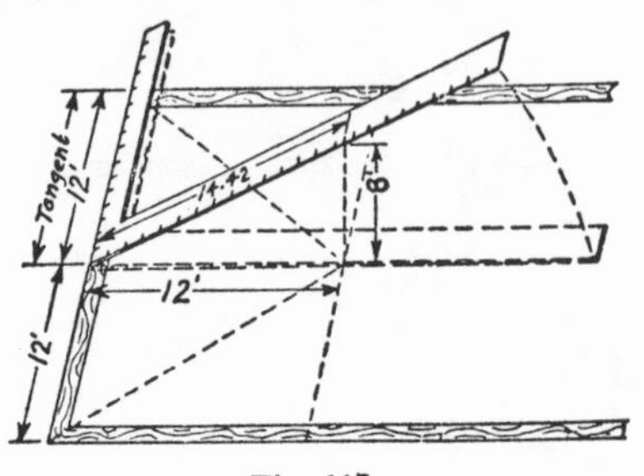

Fig. 117

117. Here the tangent and the run of the rafter are still both 12 feet, but the figures to be used on the blade of the square in order to get the edge bevel, have increased from 12 to 14.42. To obtain the edge bevel, take the tangent

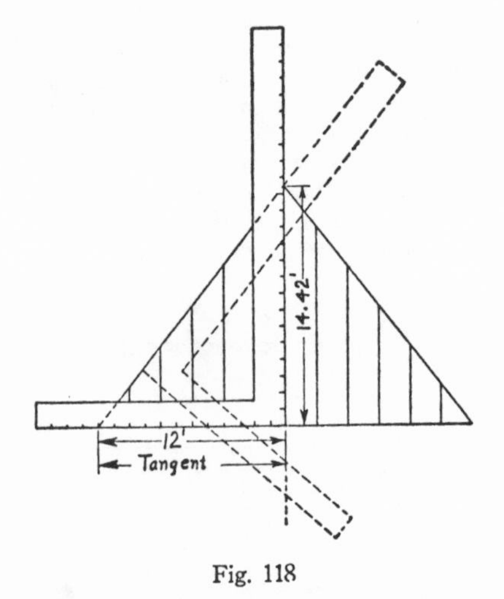

Fig. 118

on the tongue of the square, and the length of the rafter on the blade—the blade will give the bevel. While the squares in these illustrations are shown with the inches representing feet, in applying the square to obtain the cut, inches are always used, in the sense that the framing is done on a basis of a one-foot run of the roof.

Principle of Obtaining Edge Bevel.—Fig. 118 illustrates the principle on which is based the rule for obtaining the edge bevel for jack rafters. Here is shown a diagram of the end of the roof, looking straight at it, or perpendicular to the surface. The tangent is still 12 feet, but the length of the rafter is 14.42 feet, or reduced to the basis of one-foot run, we would have a tangent of 12 inches and a rafter of 14.42 inches. In practice the figures would be 12 on the tongue of the square and 14⅜, plus, on the blade—the blade of the square will give the bevel. The square, however, would be applied, as shown by the dotted-line square.

Edge Bevel for Jacks.—Fig. 119 shows the end of the roof, looking straight at it, with the rafters drawn

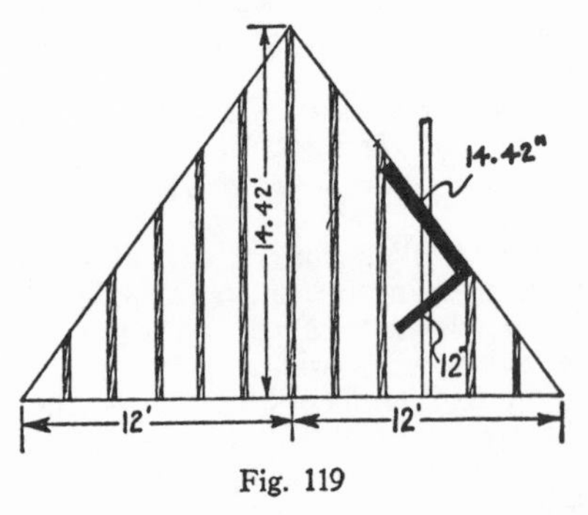

Fig. 119

in. Here the shaded square, at a larger scale, is shown applied to a rafter for marking the bevel. The tangent, or 12, is taken on the tongue of the square, and 14.42 inches is taken on the blade—the blade will give the bevel. Fig. 120 shows a side of a roof, full view, that joins a valley. The shaded square is applied here to mark the edge bevel for a valley jack, using the tangent, or 12, on the tongue, and 14.42 on the blade—the blade gives the bevel.

Obtaining Edge Bevel, Practical Way.—A practical way of getting the edge bevel for hips, valleys, or jacks,

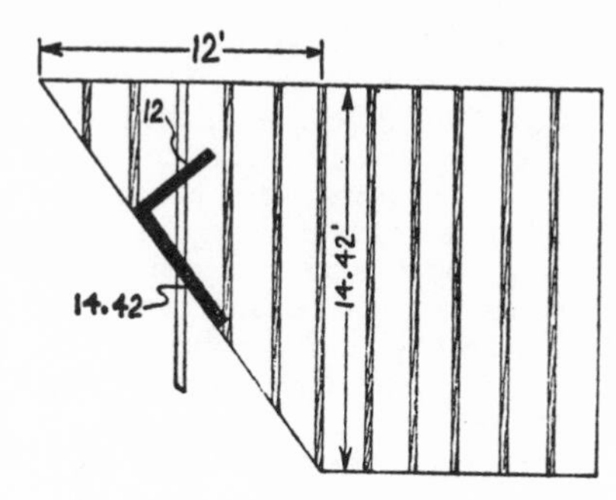

Fig. 120

is shown by Fig. 121. Here is shown a piece of rafter material with a plumb cut on it, for a rather steep roof. The first operation is to make the mark for a plumb cut on the material, as between *d* and *e*. Then apply the square as shown, making *a-b* equal to the thickness of the rafter material, and square across from *b* to *c*. Now join *c* with *d*, and the bevel *c-d* is the edge bevel you want.

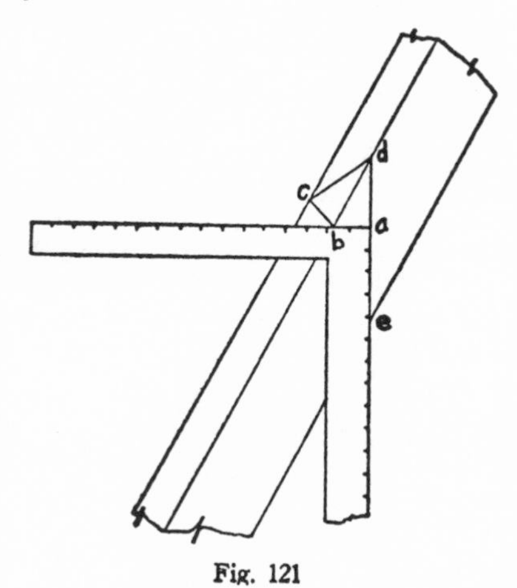

Fig. 121

Edge Bevel, Easy Way. — Another practical way of obtaining the edge bevel of hips, valleys, and jacks, is shown by Fig. 122. On a piece of rafter material, cut a foot or seat cut of a rafter, as shown by the drawing. Then with the square make a 45-degree mark on the cut, as from *a* to *b*. From point *a* square up to *d*, and from *b* square up to *c*. Now join *c* with *d*, and *c-d* is

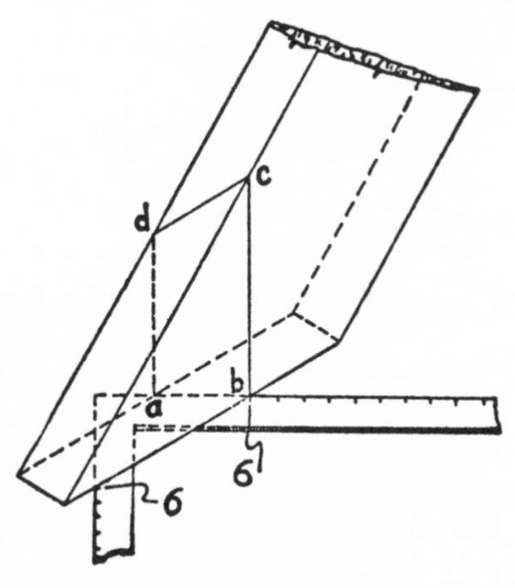

Fig. 122

the bevel you want. The square, as shown, is out of proportion, due to the perspective view, but the figures show that 6 and 6 were used in making the 45-degree mark on the foot cut. The experienced roof framer would probably set a bevel square to a 45-degree angle and use it for making the mark.

The two practical methods just explained, of obtaining edge bevels, are accurate, that is, if the workman is accurate in handling the square and in making the marks and cuts.

Chapter 14

DIFFERENCE IN LENGTHS OF JACKS

Difference in Lengths of Jack Rafters.—The difference in the lengths of jack rafters, both valley jacks and hip jacks, is based on the run and the rise of the common rafter. In case of a hip roof in which the common rafter has a 9-inch rise to the foot run, the length of the rafter for each foot run would be 15 inches. Then, if the jack rafters were spaced 12 inches on center, the difference in the lengths of the jacks would be 15 inches. If the spacing were 16 inches on center, then the difference in length would be 20 inches, or if the commonly used 2-foot spacing were used, then the difference would be 30 inches. Using a 9-inch rise for a foot run, simplifies the whole matter, because this is the simplest pitch in roof framing. So far as the cuts are concerned, they can, with a few exceptions, all be made without the use of fractions.

Difference in Length for Third Pitch Jacks.—Fig. 123 shows the square applied, using 12 on the blade and 8 on the tongue, which is the run and rise of

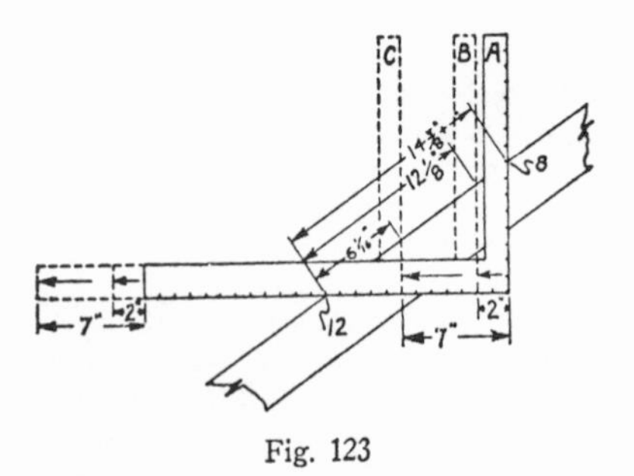

Fig. 123

a third pitch common rafter. In case the rafters for a hip roof are spaced 12 inches on center, the difference in the lengths of the jacks would be the diagonal distance between 12 and 8, or $14\frac{3}{8}$ inches, plus. The exact figure is 14.42 inches. If the square were pulled back 2 inches, or from position *A* to position *B*, the diagonal distance would be $12\frac{1}{8}$ inches, as shown, and would represent the difference in the lengths of jacks spaced 10 inches on center. But if the square were pulled back from position *A* to position *C*, then the spacing would be 5 inches, and the difference in the lengths of the jacks would be $6\frac{1}{16}$ inches. The spacings used in this illustration are rarely used in practice.

Commonly Used Spacings.—Fig. 124 shows the square applied, using 12 on

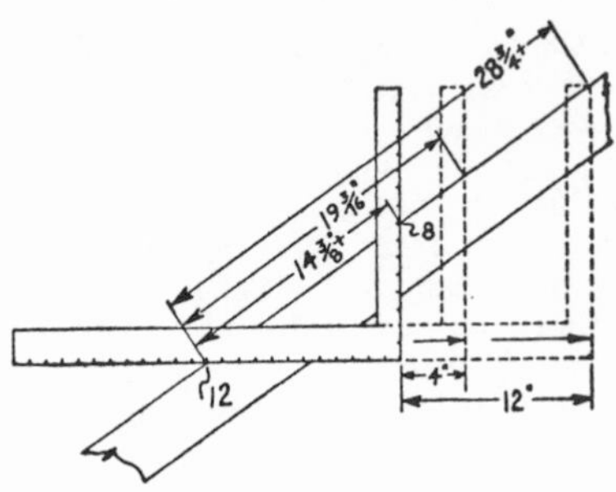

Fig. 124

the blade and 8 on the tongue. Here as in the previous figure, the difference in the lengths of jacks, for 12-inch spacings would be $14\frac{3}{8}$ inches, plus. But if the spacing were 16 inches, then the square would be pushed forward, or to the right, 4 inches, which as shown, would make the difference in the lengths of jacks $19\frac{3}{16}$ inches. Or in case the square is pushed forward 12 inches, making the spacing 2 feet, then the difference in the lengths of the jacks would be $28\frac{3}{4}$ inches, plus, as shown by the upper figures.

Rises above Eight Inches.—Because 16 inches is the standard length of the tongue of a square, the method of obtaining the difference in the lengths of jacks for 2-foot spacings, shown in the previous figure, can only be used conveniently when the rise per foot is 8 inches or less. When the rise is above 8 inches, then it is better to take two steps for obtaining the difference in the lengths of jacks for 2-foot spacing.

This is illustrated by Fig. 125. Here the square is applied, using 12 on the blade and 9 on the tongue, which as shown, would have a rafter length of 15 inches. By moving the square from position *A* to position *B*, you will have

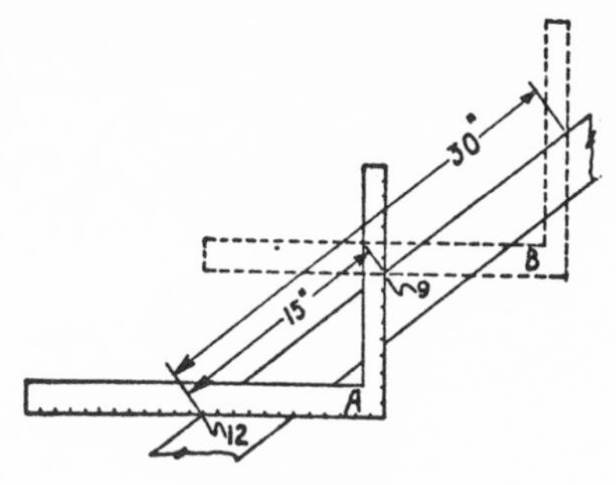

Fig. 125

two steps, which will give you the difference in length of jacks spaced 2 feet, or 30 inches. It will be observed that the difference in length of jacks for a regular hip or valley roof, is always the length of the rafter for a run, equal to the distance of one space from center to center. This is not true in irregular plan or in irregular pitch roofs.

Irregular Plan.—Fig. 126 shows a square applied to a sharp corner of an

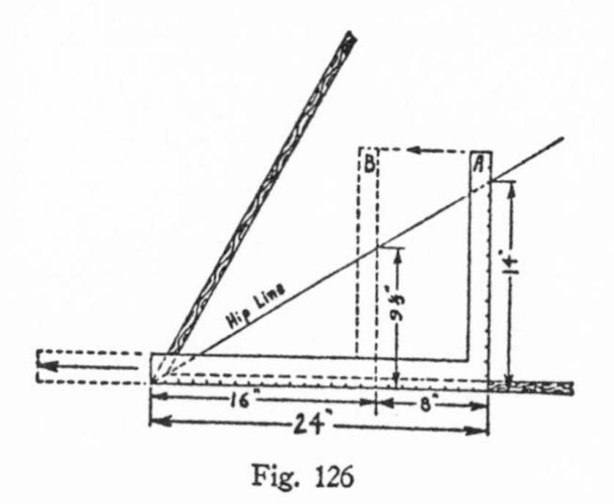

Fig. 126

irregular plan of a hip roof. The plan in this and other cases is drawn to a much smaller scale than the scale used for the square. The purpose here is to obtain the run of the jack rafter for a single space. For instance, if the rafters were spaced 2 feet on center, then the run of the jack rafters for a single space would be 14 inches, as shown to the right of the square marked *A*. Now if the square were pulled back from position *A* to position *B*, or 8 inches, it would give the run for a 16-inch space, or as shown, 9⅓ inches. The reason it is necessary to obtain the run of the jacks for a single space is that the difference in the lengths of the jacks is the same as the length of the rafter for the relative run of a single space. This is illustrated by Fig. 127, where the square is shown applied, using 12 on the blade and 9 on the tongue, which of course gives the length of the rafter for 1-foot run, or 15 inches. But the

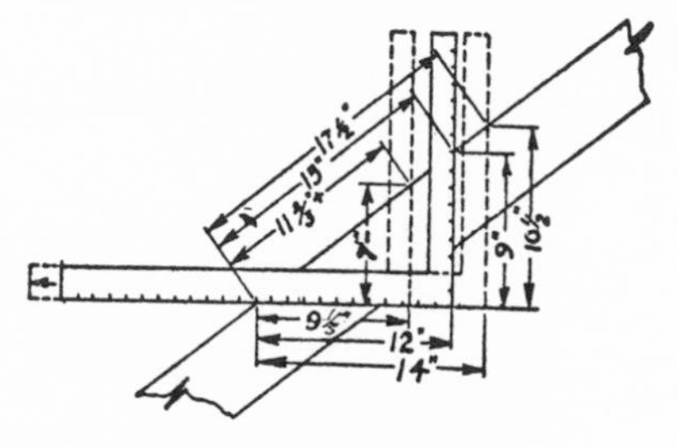

Fig. 127

runs that are shown in Fig. 126, are 14 inches for a 2-foot space, and 9⅓ for the 16-inch space. Now the drawing shows that if the square is pushed forward to 14 inches, the difference in the lengths of the jack rafters would be 17½ inches, as given in figures. But if the square is pulled back to 9⅓ inches, then the difference in the lengths of the jacks would be 11⅔ inches, plus. The drawing in Fig. 126 should be studied with the drawing in Fig. 127. The tongues, *A* and *B*, in Fig. 126 give the two respective runs shown on the blade of the square in Fig. 127, where the tongues, indicating the two positions, are shown by dotted lines. The position shown by the 12-inch run and the 9-inch rise is the starting, or base position for determining the pitch of the rafter. In irregular plan hip roofs, the jacks for each corner are to be cut in pairs, and unless the corners have the same angle, the difference in the lengths, and the cuts, must be determined for each corner separately.

Irregular Pitch.—The principle that was used in irregular plan hip roofs, will apply here. That is, the lengths of the rafter for the respective runs for a single space is the difference in length of the jack rafters. Fig. 128

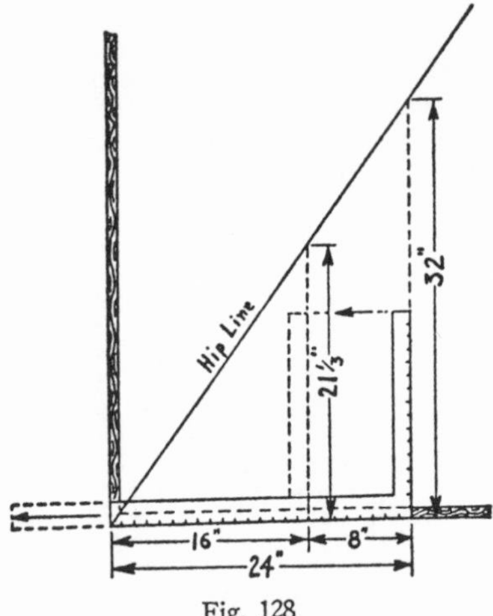

Fig. 128

shows the square applied for a 2-foot space, which gives the respective run for the jack rafter, or 32 inches. If the square is pulled back 8 inches, as shown, it would give the respective run for a 16-inch space, or as shown, 21⅓ inches. How to obtain the difference in the lengths of the jack rafters is illustrated by Fig. 129. Here the square marked *A* is applied, using 12 on the tongue of the square and 9 on the blade. This gives the pitch of the rafter. In Fig. 128 the square is applied for obtaining the run for a 24-inch space, which, as shown, is 32 inches. Hence to get the difference in the length of the jack rafters for a 2-foot spacing, the square would have to be moved from position *A* to position *C*, which shows a diagonal distance of 40 inches, or the difference in the lengths of jack rafters spaced 2 feet on center. In case the spacing were 16 inches, as also shown in Fig. 128, the run would be 21⅓ inches, so the square would have to be pulled back from position *C*, Fig. 129, to position *B*, which would result in a diagonal distance of 26⅓ inches,

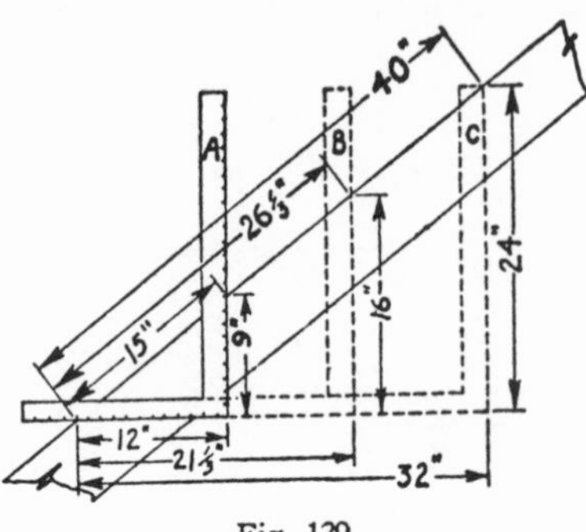

Fig. 129

or the difference in the lengths of jack rafters spaced 16 inches on center. Figs. 128 and 129 should be studied carefully, keeping in mind that the former shows how to obtain the run of single spaces, while the latter shows how to obtain the difference in the lengths of the jacks, for such runs.

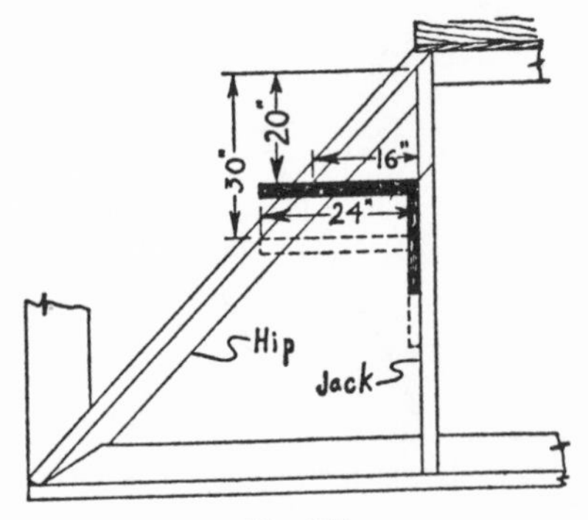

Fig. 130

A Practical Method.—Fig. 130 shows a practical method for obtaining the difference in the lengths of jack rafters. Here is shown a perspective view of a hip rafter with the longest jack rafter in place. The shaded square is shown applied for 16-inch spacing;

that is, the right angle distance from the jack rafter to the corner of the hip is 16 inches. Assuming a 12 and 9 pitch, the difference in the lengths of the jacks would be 20 inches, as shown on the drawing. The dotted-line square shows the space, or right-angle distance from the jack to the corner of the hip, as 24 inches, which on the basis of a 12 and 9 pitch would make the difference in the lengths of the jacks 30 inches, as shown. This method of obtaining the difference in the lengths of jack rafters will work on all hip or valley roofs, including irregular pitches and irregular plans.

Bevel Square for Marking. — Fig. 131 shows to the left how to transfer the edge bevel from the steel square to a bevel square. The pitch is a 12 and 9 pitch, for 12 and 15 are used for marking the bevel. To the right is shown how the pattern is marked for each length of jacks. In this case the difference in the lengths is 20 inches, indicating that the pitch is a 12 and 9

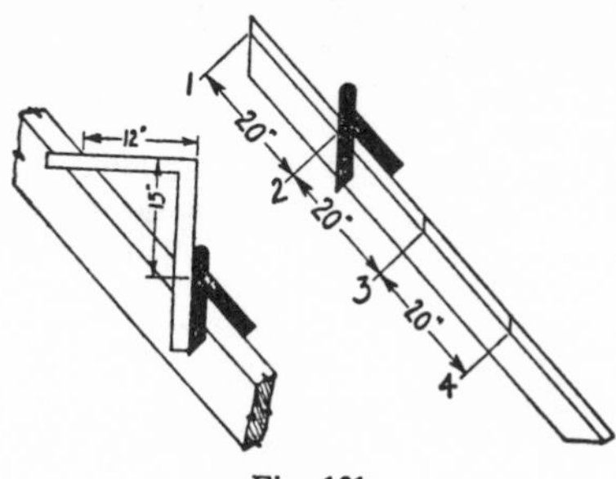

Fig. 131

pitch, spaced 16 inches on center. The edge bevel of each length should be marked on the pattern, as a guide for the roof framer in laying out the jacks.

Chapter 15

IRREGULAR PLAN ROOF

Irregular Plan.—The irregular plan roof is a sort of link between a regular roof and an irregular pitch roof. That is to say, the framing has many things in common with regular pitch roof framing, and at the same time it lays the foundation for irregular pitch roof framing, especially irregular hip roof framing. It is therefore important for the student to be sure that he understands each of the problems as he goes along, before he takes up the next one. In this way he can eliminate many of the difficulties that he might otherwise encounter.

Simple Diagram. — Fig. 132 shows the roof plan that will be used throughout this chapter as a basis. It is simple,

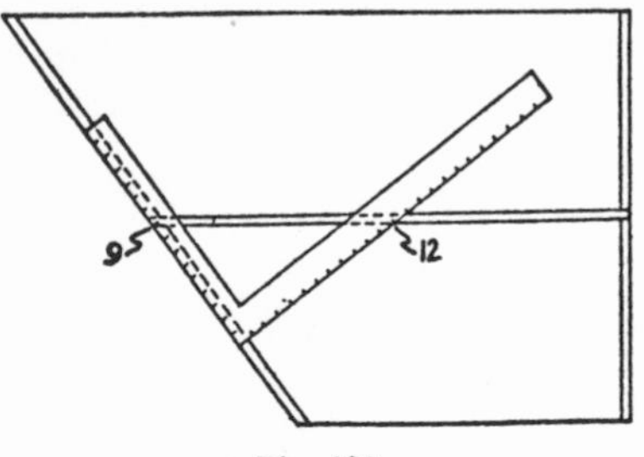

Fig. 132

and will be used largely as a diagram. The left end of the roof, as shown in plan, has a beveled shape, which is laid out by using 9 on the tongue of the square and 12 on the blade. These figures are the same as the figures to be used in marking the bevel on the left end of the ridge board. In fact, the square is shown in position for marking the left bevel of the ridge. It should be pointed out here that inches on the squares shown in these illustrations represent feet in the drawings. For example, the blade of the square is 24 inches long, or 2 feet, but the length of the blade of the square on the drawing would represent 24 feet.

The Tangent.—The word tangent is a geometrical term, and Fig. 133 shows how it is related to a circle. In actual practice, the circle is imaginary, or unseen. The roof framer is concerned with two factors in practically every cut that he marks. These factors are a point on the body and a point on the

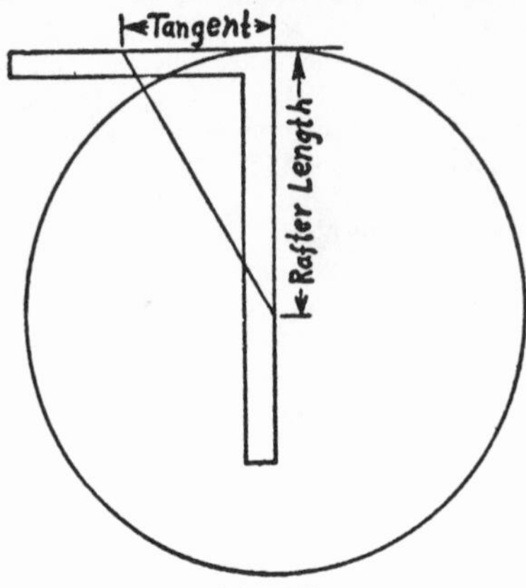

Fig. 133

tongue of the square, which represent the rafter length and the tangent. Many roof framers never use the word tangent when they lay out a roof. But whether they do or do not, the tangent is always present, and when the term

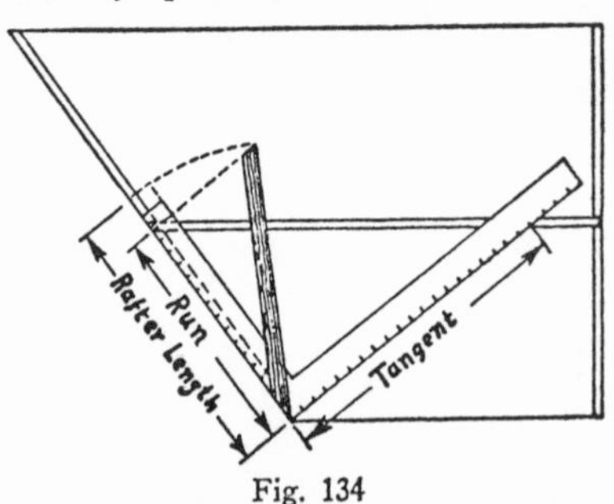

Fig. 134

is understood in the practical sense, or as it is used in roof framing, it really simplifies the work.

Edge Bevel of Irregular Gable Rafters.—Fig. 134 shows the square applied for obtaining the rafter length and tangent for marking the edge bevel of one of the irregular gable rafters. The rafter is shown, shaded, as if it were lying on the side. The curved dotted line shows how the length of the rafter is transferred to the line that contacts the edge of the tongue of the square, as shown on the diagram. Now to obtain the bevel, take the rafter length on one arm of the square, and the tangent on the other—the rafter length giving the bevel. It will be noticed that the tongue is not long enough to hold the length of the rafter. In such cases, divide both the rafter length and the tangent by the same figure, say, 2, or 3, or 4. In most cases, dividing by 2 will reduce the figures so they can be used conveniently on the square.

Edge Bevel of Jacks.—Fig. 135 shows the square applied for obtaining the points for marking the edge bevel of the jacks on the irregular end of the roof. A common rafter is shown, shaded, as if it were on the side. The dotted curved line shows how the rafter length is transferred to the blade of the square. Now, the rafter length taken on the blade and the tangent on the tongue, as shown, will give the edge bevel of the jacks—the blade giving the bevel.

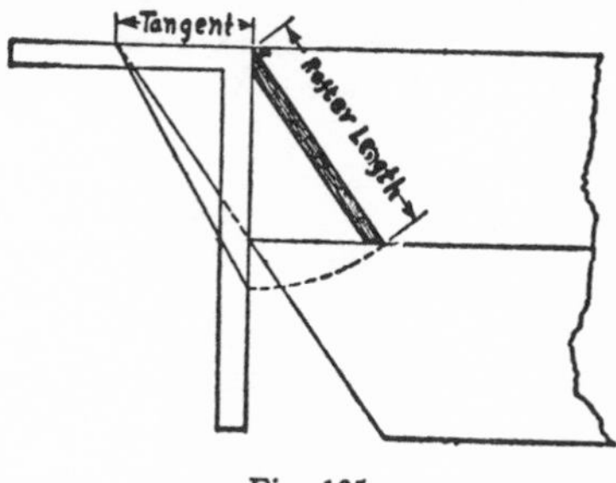

Fig. 135

Bisecting with the Square.—Fig. 136 shows how to bisect the sharp corner of the diagram with the square. While any figure on the square will do, the figure 12 is used in this case, and the two squares are applied so that the figure 12 of each square will contact the sharp corner of the diagram. A line is then drawn from the sharp corner of the drawing through the point where the tongues cross and on to the center of the drawing, as shown by dotted lines. This line represents the

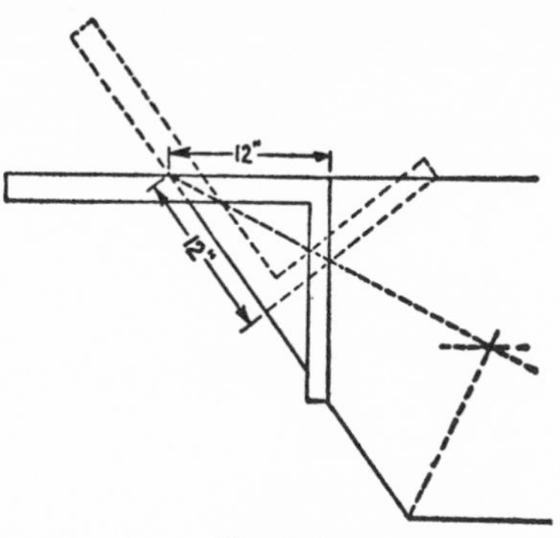

Fig. 136

run, or base line of the hip rafter. For bisecting the dull corner, measure from the corner to points *a* and *b*, say, 12 inches, as shown in Fig. 137. Then draw a line from *a* to *b*, and apply the square as shown to the left, making the tongue intersect the angle, as shown, which will bring the heel half-way between points *a* and *b*. Now apply the dotted-line square, and you have the base line of the short hip rafter.

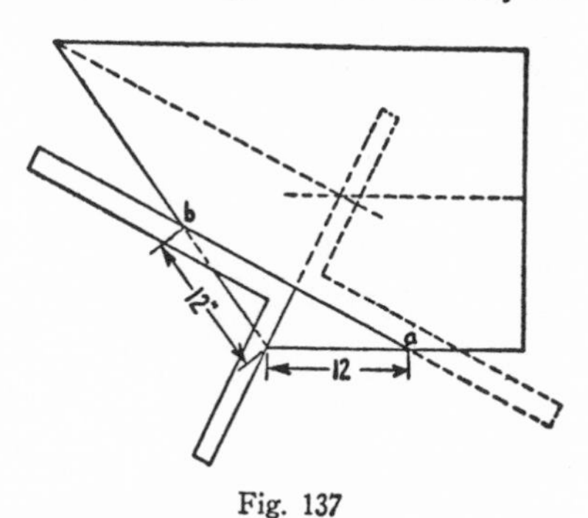

Fig. 137

Edge Bevels for Hips.—Fig. 138 shows the square applied to the base line of the long hip rafters. The hip, lying on the side, is shown shaded. The curved dotted line shows how the rafter length is transferred to the base line, bringing the rafter length down in line with the run of the rafter. To obtain the tangent, extend the beveled end line of the deck, as shown by dotted line, until it strikes the edge of the tongue to the upper left. The tangent

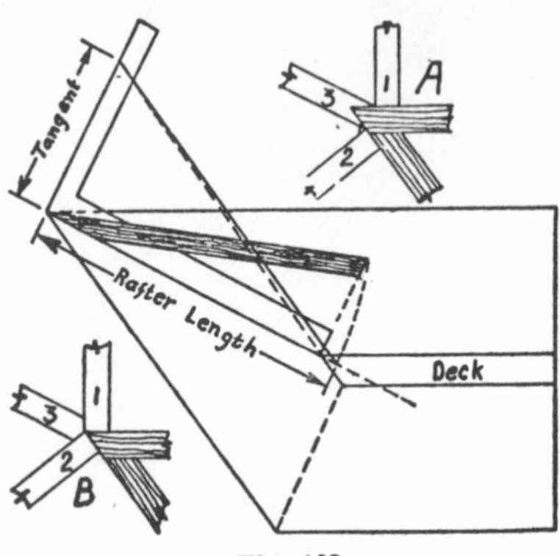

Fig. 138

is the distance between this point and the heel of the square. Using the rafter length on one arm of the square and the tangent on the other will give the edge bevels needed for the constructions shown in details *A* and *B*. In these details the parts marked *1* and *2* are common rafters joining the deck at the sharp corner. For the construc-

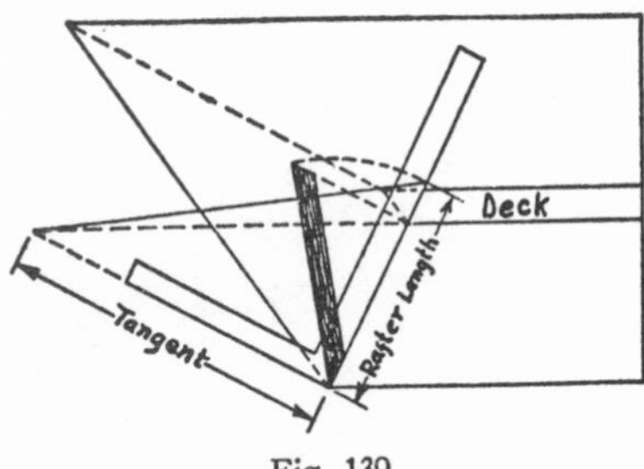

Fig. 139

tion shown at *A-3*, the rafter length will give the edge bevel of the joint that straddles the sharp corner of the deck, while the tangent will give the edge bevel of the hip for the construction shown at *B-3*.

Edge Bevel for Short Hip.—Fig. 139 shows the square applied for obtaining the rafter length and the tangent for marking the edge bevel that will fit the dull corner of the deck. In this case the rafter length will give the bevel of the joint that will straddle the corner of the deck, while the tangent will give the bevel that will fit the common rafters when they are put in place first,

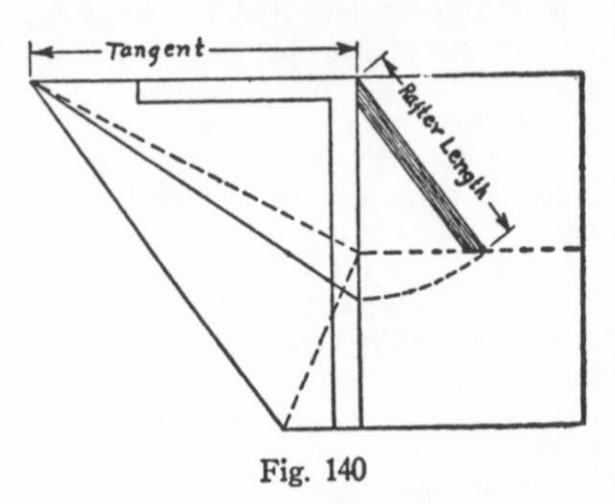

Fig. 140

contacting each other at the corner. Study this diagram with the one shown in Fig. 138. The principle is the same in both cases.

Jack Rafters and Edge Bevel.—Fig. 140 shows the square applied to the diagram for obtaining the rafter length and the tangent for marking the edge bevel of the jack rafters for the long hip. A common rafter, shaded, is shown

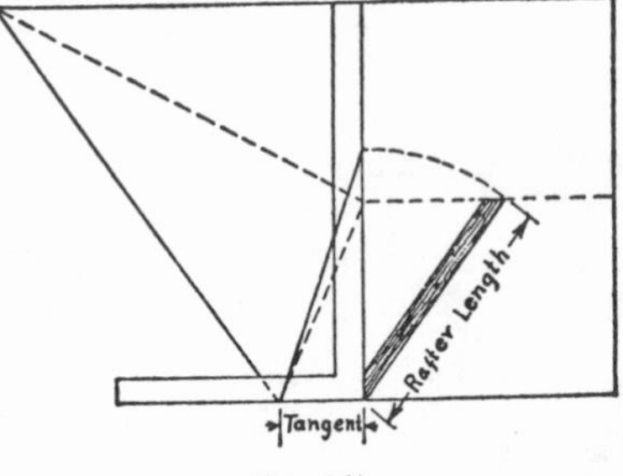

Fig. 141

on the side. The curved dotted line shows how the rafter length is transferred to the blade of the square. The distance between the sharp corner and the heel of the square gives the tangent. The rafter length, in this case, gives the edge bevel of the jacks. The square is shown with the rafter length transferred to the blade, and the tangent is on the tongue. But because the tongue is too short for the tangent, in practice, the tangent in such cases should be taken on the blade and the rafter length on the tongue. These adjustments must frequently be made, and the student should train himself to make them automatically. Fig. 141 shows how to get the edge bevel for the jack rafters that join the short hip. Here again, the principle is the same as in the other case. The student, for practice, should figure out this problem on the basis of the explanation given in Fig. 140.

Chapter 16

CUTS FOR IRREGULAR PLAN ROOFS

Other Irregular Plan Roofs.—It should be stated that the irregular plans that are used in this chapter were chosen, not because they are much in use, but because they make possible different applications of the principle involved. The truth is that the principle of irregular plan roof framing must be adapted to whatever design the roof framer finds himself confronted with, just as he must do in regular roof framing. The method used here for obtaining the points with which the edge bevels for hips, valleys, and jacks are made, will apply on any kind of hip or valley roof. The only reason that

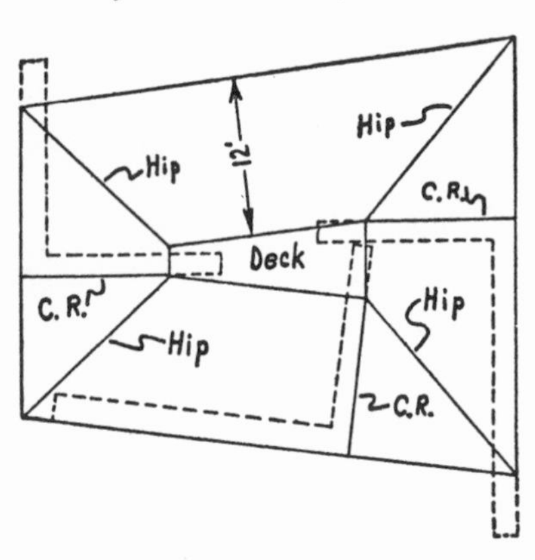

Fig. 142

the different designs are given is to show the different results that the different applications bring out. The roof framer should make diagrams, such as are shown here, letting inches on the square equal feet in the plan, or he might use some other convenient scale. This is the most practical way to obtain the points to be used on the square.

Right Angle to Plate.—Puzzling to the beginner, and sometimes the journeyman, in irregular plan roof framing, is the position in which the common and jack rafters are placed with regard to the plates. Fig. 142 shows an irregular roof plan. Here the hips are pointed out, and three one-line common rafters, presumed to be in place, are indicated with the letters, *C.R.* The three squares are in position showing that the common and jack rafters, in plan, are set at a right angle to the plates, as the tongues show.

Edge Bevel for Hips.—Fig. 143 shows the left end of the diagram shown in Fig. 142, with the square applied to the run of the upper left hip. (The scale, in these diagrams, as mentioned before, is one inch on the square equals one foot on the plan.) Now ex-

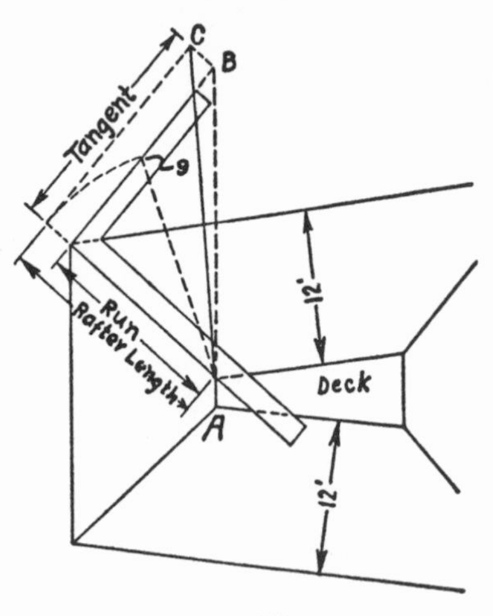

Fig. 143

tend the edge line of the tongue to point *B*, or a little beyond, and the edge line of the blade to the rafter length, as shown by dotted lines. The rise is 9 inches to the foot, which for the hip would be the hip run per foot run of the common rafter. The diagonal dotted line from 9 on the tongue to the corner of the deck, is the length of the hip rafter. With a compass, transfer this rafter length to the extended line of the blade, as shown. Draw the dotted line between *A* and *B* in line with the small end of the deck. Where this line intersects the extended line of the tongue, is the point giving the length of the tangent. Draw the dotted line from *B* to *C*—also the dotted line representing the tangent, parallel with the tongue of the square and in line with the point giving the rafter length. Complete the triangle by drawing the line from *C* to the corner of the deck. To obtain the edge bevel for the hip where it fits the deck, take the rafter length on the blade, and the tangent on the tongue—the rafter length will give the bevel. It is apparent that the tongue is not long enough to hold the tangent, so by dividing both the rafter length and the tangent by 2, the figures will be reduced so that the square will hold them.

Fig. 144 shows the same end of the diagram, but the application of the

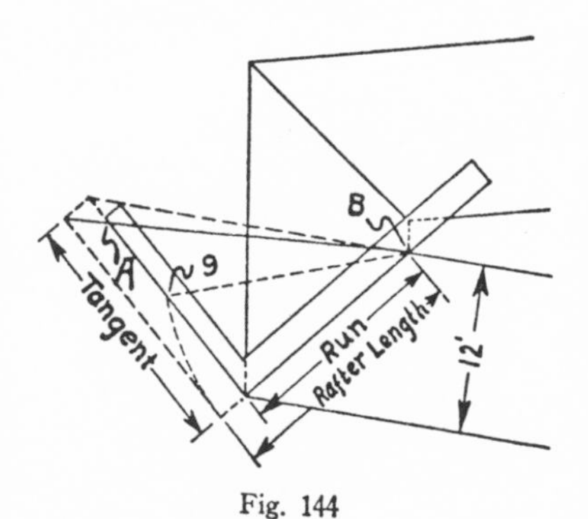

Fig. 144

square is different. Here the side line of the deck is extended, as shown by dotted line, until it contacts the extended tongue line. The rafter length is transferred with the compass, as indicated by the part-circle. The other lines are then drawn in as explained in Fig. 143. Now the rafter length and the tangent will give the edge bevel for the hip. The same results can be obtained by taking the run on the blade of the square, and point *A* on the tongue—the blade will give the bevel. Figs. 143 and 144, should be studied and compared, for the principle involved in all of these illustrations, is explained more fully here than it is in the following examples.

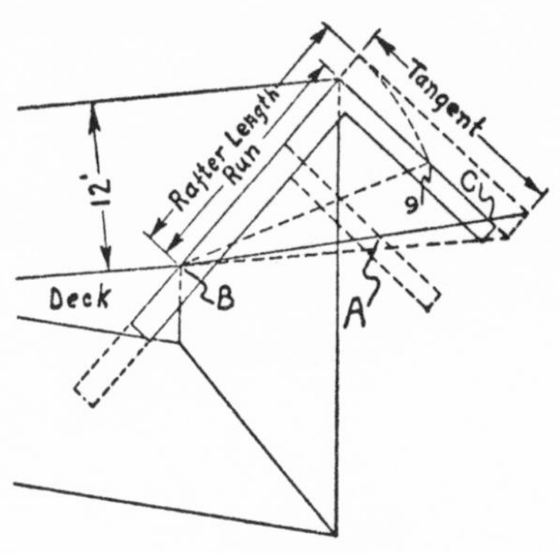

Fig. 145

Fig. 145 shows the wide end of the roof diagram, with the square applied to the hip run. Here again, the side line of the deck is extended until it contacts the extended tongue line. The rafter length is transferred with a compass to the extended blade line, as indicated by the dotted part-circle. The rafter length and the tangent will give the bevel. The edge bevel can also be obtained by taking the run on the blade, and point C, on the tongue. In both cases the blade will give the edge bevel. The figures can be made still more convenient, by pulling the square back until the figure 12 comes to point *B*, then 12 on the blade and point *A* on the tongue, shown by dotted lines, will give the bevel—mark along the blade.

Edge Bevel for Jacks.—Fig. 146 shows the wide end of the roof plan,

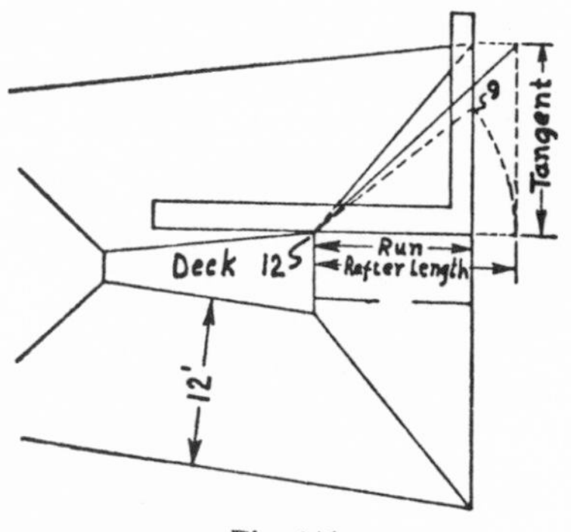

Fig. 146

with the square applied for obtaining the points with which the edge bevel for the jacks is marked. The diagonal distance between 9 and 12 is the rafter length, as shown by dotted line. This length is transferred with the compass to the extended blade line, as indicated by the dotted part-circle. The tongue is long enough in this case to take the tangent. The other lines are drawn in as before. Now the rafter length on the blade, and the tangent on the tongue will give the edge bevel for the jacks. The same bevel can be obtained by taking the run on the blade and the point where the diagonal line crosses the edge of the tongue. The blade will give the bevel.

Edge Bevel for Valleys. — Fig. 147 shows a roof plan with a dull angle,

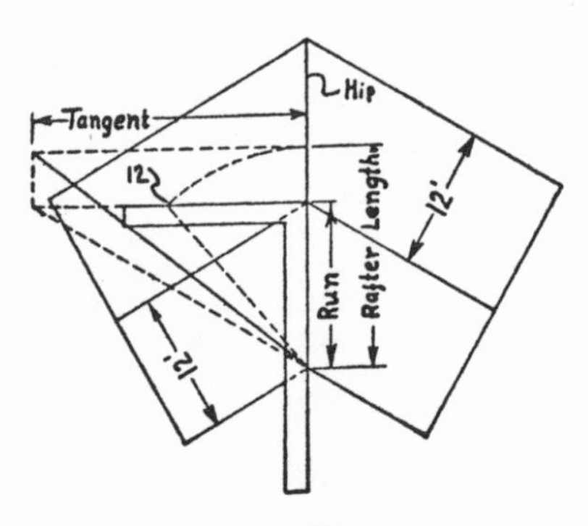

Fig. 147

where a valley rafter is to be placed. The side of the roof is extended, as shown by dotted line, until it contacts the extended tongue line, which is also shown by dotted line. In this case the rise shown is 12 inches. The length of the rafter is transferred from this point to the extended blade line, as shown by the dotted part-circle. Now the rafter length and the tangent will give the edge bevel. The rafter length giving the bevel. In this case it is again necessary to divide both the rafter length and the tangent by 2, so that the figures can be taken on the square.

Edge Bevel for Valley Jacks.—Fig. 148 shows how to obtain the edge bevel for valley jacks joining the valley rafter dealt with in Fig. 147. The rise is 12 inches, making it a half pitch roof. Transfer the rafter length with a com-

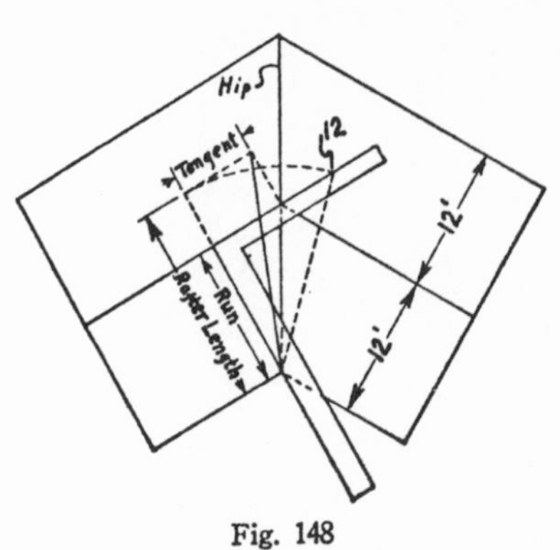

Fig. 148

pass, as shown by the part-circle. The rafter length and the tangent taken on the square will give the bevel. Mark on the rafter length.

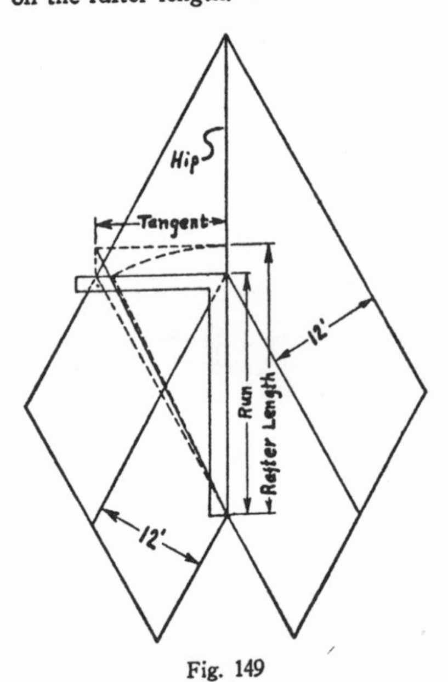

Fig. 149

Edge Bevel for Sharp Angle Valleys. —Fig. 149 shows how to obtain the points for marking the edge bevel of valley rafters for a sharp angle, such as shown. Transfer the rafter length with a compass, as shown by the part-circle. The rafter length and the tangent will give the edge bevel — the former giving the bevel.

Edge Bevel for Valley Jacks.—Fig. 150 shows how to obtain the points for

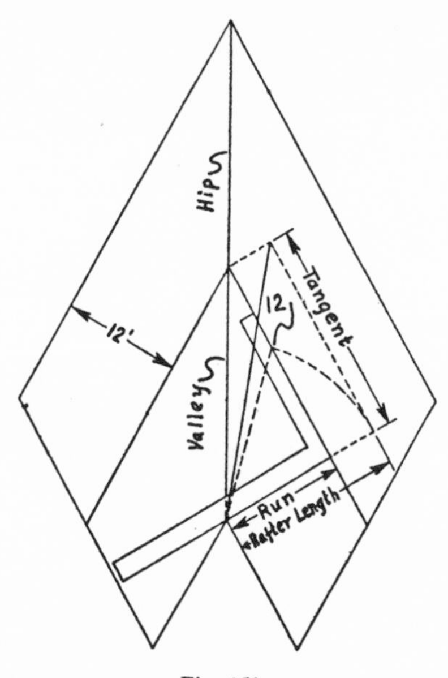

Fig. 150

marking the edge bevel for the valley jacks joining a sharp angle valley rafter. Transfer the rafter length, as shown by the dotted part-circle. The rafter length and the tangent taken on the square will give the bevel. Mark on the rafter length. The run of this roof diagram is 12 feet, which is true of all the other roof diagrams shown in this chapter. Also, 1 inch on the square equals 1 foot on the drawing, as mentioned before.

Reducing Figures.—Fig. 151 shows the principle of obtaining the points for marking the edge bevels in this chapter, applied to a regular hip roof with a 12-foot run and a 12-foot rise. (To

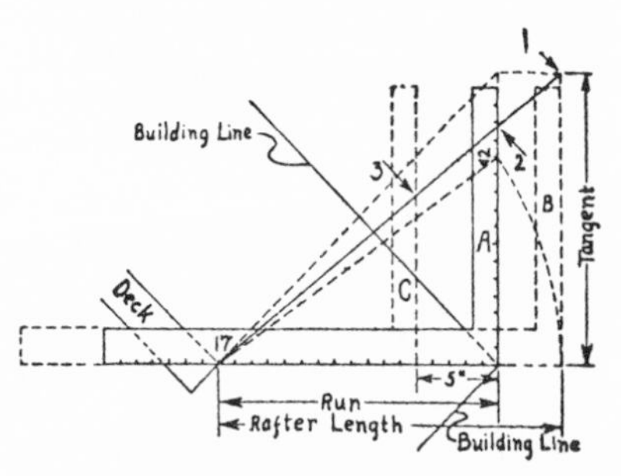

Fig. 151

get the right viewpoint, notice the deck and the building lines, as indicated on the drawing.) Now if the square were pushed forward to position *B*, the rafter length and the tangent (the distance between the upper arrow and the heel of the square) will give the edge bevel. In position *A*, 17 and point *2*, will give the bevel, while in position *C*, 12 and point *3* will give the edge bevel. In each of these cases mark on the blade. The three positions give the same bevel.

Chapter 17

IRREGULAR PITCH ROOF FRAMING

Irregular pitch roof framing is no more difficult than regular pitch roof framing. It is true that the roof framer must be on his guard, so as not to become confused, but the principle is the same. That is also true of irregular plan roof framing. The reason so many workmen find these two branches of roof framing difficult, is that they do not fully understand the principle of regular hip roof framing. The rule, that the run and the length of the rafter will give the edge bevel for hips, valleys, and jacks, increases the confusion. While this old rule is easy to remember and will work in regular roof framing, it is an incorrect rule. The correct rule is this: *The tangent and the rafter length will give the edge bevel for hips, valleys, and jacks—the rafter length giving the bevel.* This rule covers regular hip and valley roofs, as well as irregular hip and valley roofs, both in plan and in pitch. By using the word tangent, instead of the word run, the rule becomes applicable to all hip and valley roofs. When the use of the tangent is thoroughly understood, the roof framer will have no more difficulty in framing the irregular hip and valley roofs than he has in framing the regular hip and valley roofs. It is suggested, however, that diagrams be made for irregular hip and valley roofs, on the order of the diagram shown in these chapters.

The Run of Irregular Hip Rafters. —Fig. 152 gives a plan of an irregular hip roof, 28 feet by 40 feet, with a deck 4 feet square. The run of the sides is 12 feet, while the run of the ends is 18 feet. This is shown by the shaded square, which is in position for obtaining the hip run. The tongue holds the short run, 12 feet, while the blade holds the long run, 18 feet. The diagonal distance between 18 and 12 on the square, is the hip run, as shown on the diagram. (In regular hip roof framing it is the diagonal distance between 12 and 12, because the runs are the same.)

It should be remembered that in all of these diagrams, inches on the square represent feet in the diagram. In this connection the suggestion is repeated, that the roof framer make a diagram

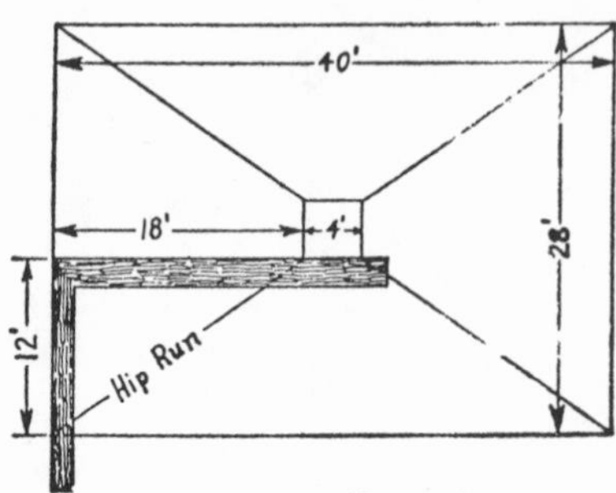

Fig. 152

of the roof in hand, using a convenient scale, say, 1 inch equals 1 foot. Then by applying the square to the diagram, as shown in these lessons, he can get

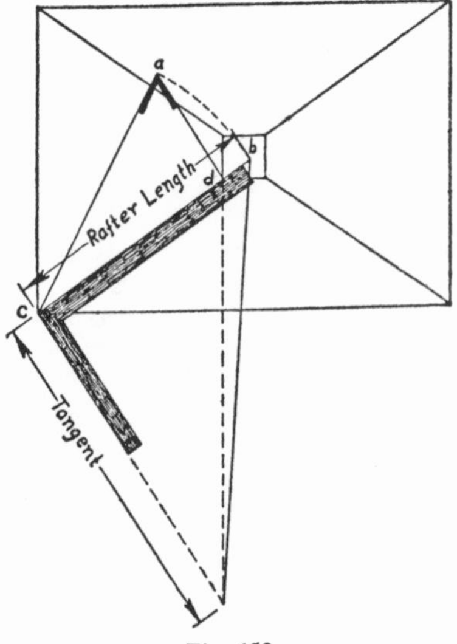

Fig. 153

the various bevels and cuts of the rafters that he is framing.

Edge Bevels of Hips. — Fig. 153 shows the blade of the square applied to the run of one hip. The lines, *c-d, d-a,* and *a-c,* respectively represent the run, the rise, and the hip rafter. This triangle, pivoted on the run, is shown as if it were lying on the side. The rise is 12 feet. The shaded bevel at *a,* gives the plumb cut of the hip rafter. With the compass set at *c,* transfer point *a* to *b,* as shown by the dotted part-circle. This brings the rafter length, *c-a,* in line with the hip run, *c-d,* and on to *b.* Now the edge bevel that will fit the ends of the deck is obtained by taking

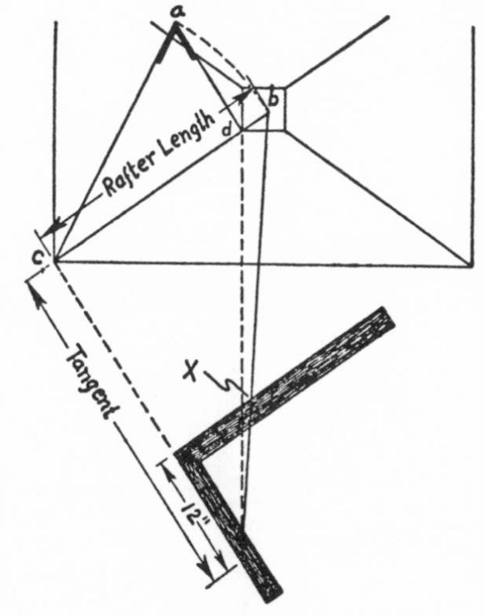

Fig. 154

the tangent and the rafter length on the square — the rafter length giving the bevel. It is obvious that the square is not large enough to hold the tangent on one arm and the rafter length on the other, so the two distances should be divided by 2, which will give a reduced tangent and a reduced rafter length. With these reduced points taken in the most convenient way on the square, the edge bevel can be marked —the arm on which the reduced rafter length is taken gives the bevel.

Fig. 154 shows in part, the roof plan shown in Fig. 153 and the diagram for obtaining the points to use for marking the edge bevel of the hip rafters to fit the ends of the deck. In this case the tongue of the square is applied to the tangent line. To get the edge bevel, use 12 on the tongue of the square and the point where the diagonal line crosses the edge of the blade, or point X. The blade gives the bevel.

Fig. 155 shows the blade of the square applied to the run of one hip,

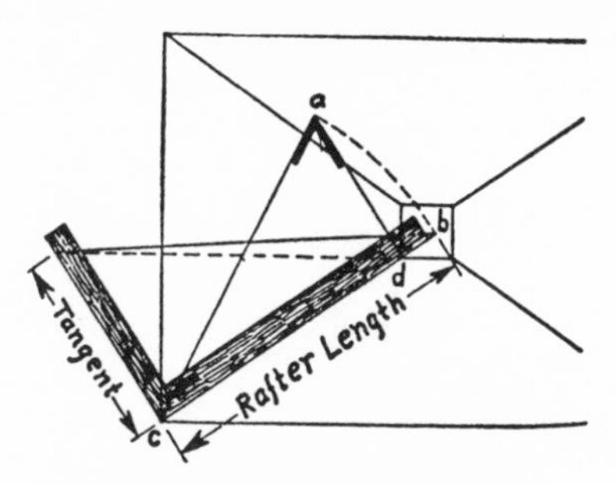

Fig. 155

to obtain the points on the square for marking the bevel that will fit the side of the deck. The triangle, *a-c-d,* is the same as in the previous diagram. The rafter length is transferred with a compass from *c-a* to *c-b,* as indicated by the dotted part-circle. The tangent and

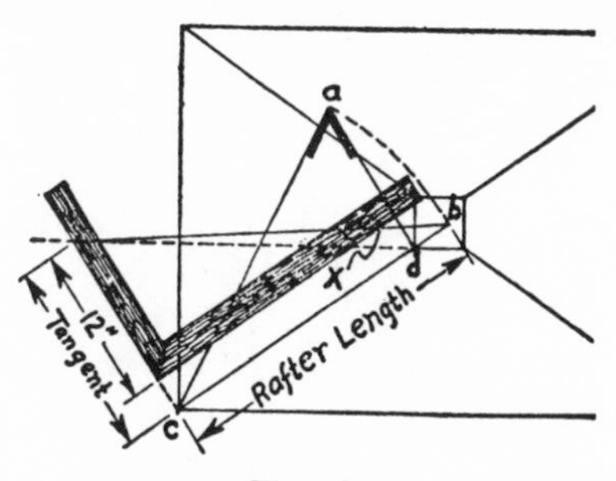

Fig. 156

the rafter length taken on the square will give the edge bevel—mark on the arm that holds the rafter length. Again, the tangent and the rafter length should be reduced by dividing both by 2.

Fig. 156 shows the diagram shown in Fig. 155, but here the square is applied on the tangent line, as shown. To mark the bevel, take 12 on the tongue of the square and point X on the blade —the blade giving the bevel. The principle here is the same as in Fig. 154.

Fig. 157 shows the square applied to the timber for marking the two bevels necessary to make the rafter straddle the corner of the deck. The shaded square is applied, using 12 and point

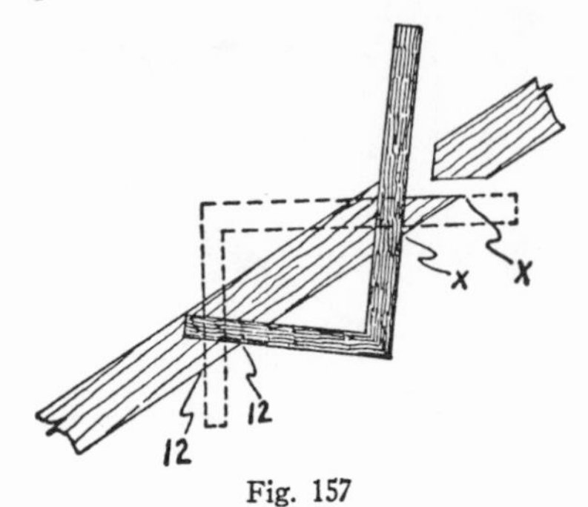

Fig. 157

X. Point X was found by applying the square as shown in Fig. 154. The dotted-line square is also applied by using 12 and point X, but as shown by the application of the square in Fig. 156. Compare and study the two applications of the square in Fig. 157, with the applications of the square shown in Figs. 154 and 156.

Edge Bevels for Jacks. — Fig. 158 shows the square applied to get the

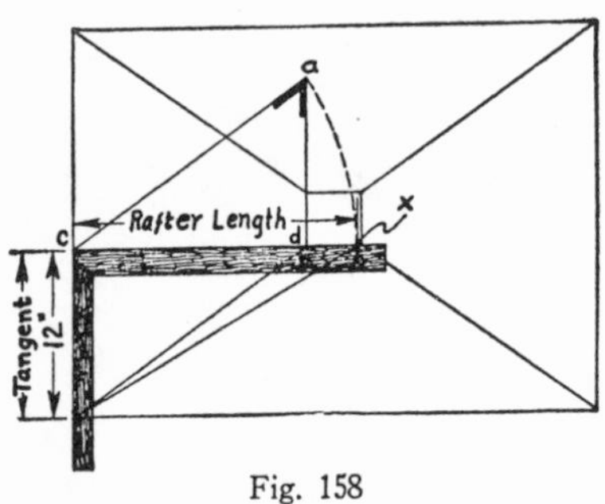

Fig. 158

points for marking the edge bevel of jacks for the ends of the roof. Here the lines, *c-d, d-a,* and *a-c,* show respectively, the run, the rise, and the length of the rafter as if the rafter were on its side. The rafter length, *c-a,* is transferred to the run line, *c-X,* with the compass, as shown by the dotted part-circle. Now the tangent and the rafter length will give the edge bevel—the rafter length giving the bevel. In other words, 12 on the tongue and

Fig. 159

point X on the blade will give the edge bevel—the blade giving the bevel. Fig. 159 shows the square applied to the rafter timber—mark along the blade.

Fig. 160 shows the square applied to get the points for marking the edge

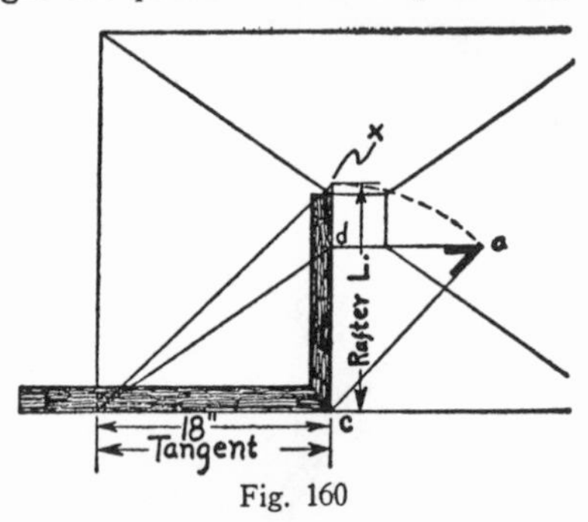

Fig. 160

bevel of the jacks for the sides of the roof. The tangent and the rafter length will give the edge bevel — the rafter length giving the bevel. Again, the square is too small to hold the tangent

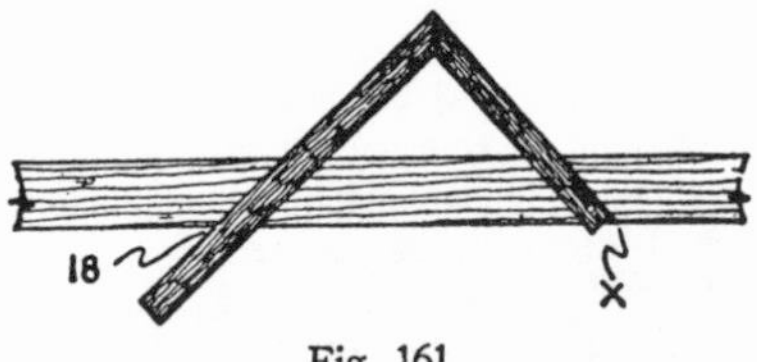

Fig. 161

and the rafter length, so the two distances should be reduced as explained before, and taken in the most convenient way on the square. Fig. 161 shows the square applied to the rafter material, using 18 and point X. Mark along the tongue.

Chapter 18

CUTS FOR IRREGULAR VALLEY RAFTERS

The Use of Diagram, Practical.—In framing any irregular roof, the use of a diagram is recommended. Such a diagram should be drawn to some convenient scale. Perhaps the most practical scale is the one used in these chapters, in which 1 inch on the square equals 1 foot in the diagram. But it should not be presumed that this is the only practical scale that can be used. If the roof is rather large, then a smaller scale is more convenient, and for that reason more practical. If the roof is small, a convenient larger scale might prove to be more practical than the one suggested here.

Rule for Framing Hips, Valleys, and Jacks. — If the use of the tangent is clearly understood, as it is used in roof

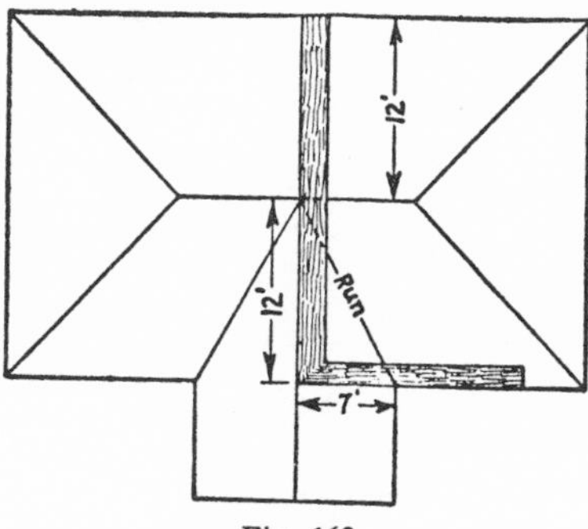

Fig. 162

framing, then any hip or valley roof, regular, irregular plan, and irregular pitch, can be framed by the following rule:

The diagonal distance of the two full runs that intersect at the hip or valley, is the run of the hip or valley, whichever it might be. The edge bevel of any hip, valley, or jack is obtained by taking the tangent on one arm of the square, and the rafter length on the other, the latter giving the bevel. The run and the rise of any rafter taken on the square, will give the level and plumb cuts.

The Valley Run. — Fig. 162 shows the square applied to a diagram of a roof, which has two irregular valleys, for obtaining the run of the valleys. Here the run of the main roof is 12 feet, and the run of the secondary roof is 7 feet. The diagonal distance of these two runs, as shown on the diagram, is the run of the valleys.

Edge Bevels for Valley Rafters.— Fig. 163 shows the square applied to the valley run for establishing the points to be used on the square for marking the edge bevel that will fit the ridge of

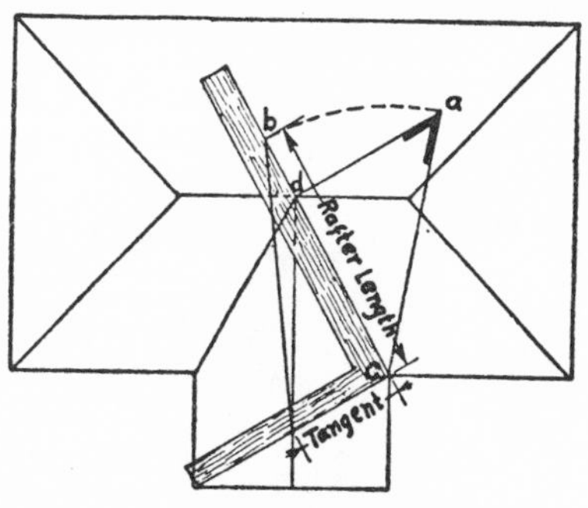

Fig. 163

the secondary roof. The sides of the triangle, *c-d, d-a,* and *a-c,* represent respectively, the run, the rise, and the rafter length, shown as if lying on the side. The dotted part-circle from *a* to *b,* shows how the rafter length has been transferred from *c-a,* to *c-b.* The tangent, as shown, is the right-angle distance from the toe of the valley rafter to the center of the secondary roof. Now the tangent and the rafter length, as shown, will give the edge bevel of the valley rafter that will fit the ridge of the secondary roof—the rafter length giving the bevel.

The bevel to fit the ridge of the main roof, is obtained as shown by the diagram in Fig. 164. Here again the valley rafter is on the side, as shown. The shaded bevel at point *a,* is the bevel for the plumb cut of the valley. The rafter length has been transferred, as explained in the other diagram, and as shown here by the dotted part-circle, between *a* and *b*. To get the edge bevel, take the tangent and the rafter length on the square—the latter will give the bevel. As can be seen, the square is not large enough to take the tangent and the rafter length,

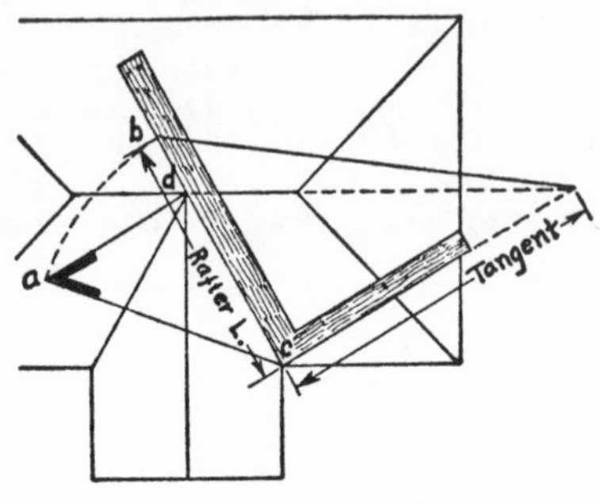

Fig. 164

so these must be reduced. A good way to do this is shown by Fig. 165. Here the shaded square is pushed to the right on the tangent until 12 on the tongue comes to the point of the triangle, as shown. Then 12 inches on the tongue represents the reduced tangent, and the

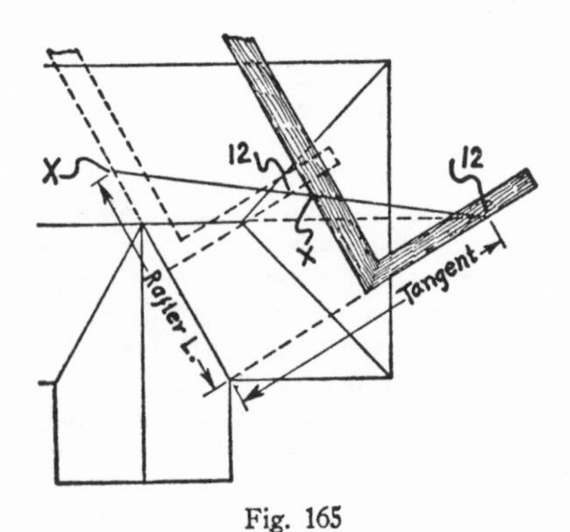

Fig. 165

distance from the heel of the square to point *X* represents the rafter length. The same results can be obtained by pushing the square up on the valley run, as shown by the dotted-line square, until 12 on the tongue intersects the diagonal line, as shown. Then 12 on the tongue and point *X* on the blade will give the edge bevel—the blade giving the bevel. The rafter length and the tangent can also be reduced, as mentioned a number of times in previous chapters, by dividing both distances by 2.

Edge Bevels of a Valley.—Fig. 166 shows two applications of the square for marking the edge bevels for a valley rafter. Here the application numbered *2,* is the same as the one found in Fig. 163, while the one numbered *4,* is the

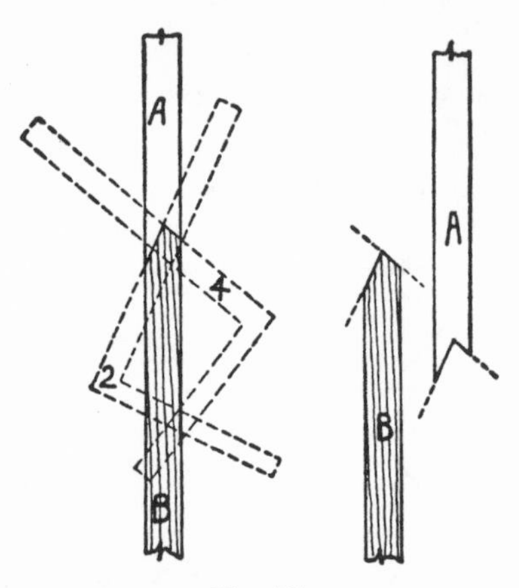

Fig. 166

same as either of the two applications shown in Fig. 165. Study the three drawings. The parts of the main drawing, Fig. 166, marked *A* and *B,* are shown to the right cut in two, again marked *A* and *B*. At the bottom of *A,* the cut shows what it would be like in case the rafter were a hip straddling the corner of a deck, while the upper part of *B* shows the cut of a valley rafter that is to fit into an angle of two ridges, as shown in Fig. 164. The dotted lines respectively indicate the corner of a deck and an angle of two ridges.

Edge Bevels of Jacks.— Fig. 167 shows the square in position for obtaining the points to be used for marking the edge bevel of the valley jacks of the main roof. Here the common rafter is shown as if it were on its side. The rafter length, *c-a* as shown by the dotted part-circle, has been transferred to *c-b*. Now the tangent

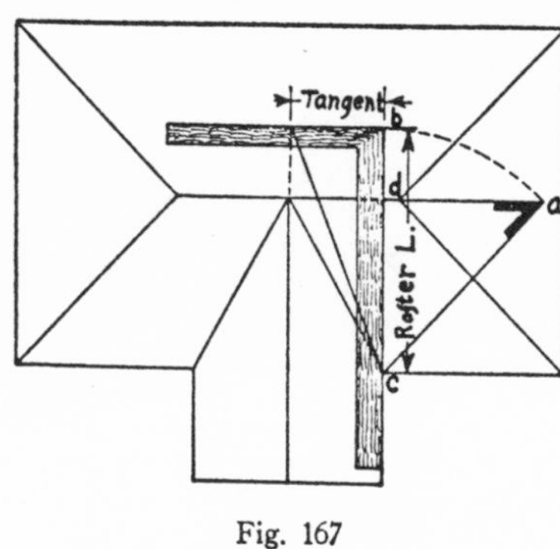

Fig. 167

and the rafter length will give the edge bevel of the jacks. The application of the square to the rafter material is

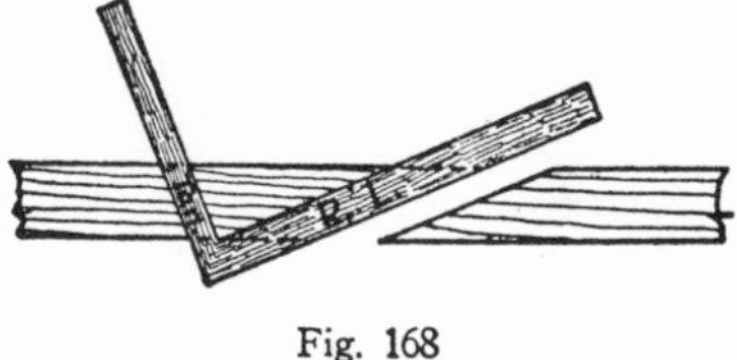

Fig. 168

shown by Fig. 168—the blade giving the bevel.

How to get the points for marking the edge bevel of the jack rafters for the secondary roof is shown by Fig. 169. Here again, the rafter length, *c-a* is transferred to *c-b*, as indicated by the

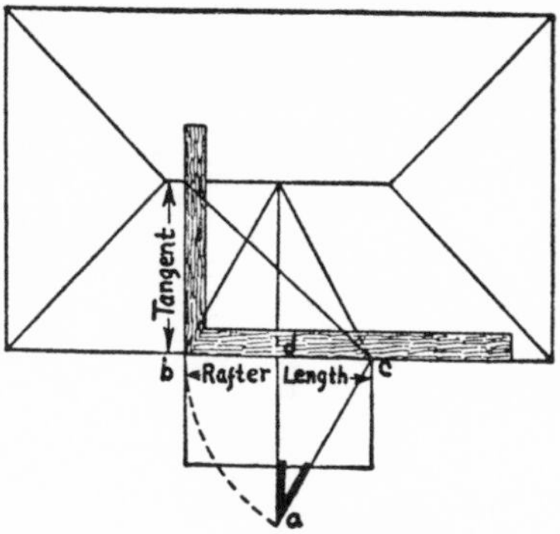

Fig. 169

dotted part-circle. Now the tangent and the rafter length will give the edge bevel—the rafter length giving the bevel.

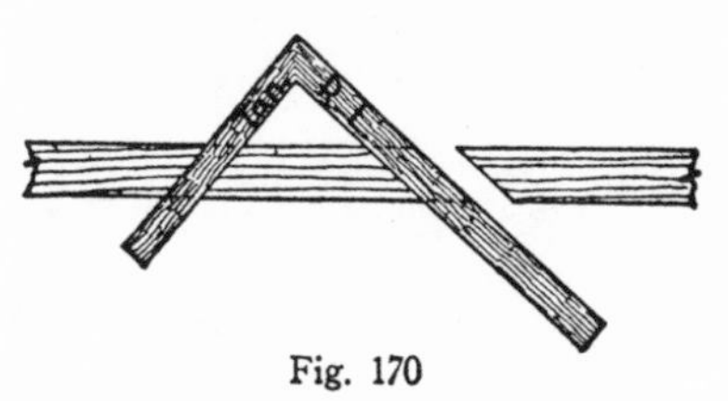

Fig. 170

Fig. 170 shows the square applied to the rafter material—the blade giving the bevel.

Chapter 19

SHEETING AND PLANCIER BEVELS

While the sheeting bevels for ordinary roofs are rarely marked with the square, there are times when this must be done. This is especially true on roofs that must have a perfectly straight hip line. It is also true with planciers, which must be cut to fit, and

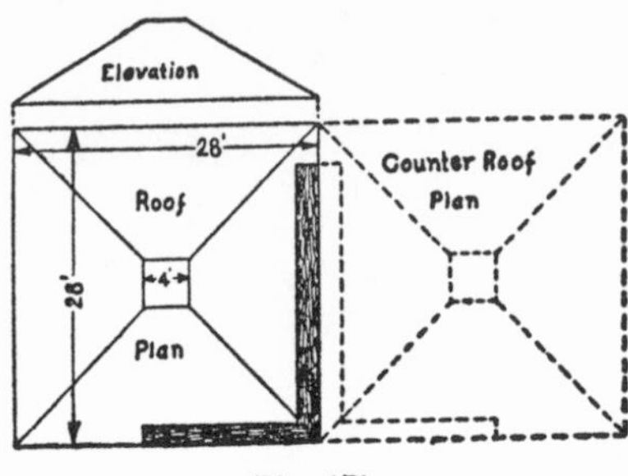

Fig. 171

on hopper work of all kinds—for hoppers and hip roofs are framed on the same principle, but in upside down order.

Hip Roofs.—Fig. 171, to the upper left, shows an elevation of a rather flat

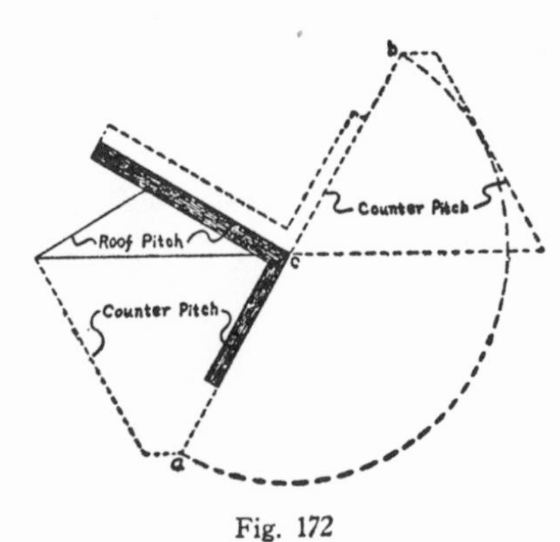

Fig. 172

hip roof. Directly under it is shown a roof plan, which is 28 feet by 28 feet, with a 4x4-foot deck. To the right, by dotted lines is shown the same size plan of the counter roof, a sort of imaginary roof. The shaded and dotted-line squares indicate that the angles of these plans are perfect right angles, which is necessary to make any regular hip roof framing work out right.

The Counter Pitch.—Fig. 172 shows how the counter roof plan, or rather, the counter pitch, as shown in this diagram is obtained. Here the shaded square is in position for laying out the counter pitch, directly under the roof, which is shown by dotted lines. To bring this up in line with the roof shown to the left, set the compass at the heel of the square, or *c*, and with it swing point *a* around to point *b*, bringing with it the hopper-shaped outline, as shown. The dotted-line square shows that the pitch in this position

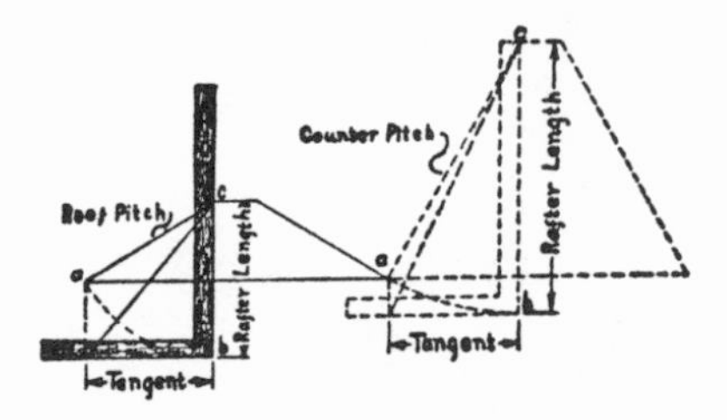

Fig. 173

would have the same relationship to the pitch of the real roof as it had before, but in reverse order.

Obtaining Points for Bevels.—Fig. 173 to the left, shows a cross section of the roof shown in Fig. 172. What we want to find is the points for marking the face bevel for the sheeting. To do this, transfer the rafter length, *c-a*, to *c-b*, as shown by the dotted part-circle. Then apply the square in the position shown. Now to get the face bevel for the sheeting, take the rafter length on the blade and the tangent on the tongue —the tangent giving the bevel. The process is exactly the same on the counter pitch roof, shown to the right. The rafter length and the tangent, as shown, will give the face bevel of the sheeting. For the edge bevel of the sheeting for the roof to the left, take the face bevel of the sheeting for the roof to the right, or vice versa.

Bevels for Jacks and Sheeting.—Fig. 174 shows the flat hip roof shown in previous diagrams, as if it were separated at the hips and flattened out on

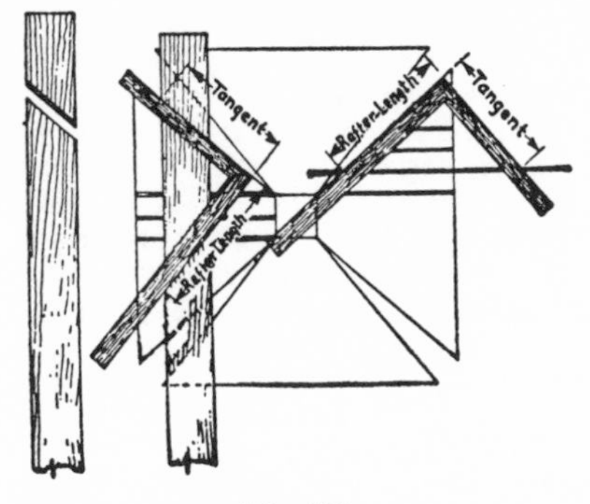

Fig. 174

a level floor. The two squares shown here are applied with the rafter length and the tangent shown in the diagram to the left in Fig. 173. The square to the left is applied to a sheeting board for marking the face bevel—the tangent giving the bevel. The same points

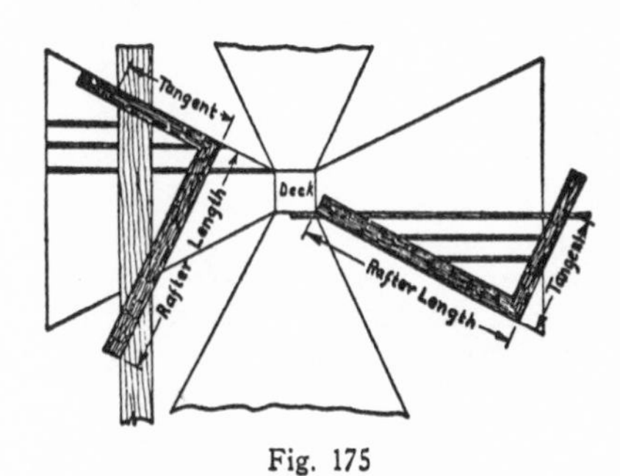

Fig. 175

are used on the square shown to the right, for marking the edge bevel of the jack rafter, but the rafter length, not the tangent, gives the bevel. The sheeting board cut to the proper bevels is shown to the extreme left.

Fig. 175 shows the counter pitch roof (two sides in part) flattened out. The square to the left is applied to a sheeting board, using the rafter length and the tangent, shown to the right in Fig. 173. The tangent, again, gives the face bevel for the sheeting. The same rafter length and tangent are used in applying the square to the right, for marking the edge bevel of the jack rafter, but the rafter length, not the tangent, gives the bevel.

Two applications of the square are shown by Fig. 176. The upper one gives the face bevel of the sheeting for the hip roof shown in this chapter, and the edge bevel on the counter pitch; while the bottom one gives the edge bevel of the roof sheeting and the face bevel of the counter pitch. Which is the same as saying that if you know how to determine one bevel of the

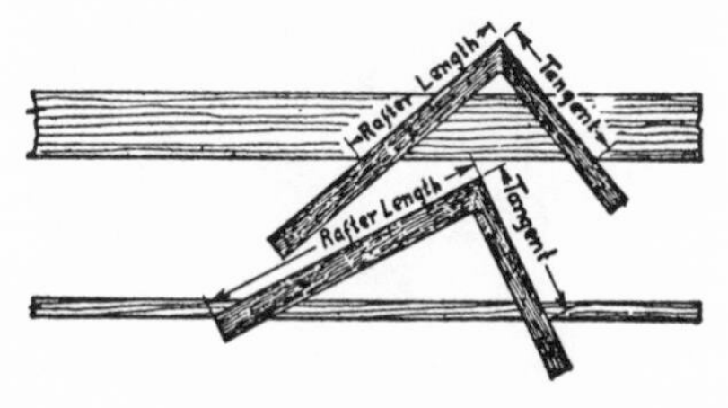

Fig. 176

roof sheeting, you can get the other bevel, by determining the same bevel of the counter pitch. Study Figs. 173, 174, 175, and 176.

Cuts for Sheeting and Plancier.—The upper drawing of Fig. 177 shows a cross section of the counter pitch roof shown in Figs. 172 and 173, at a little larger scale. In this drawing the roof is shown with a cornice and a plancier board. The inside of the roof is shown lined with boards. Now the bevels for the lining and for the plancier boards are exactly the same as those for the sheeting, shown in Fig. 175. The center drawing, Fig. 177, shows the square, numbered *5,* applied to a plancier board for marking the face bevel, which, in reverse order, is the same as the application shown for the face bevel of the sheeting in Fig. 175. The bottom drawing shows the square, numbered *4,* applied for marking the edge bevel of the plancier board, which, in reverse order, is the same as the application shown in Fig. 174, for marking the face bevel of the sheeting. It should be remembered here, that the

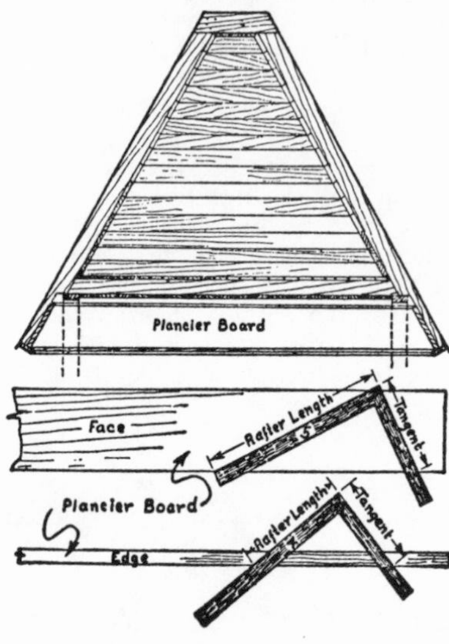

Fig. 177

face bevel of the sheeting for one roof is the edge bevel of the sheeting for the other roof, or vice versa.

Face and Edge Bevels of Sheeting. —Why the face bevel of the sheeting for any hip roof is the same as the edge bevel of the sheeting for the counter

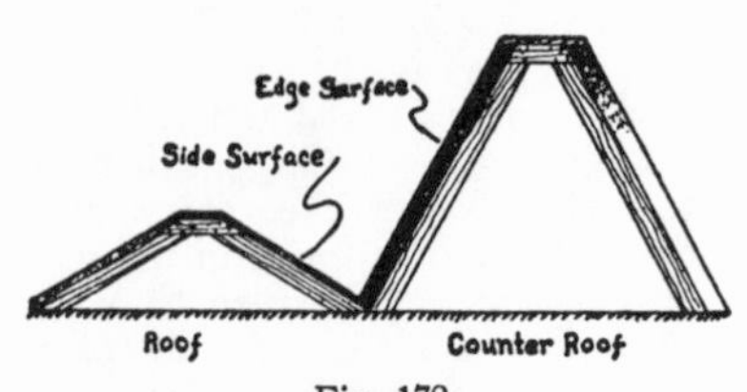

Fig. 178

roof, or vice versa, is illustrated by Fig. 178. Here the low pitch roof has the sheeting on in the regular way, but the counter pitch roof, to bring out the point, has the sheeting on edgewise. Now if the sheeting were put on the low pitch roof as shown, and on the counter pitch roof also as shown, then the edges of the boards in both roofs would have the same bevel—the face bevel of the boards would also be the same in both roofs. Therefore, if the sheeting is put on in the regular way on both roofs, the face bevel of one roof, speaking of the sheeting, would become the edge bevel of the other. Think that through.

Butt Joint of Sheeting. — Fig. 179 shows a diagram of the roof and counter roof that has been used throughout this chapter, illustrating how to get the edge bevel of sheeting for a butt joint. The tongue of the square here is applied to the counter pitch of the common rafter, from the comb down to-

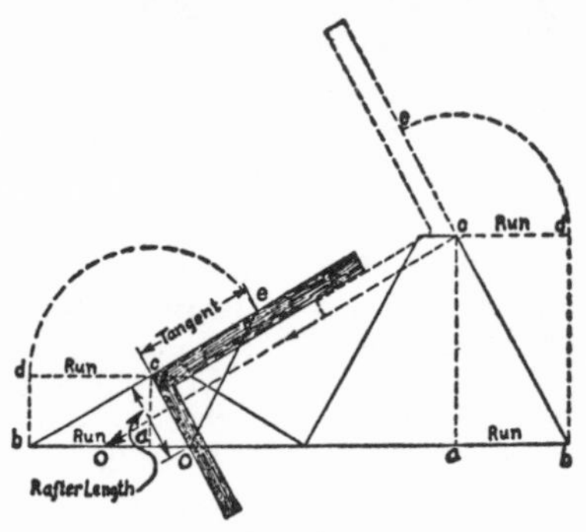

Fig. 179

ward the base line. In this instance the counter pitch from *c* to *o*, becomes the rafter length, as shown. The run, *b-a,* is raised to the level of the top of the common rafter, *d-c,* and then with a compass transferred to the blade of the square, as shown by the dotted part-circle, which becomes the tangent. Now the tangent and the rafter length will give the edge bevel for the butt joint—the tangent giving the bevel. (The reference letters in both the pitch and the counter pitch, refer, relatively to the same points, hence the explanation will apply to both pitches. It should also be mentioned that the rafter length shown in the diagram for the counter pitch roof, runs through the diagram of the original roof.)

Chapter 20

IRREGULAR ROOF PROBLEMS

Difference in Height of Walls.—The walls supporting an irregular pitch hip roof with a cornice must be built to accommodate the different pitches of that roof. For instance, you have an irregular pitch hip roof, in which a one-fourth pitch and a one-half pitch are used, with the cornice overhanging one foot. The difference in the height of the walls would be the difference found in the two rises for the width of the cornice, or in this case, for a one-foot overhang, it would be 6 inches.

Irregular Hip Roof Plan.—Fig. 180 shows an irregular hip roof plan, on which the problems of this chapter are based. Two dotted-line squares are

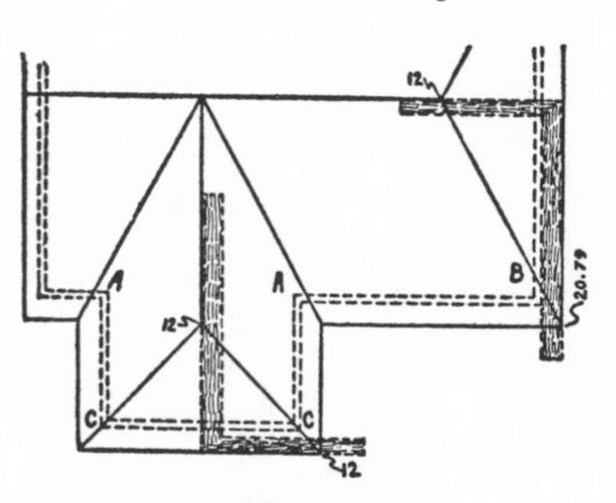

Fig. 180

shown applied to the drawing. The one to the bottom left, is applied to a part of the roof that has regular hips, while the square to the right is applied to the plan where the hip roof is irregular. The dotted lines show the relationship of the outside walls to the roof. At *A* and *A* is shown how the seats of the valleys come to one side of the angle. At *B* is shown how the seat of the hip is offset from the corner, while at *C* and *C* the hips come directly over the corners, because the pitch of this part of the roof is regular.

Run, Rise and Rafter.—Fig. 181 shows the same roof plan shown in Fig. 180, giving the run, the rise, and the rafter of, *A*, an irregular valley, *B*, an irregular hip, and *C*, a regular hip. The difference in the height of the walls is pointed out at *a* and *b* with double indicators. Study the drawing.

Details of Difference in Height of Walls.—Fig. 182 shows a dotted-line square applied to the tail of a common

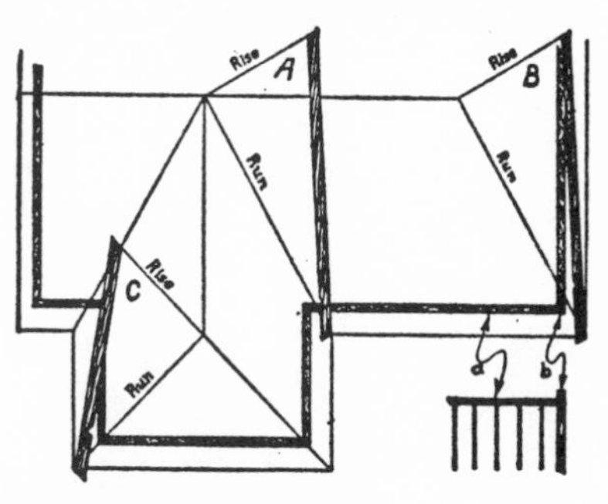

Fig. 181

rafter of the steep pitch shown in Fig. 181, using 12 on the blade and the rise of the pitch on the tongue. The tail of the common rafter for the lower pitch is shown by dotted lines. Here again, 12 is used on the blade of the square and the rise of this pitch on the tongue,

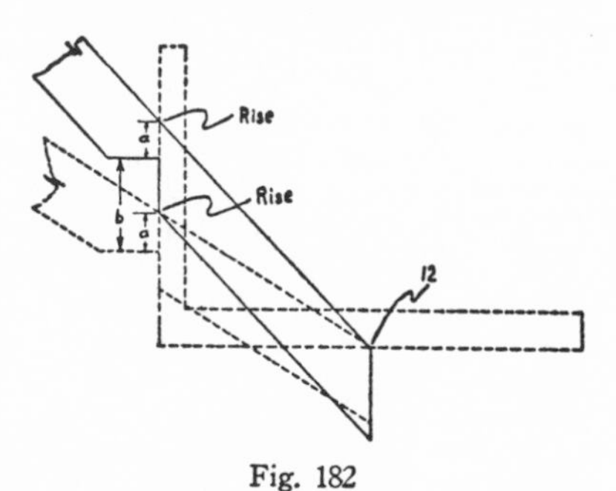

Fig. 182

as shown. The points that should be watched are the distances *a* and *a*, which must be the same. The distance marked *b*, is the difference in the height of the walls. The top of the rafter tails must meet at point 12 on the blade of the square, as shown.

Fig. 183 shows a little different way to obtain the same results. Here a detail of the cornice is given of each of the pitches. The distances at *a* and *a*, again must be the same. Two ways of getting the difference in the height of the walls are shown at *b* and *b*. Points *1* and *2* must be on the same level, as indicated by the dotted line. Study and compare Figs. 182 and 183. They deal with the same problem.

Seat and Tail of Irregular Hip Rafter.—Fig. 184 shows a detail in plan of the irregular hip rafter tail, shown to the right in Fig. 181. The corner of

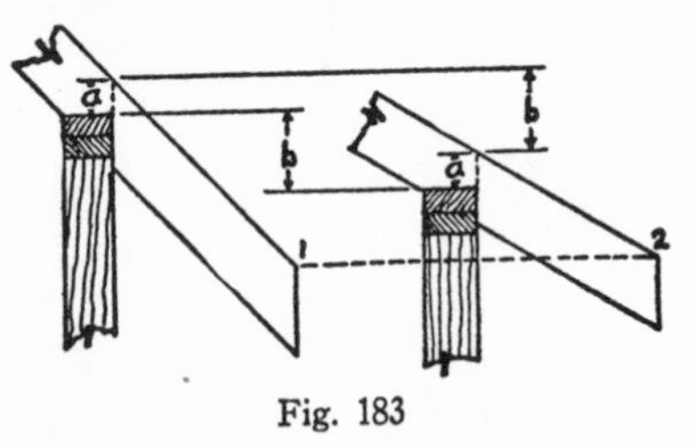

Fig. 183

the cornice is shown off center on the tail. The dotted-line square shows why this is so—the two bevels must intersect the side corners of the tail exactly square across from each other. This keeps the two upper corners of the tail cut on the same elevation. If the corner of the cornice would center the tail of the hip, the two side corners would come at different elevations.

How to obtain the points for marking the edge bevel for the hip seat cut, is shown by Fig. 185. The diagram shows a right-angle triangle representing the run, rise, and length of rafter (tail) as if it were lying on the side. The side wall line is extended, as shown by dotted line, giving the tangent point on the tongue of the square. The rafter length is transferred with the compass to the seat line, as shown by the part-circle. Now the rafter length taken on the blade of the square and the tangent on the tongue will give the edge bevel of the seat—the rafter length giving the bevel. Because the rafter length, as shown, is longer than the blade, both the rafter length and the tangent should be divided by 2,

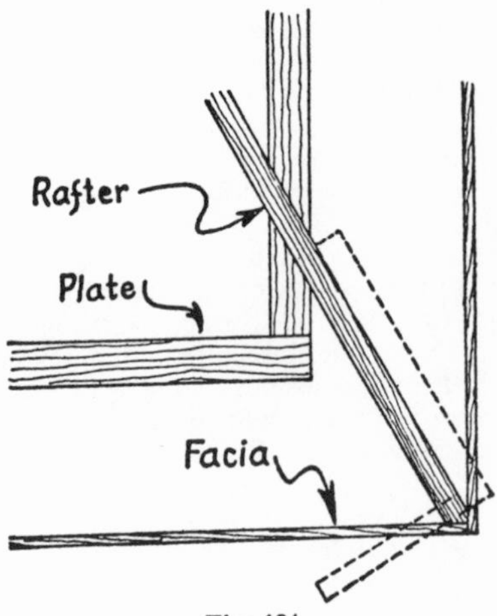

Fig. 184

which will give a reduced rafter length and tangent to use on the square.

Fig. 186 shows a detail in plan, of the irregular hip tail, giving three views of the seat cut. At *A* the tail is shown in place; at *B* it is shown on the side, giving a view of the seat and tail

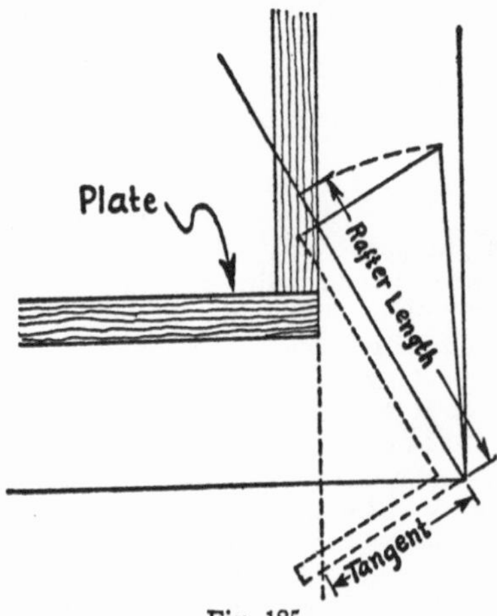

Fig. 185

cuts; at *C* is a bottom view, looking straight at it, and at *D* is the other side view. For instance, if you will imagine that what is shown at *D* is the right side view of position *A*, you will have the right idea. The cornice is indicated by dotted lines.

Seat and Tail of Irregular Valley Rafter. — Fig. 187 shows a detail in plan of the valley rafter tail, shown at

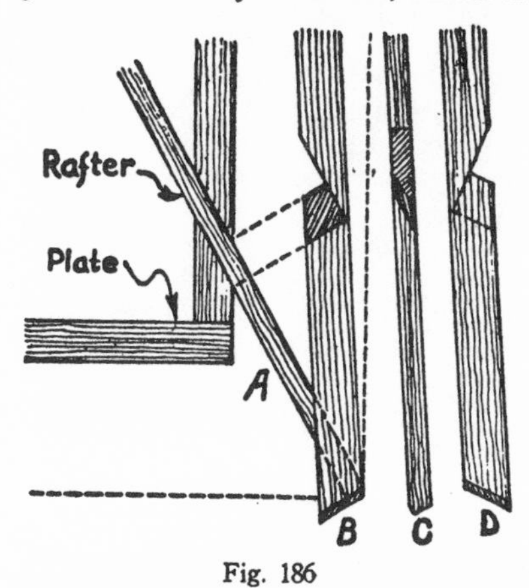

Fig. 186

the center in Fig. 181. If this detail is compared with the one shown in Fig. 184, it will be seen that the valley tail

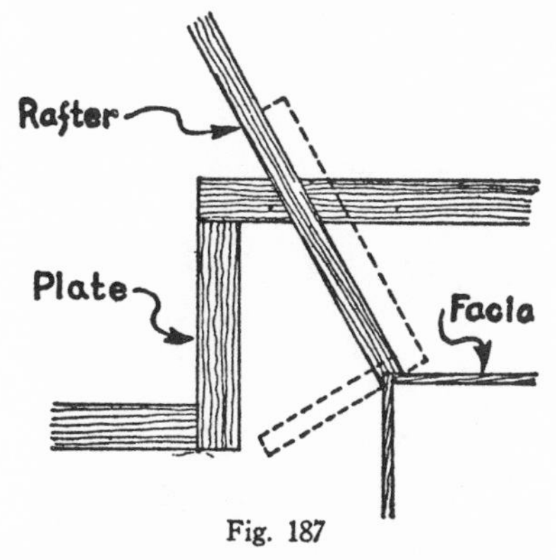

Fig. 187

cut and the hip tail cut are exactly in reverse order. In both instances, however, the tail cuts intersect the side corners square across from each other, as the dotted-line squares show.

Fig. 188 shows a right-angle triangle representing the run, rise and rafter of the cornice, as if lying on the side. In order to get this, the side of the wall under the steep pitch must be extended, as shown by dotted line, from *c* to *d*. Then make the rise equal to the rise of the steep pitch for the width of the

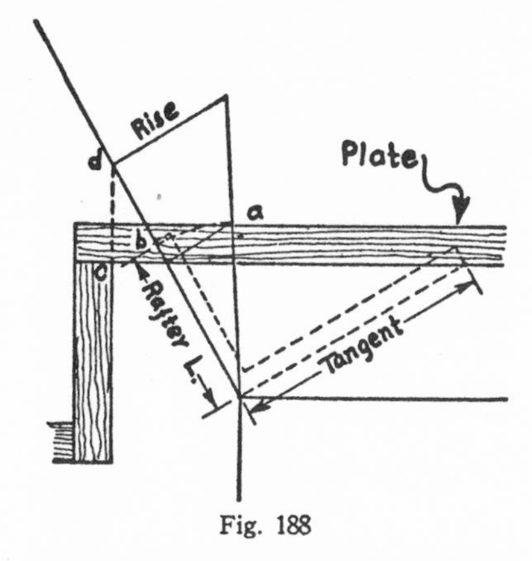

Fig. 188

cornice, in this case a foot run, and draw in the rafter line. Now where the run line crosses the outside line of the wall, draw a perpendicular line, to point *a*. With a compass set at the toe, transfer the rafter length from point *a* to point *b*. Now the tangent and the rafter length, as shown, will give the edge bevel of the valley seat—the rafter length giving the bevel.

Fig. 189 is a detail in plan, showing the valley tail and three views of the seat and tail cuts. At *A* the valley is shown in position, at *B* is shown a side

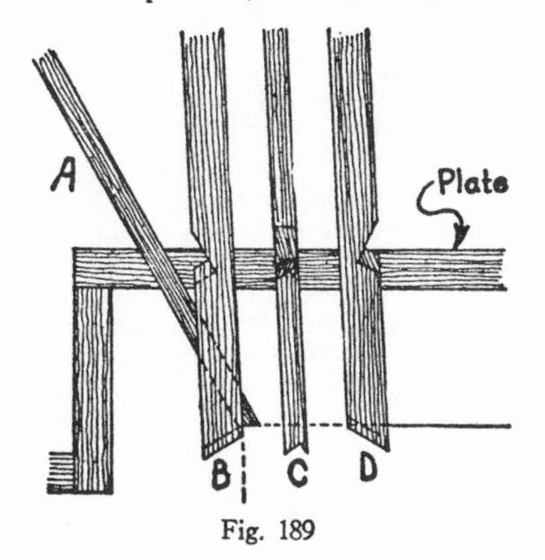

Fig. 189

view, at *C* is a bottom view, looking straight at the rafter, and at *D* is the other side view. Now imagine that you are rolling what is shown at *D*, back to *C*, *B* and then to position *A*, and you will have a good idea of the three views.

Chapter 21
BACKING HIPS AND VALLEYS

Every roof framer should know how to back hip and valley rafters. In practice, however, the backing is usually omitted. This is especially true when the rafters are made of what is called 2-inch stuff. But since no one can tell in advance what the requirements will be on the next job, the roof framer should be prepared to master anything that he might be called on to do, and that includes backing hip and valley rafters.

Rule for Backing. — A simple rule for obtaining the bevel for backing hips and valley is: *Take the length of the rafter on the blade of the square and*

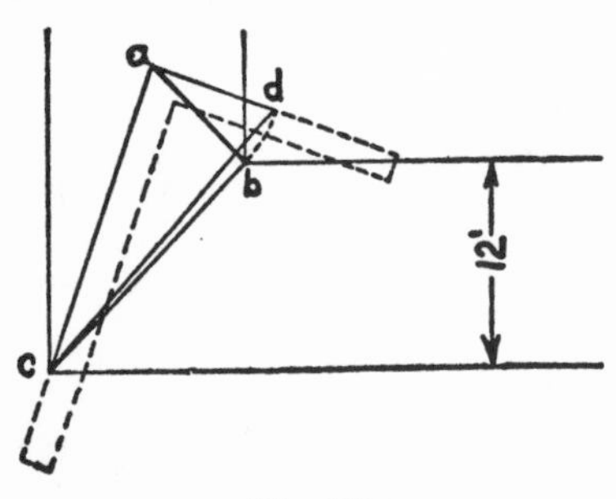

Fig. 190

the rise on the tongue—the tongue will give the bevel. In the diagram shown in Fig. 190, inches on the square, represent feet on the drawing. The diagram shows one corner of a hip roof. The hip rafter is shown in triangular form, as if it were lying on the side, in which *c-b* is the run; *b-a,* the rise, and *a-c,* the rafter. Now set the compass at point *a,* and make *a-d* equal to *a-b,* as indicated by the dotted part-circle. The square, as shown, is applied to the right-angle triangle, *c-a-d,* in which *c-a* is the rafter length; *a-d,* the rise, and *d-c* the diagonal distance, or hypotenuse. The run of the common rafter in the diagram, Fig. 190, is given as 12 feet, and assuming that the rise is 8 feet, it will be an easy matter to find the figures to be used on the square. The run of the hip would be 17 feet, minus. Now the diagonal distance between 17 and 8 would be approximately 18¾ inches. Then 18¾ on the body of the square, and 8 on the tongue, would be the points to use for marking the bevel for the backing. Fig. 191

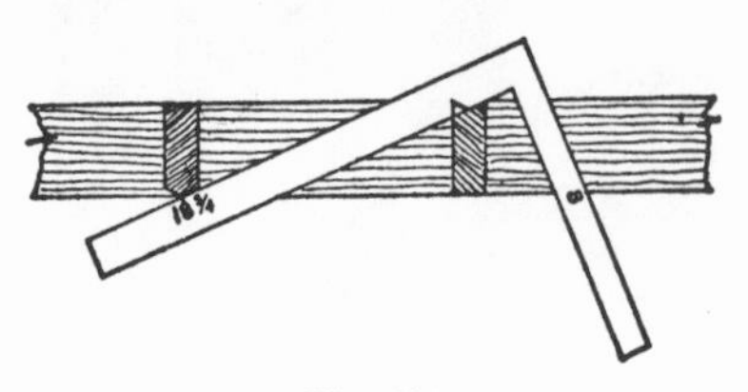

Fig. 191

shows the square applied to a piece of rafter material for marking the backing bevels, to the left, of a hip, and to the right, of a valley. These bevels, which are the same should be taken on a bevel square, with which the marking can be done conveniently.

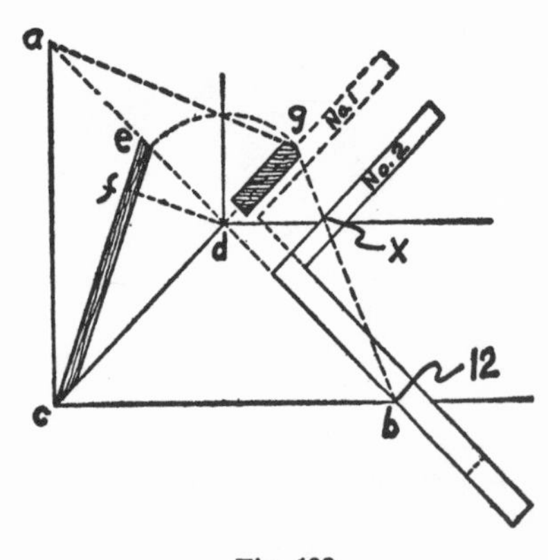

Fig. 192

Backing Hips and Valleys.—Fig. 192 shows another way to get the bevel for backing hips and valleys. The diagram again shows a corner of a hip roof with the hip rafter lying on the side. Here *c-d* gives the run; *d-e,* the rise, and *e-c* the rafter length. Now draw *f-d* at a right angle to the rafter line, and with a compass set at *d*, make *d-g* equal to *d-f*. At right angles to *c-d*, draw *a-b*, crossing point *d*. Join points *a* and *b* with point *g*, as shown by dotted lines. The bevels for the backing, which are again the same, are shown on the enlarged cross section of the hip at point *g*. How to get the points to be used on the square is shown by the applications of squares No. *1* and No. *2*. For application No. *1*, shown in part by dotted lines, take the base of the triangle on the body, and the altitude on the tongue, the tongue giving the bevel. For application No. *2*, take the base figure, 12, on the body of the square, and point *X* on the tongue—the latter gives the bevel.

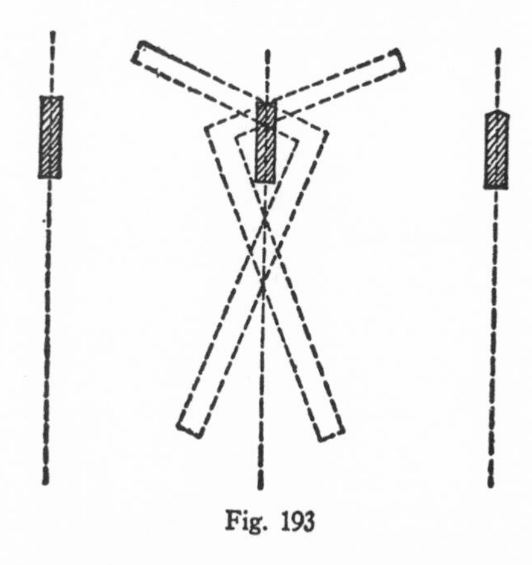

Fig. 193

Fig. 193 shows, to the left, a cross section of a rafter timber with a dotted line through the center. At the center, by dotted-line squares, is shown how the square is applied to the center line to get the two bevels for the backing. The points used on these applications are the same as those shown by square No. *2*, Fig. 192. To the right the backing is shown with the square removed.

Backing for Irregular Pitch Hips and Valleys. — Fig. 194 shows a diagram of an irregular pitch hip rafter on the side, represented by the triangle *c-b-a*. Now draw *h-i* at a right angle to *c-a*, and *d-e* at right angles to *c-b*, crossing point *i*. With a compass set at point *i*, make *i-g* equal to *i-d*, and *i-f* equal to *i-e*. Join point *h* with point *g*, and also with point *f*. The bevels for the backing will be found on the cross section of the rafter shown at *h*. The application of the square to get the bevels, is the same as explained in Figs. 192 and 193.

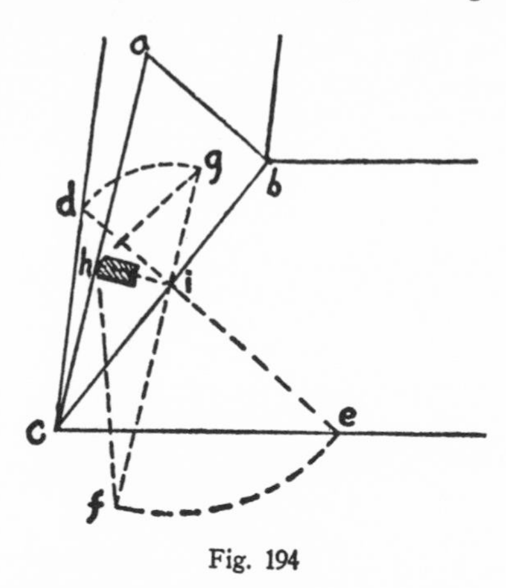

Fig. 194

Practical Backing Method. — At *A*, Fig. 195, is shown a plan in part, of the toe of a hip rafter, and by the dotted lines the two toes of the valley, marked *X, X*. The main drawing, marked *B*, is a side view of the bottom part of a hip rafter. Now draw the two perpendicular dotted lines from the plan, *A*, to the toe of the side view, *B*. The distance between these two lines is the distance from the point of the foot cut to the beginning of the dotted line that runs parallel with the pitch of the rafter. This line gives the depth

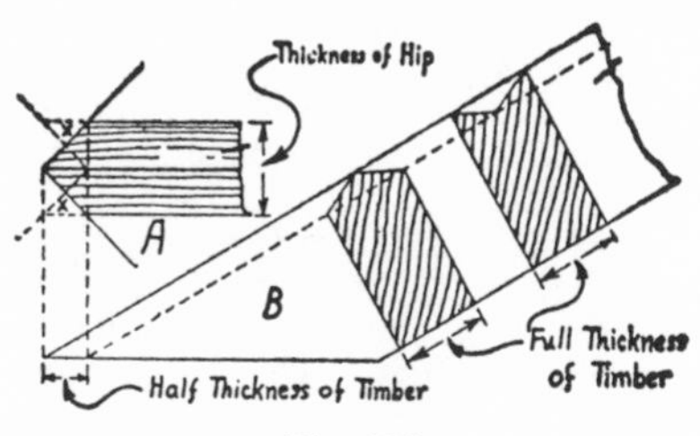

Fig. 195

of the backing, as shown by the two cross sections on the main drawing, one of which is for a hip and the other is for a valley. The bevels are the same, and should be taken on a bevel square with which the marking can be done.

A similar diagram is shown by Fig. 196. Here the problem is to get the bevels for backing irregular pitch hip and valley rafters. At *A* is shown a plan of a corner, in part, giving the toe

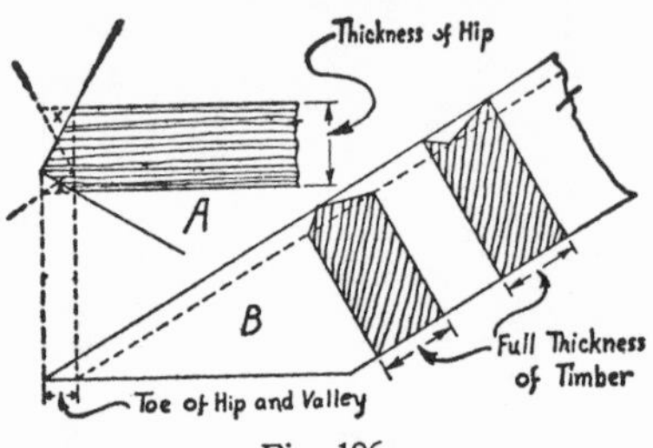

Fig. 196

of the hip rafter, and by dotted lines, the toe of the corresponding valley rafter, marked *X, X*. The two perpendicular dotted lines that run from the toe of the rafter plan, marked *A,* to the toe of the side view of the rafter, marked *B,* give the distance from the point of the foot cut to the beginning of the dotted line that runs parallel with the rafter pitch. This line marks the depth of the backing, as shown by the two cross sections. One of these is for a hip rafter and the other is for a valley rafter. Compare and study Figs. 195 and 196.

Different Backing for Different Angles.—Fig. 197 shows, to the left, four plans in part, of four corners, each having a different angle. The toes of the backed hips are shown in plan, somewhat shaded. The dotted lines that run parallel with the pitch of the rafter shown in part, to the right, give the depth of the backing for the four different hip rafters, shown in part to the left. The figures just below the foot cut of the rafter to the right, show the same distances on the foot cut, as are

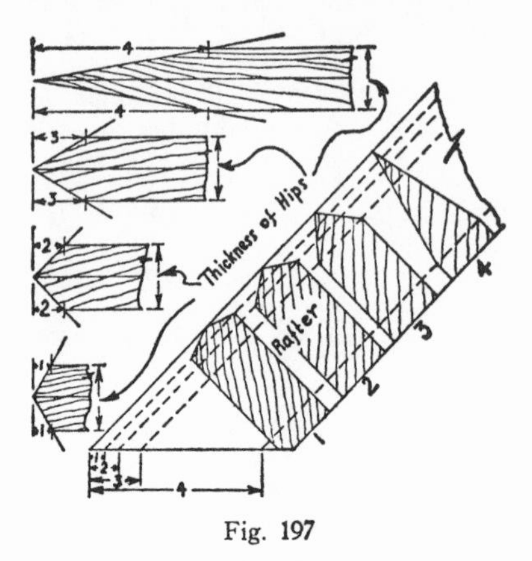

Fig. 197

shown on the different corners, and numbered correspondingly the same. The distance at *1,* directly under the foot cut, is the same as the distances *1, 1,* shown on the plans to the left; the distance *2,* is the same as the distances *2, 2,* and *3* is the same as *3, 3.* The distance *4* goes to the extreme, and is the same as the distances *4, 4.* The angle for this hip is very sharp, and is rarely found in practice. It is given here merely to give a definite contrast between it and the other angles. For the corresponding valleys, the backing is the same, but in reverse order.

Chapter 22

DIFFERENCE IN LENGTH OF GABLE STUDDING

Framing and setting gable studding is a part of roof framing; at any rate, the studding join the roof and the cuts that make the joints are roof framing cuts. The problem of finding the difference in the lengths of the studding involves the pitch of the roof. For instance, if the roof has a one-third pitch or, say, 8 inches rise to the foot run, then the difference in the lengths of the studding would be 8 inches for every foot of distance in the space. If the studding were spaced 1 foot on center, then the difference would be 8 inches, if they are spaced 16 inches, then the difference in the lengths would be 10⅔ inches, or if the gable studding were spaced 2 feet on center, then the difference in the lengths of the stud-

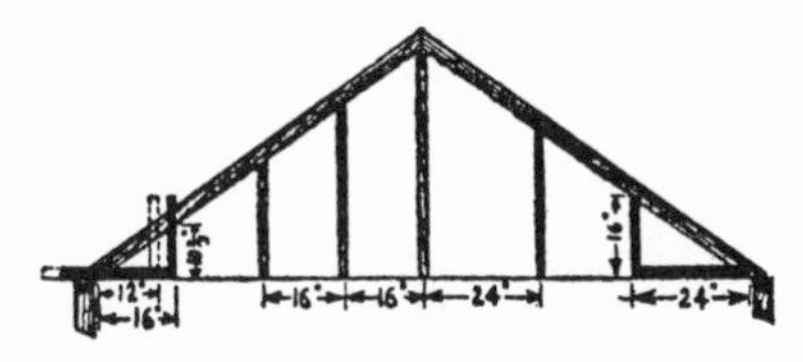

Fig. 198

ding would be twice 8, or 16 inches. How to obtain these differences in the lengths of the studding with the square is shown to the left and to the right of Fig. 198. To the left the square is applied to the rafter so as to intersect the bottom edge of the rafter for a 16-inch space, which shows a rise of 10⅔ inches. If the square were pulled back to the position shown in part by dotted lines, or as shown for a 12-inch space, then the difference in the lengths of the studding would have to be the same as the rise per foot run, or 8 inches. To the right, the square is shown applied to the bottom edge of the rafter for a 24-inch space, which shows that the difference in the lengths of the studding would be 16 inches.

Marking Bevel.—Fig. 199 shows at *A* how the bevel for the cuts is marked with a bevel square. At *B*, by dotted lines, is shown how a double bevel is marked with the square for the center studding. This marking can also be done with a bevel square. At *C* the square is applied for obtaining the bev-

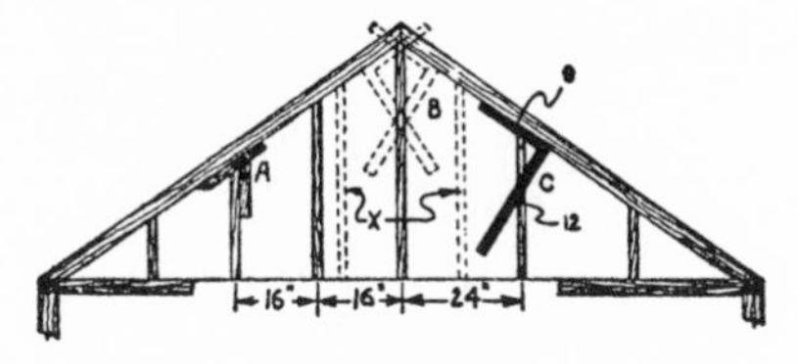

Fig. 199

el for a common studding, using the figures 12 and 8 on the square, which are the same as those used for framing the common rafters. In setting the bevel square, the bevel is first marked on a piece of timber, and then the bevel square is set accordingly.

Framing Studding.—Fig. 200 shows details of the gable studding in a little larger scale. At *A* are shown two views of the center studding, an edge view and a side view. This studding is cut ready to be put in place. Before the center studding is set, it is marked and used for a pattern with which the common studding are marked and cut. The difference in the length is shown as be-

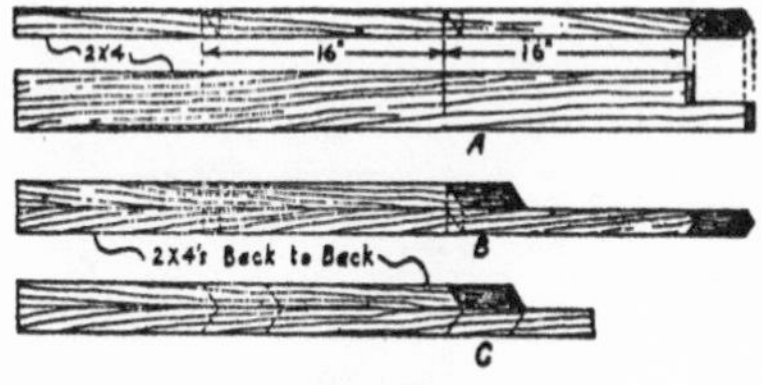

Fig. 200

ing 16 inches, which means that the spacing is 2 feet from center to center. At *B* is shown how the length of the first common studding is obtained, while at *C* is shown how the second studding of the pair is marked. How the shortest pair is marked, is shown by dotted lines at the center of the pair marked *C*. Study the drawings and compare them with the 2-foot spacing in Fig. 199.

Cutting in Pairs.—Fig. 201 shows to the left a detail giving two applications of the square for marking the double

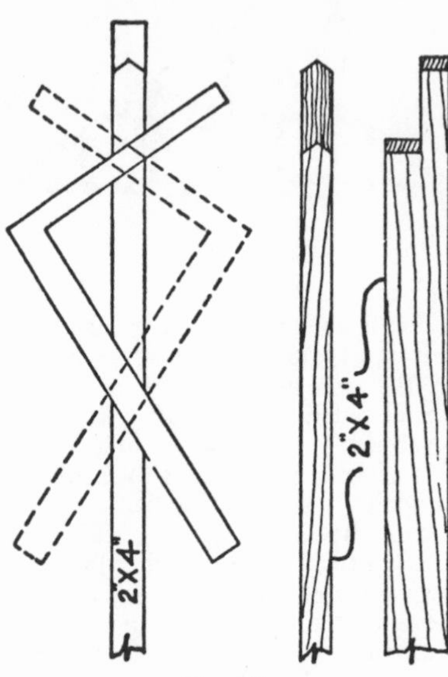

Fig. 201

bevel for the center studding, as shown in Fig. 199. To the right in detail, are shown two views of the top cut of the same studding. Fig. 202 gives details of three pairs of gable studding when they are all cut and set in pairs. (The

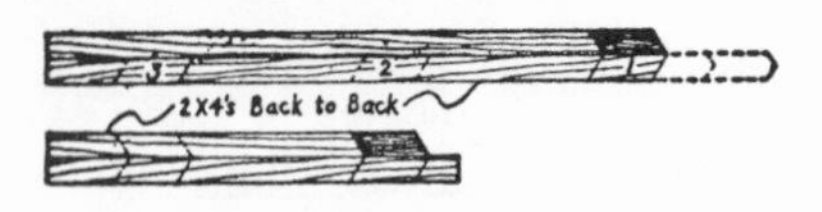

Fig. 202

dotted lines to the right show how much longer the center studding is than the longest pair, when the center studding is omitted.) The upper detail shows the pair of studding, back to back, which is shown by dotted lines in Fig. 199, and pointed out with indicators at *X*. It will take three pairs of studding spaced 2 feet on center for this gable. The different lengths are marked on the upper detail at *1, 2,* and *3*. The bevels of the cuts are shown by dotted lines. The bottom drawing shows the second pair back to back, one is cut and the other is marked by it, ready to be cut. The top of the short pair, is indicated by dotted lines to the left of the bottom drawing. How to get the bevel for the cuts is shown by the detail to the left, in Fig. 203. Here a square is applied to a timber, using 12 and 8, which are the same as the

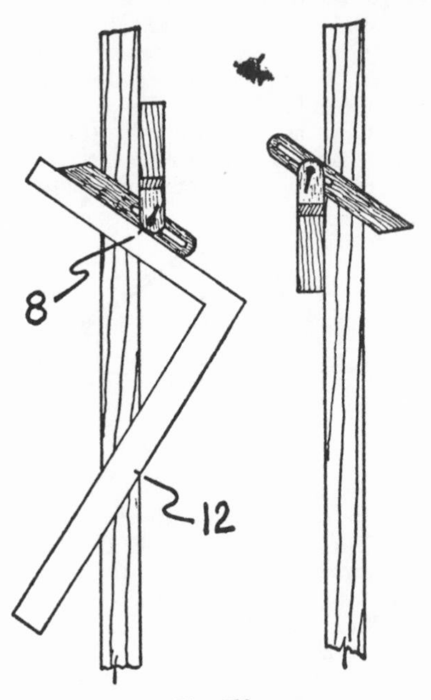

Fig. 203

figures used in framing the common rafters. This bevel is transferred to the bevel square, as shown, and used for marking the bevels for the other studding.

Studding Between Two Pitches.—Fig. 204 shows a half pitch gable roof set over a one-sixth pitch gable roof. This combination is used here for convenience, to show how to figure the difference in the lengths of the gable studding when two pitches, as shown, are involved. The right half of the drawing is in diagram form. The studding are spaced 2 feet on center. The difference in the lengths of the studding for the half pitch would be 2 feet, but since it is set over a one-sixth pitch, there would have to be deducted from 2 feet, the rise of the one-sixth pitch for a 2-foot run, or 8 inches, which would leave the length of the studding only 16 inches, as shown in figures to the left of the studding line. The next studding would be 4 feet for the half pitch, but would have to be cut 16 inches shorter, leaving it only 32 inches long. The center studding would be 24 inches shorter than for a half pitch, leaving it only 48 inches long. These figures are all shown on the diagram to the right, which should be studied. The square to the left is shown applied to a studding, using 12 and 4, to obtain the bevel for the cut that fits the one-sixth pitch. These figures are the same as would be used in framing a one-sixth pitch roof. To the left, by dotted lines is shown exactly the same problem, excepting that the one-sixth pitch is above. The difference in the lengths of the studding would be the same as we found in the problem shown by the drawing to the right. The dotted-line square is shown applied for obtaining the bevel for the studding where they join the one-half pitch roof, using 12 and 12 on the square. Study this drawing and think it through.

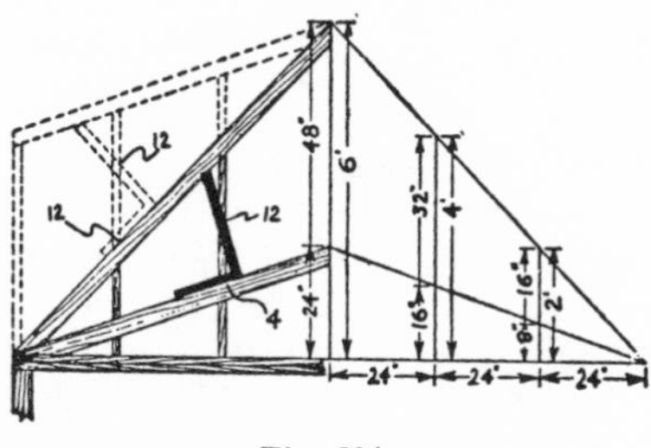

Fig. 204

Difference in Lengths of Gable Studding.—Fig. 205 shows a one-line drawing, or rather, a diagram, dealing with a still different problem in finding the difference in the lengths of the gable studding. Here we have a gable of a third pitch roof, but the base line on which the studding rest, slopes 2 inches to the foot, as shown. The gable is separated from the rest by a dotted line. In this case, working from right to left, if the studding were spaced 12 inches on center, the difference in the lengths of the studding would be 8 inches for the pitch of the roof, and 2 inches for the slope of the base, or 10 inches. How to apply the square for obtaining this, is shown to the upper right. The angles on the detail are the same as those on the diagram, as pointed out by the double indicator. If the spacing is 2-feet on center, then the difference in the lengths of the studding would be 16 inches for the roof, and 4 inches for the slope, or 20 inches. But when you pass the center studding the roof slopes in the opposite direction, which makes a difference in finding the difference in the lengths of the studding. The diagram to the left, shows the studding spaced 16 inches on center. Starting with the center studding, which is 48 inches, plus 12 inches, or 60 inches long, the next studding would be 10⅔ inches shorter than 48 inches, for the roof, or 37⅓ inches, to which would have to be added the increase for the slope in the base line, or 14⅔ inches. The same process is continued until all the lengths have been found. The problem is rather simple and the student is asked to study the diagram, working it out for both the 16-inch and the 2-foot spacings.

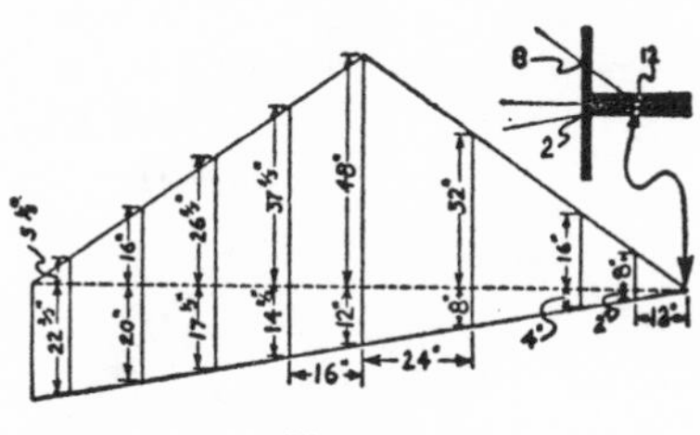

Fig. 205

Chapter 23

VARIOUS ROOF FRAMING PROBLEMS

Making Cuts.—There are different kinds of cuts that carpenters often have to make, which are not covered in ordinary roof framing, although, most of them are roof framing cuts. Any kind of timber that joins a roof must be cut to fit the roof it joins, and that makes it necessary for the workman to know how to obtain the cut with the steel square. If he can't do that, then he will have to resort to the cut-and-fit method of making such joints.

Cuts for Dormer Rafters.—Fig. 206, to the right, shows a pair of rafters for a half pitch roof in place, with a dutch dormer to the left. The problem here is how to get the cut for the dormer rafters when they join the sheeting of the main roof. As shown on the drawing, the dormer rafters have a 30-degree slope, or pitch, and the main roof, as already stated, is half pitch. There are different ways to get this cut, but a simple way is shown to the upper left. Draw the horizontal line, *A-B*. Then draw the rafter timber, as shown, at a 30-degree angle. Now apply the square to the horizontal line, using 12 and 12. These figures are used because they will bring the blade of the square to a half pitch slope. The cut for the dormer rafters that will fit the sheeting of the main roof, can be made by using 12 on the blade of the square, and point *X* on the tongue, the blade giving the cut. It is not necessary to draw the rafter timber, as shown here. All that is needed is a line to the pitch of the dormer roof, as shown in a little larger scale in Fig. 207. To make this diagram, draw the horizontal line first. Then draw line *C-D* to the pitch of the dormer roof, in this case 30 degrees, as shown by the drawing. Now apply the square to line *A-B*, using 12 and 12 and making 12 on the body come to the crossing of the two lines. Then 12 on the body and point *X* on the tongue will give the cut. This principle will work on any other pitch, by making line *C-D* slope to the pitch of the dormer roof, and bringing the blade of the square to the pitch of the main roof.

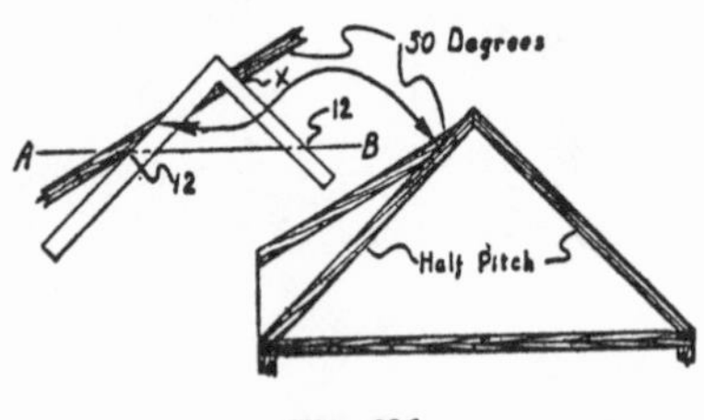

Fig. 206

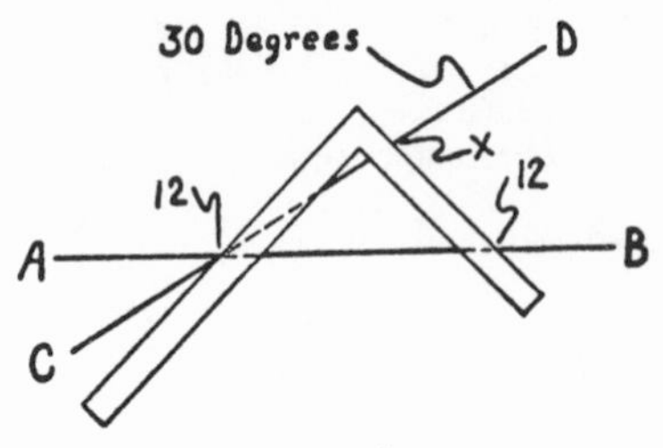

Fig. 207

Cuts for Joining Pitch Roofs.—Fig. 208 shows a diagram of a roof with a half pitch to the right, and a 30-degree pitch to the left. The main building, however, is to be altered so as to give it a full half pitch gable roof, as indicated by the dotted line. The problem here is to obtain the foot cut for the rafter, where it fits the 30-degree roof.

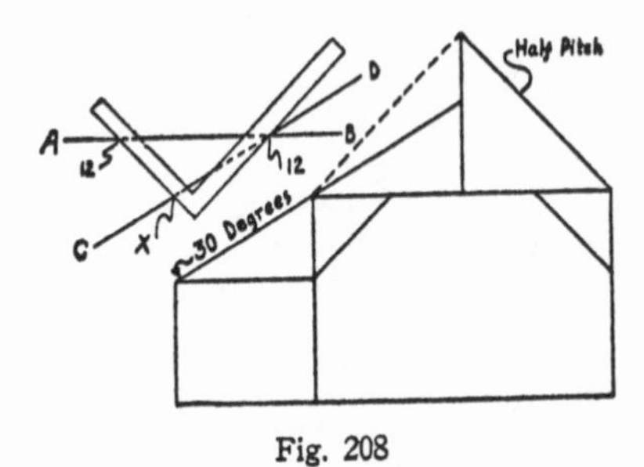

Fig. 208

As the diagram to the left shows, draw line *A-B*. Also draw line *C-D* to the slope of the roof to the left, which is 30 degrees. Then apply the square as shown, using 12 and 12, the figures that give the pitch of the main roof. Now 12 on the body and point *X* on

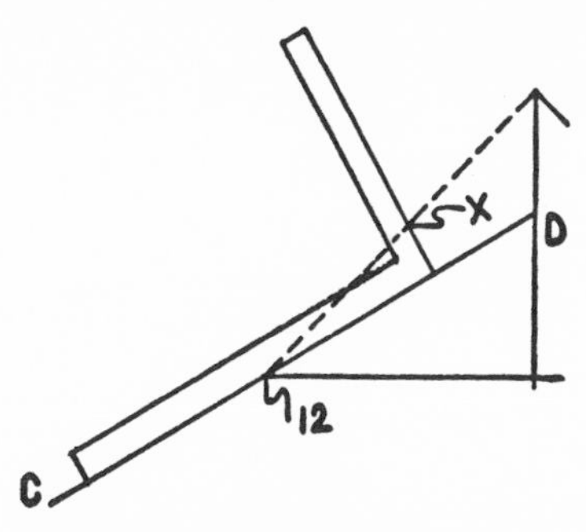

Fig. 209

the tongue will give the cut. Mark along the blade of the square. The same results can be obtained by making a diagram as shown in Fig. 209. Here only a part of the roof shown in Fig. 208 is used. Line *C-D* represents the 30-degree roof and the dotted line represents the rafter that is to be framed. Proceed by applying the square in the position shown, bringing

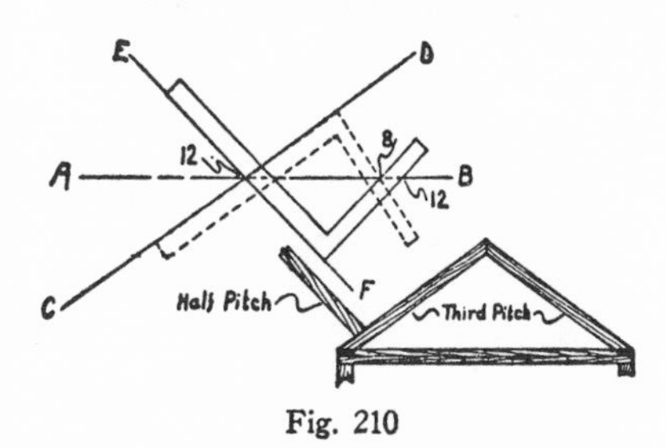

Fig. 210

12 on the body to the toe of the dotted-line rafter. Now 12 on the body of the square, and point *X* on the tongue will give the cut. Mark along the body of the square. This drawing, made to a little larger scale, should be compared with the diagram to the left in Fig. 208.

One Pitch Joining Another Pitch.— Fig. 210 shows a cross section of a third pitch roof, to the right, which is joined by a half pitch rafter, shown in part to the left. The drawing to the upper left shows how to make the diagram for obtaining the cut. Draw line *A-B,* as shown. Next draw line *C-D,* giving it the slope of the main roof, as indicated by the dotted-line square,

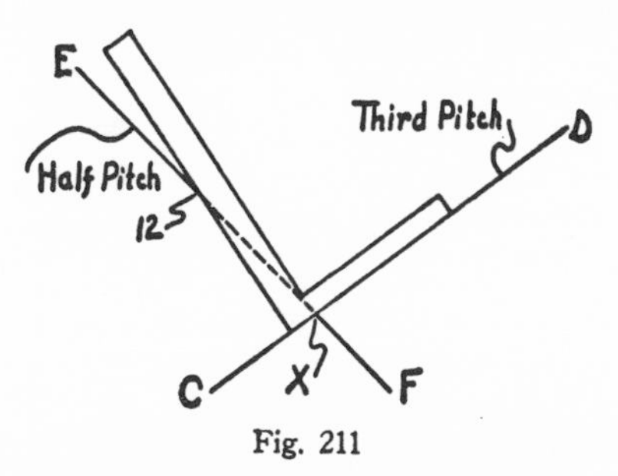

Fig. 211

which is applied, using 12 and 8. Then draw line *E-F* with the square, using 12 and 12, so as to give it a half pitch. Now turn to Fig. 211. This diagram is drawn to a larger scale, and shows the square applied to line *C-D* and line *E-F* in such a manner that it will give the cut for the half-pitch rafter where it joins the third-pitch roof. To mark

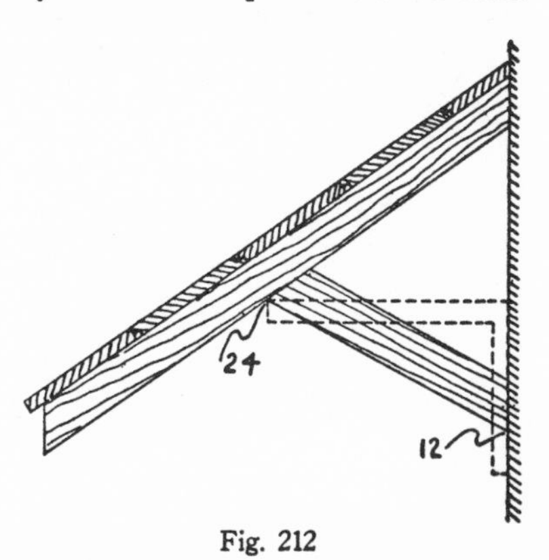

Fig. 212

the cut on the rafter material, use 12 on the body of the square and point *X* on the tongue, as shown. Mark along the tongue.

Cuts for Braces.— Fig. 212 shows how to get the cut for a brace where it joins the main building, at the same time locating the point of the cut where it joins the rafter. In this case, 24 on the body of the square and 12 on the tongue are used, so that one operation will do the job. The same results can be obtained by using 12 on the body of the square and 6 on the tongue, but then two steps must be taken. Fig. 213

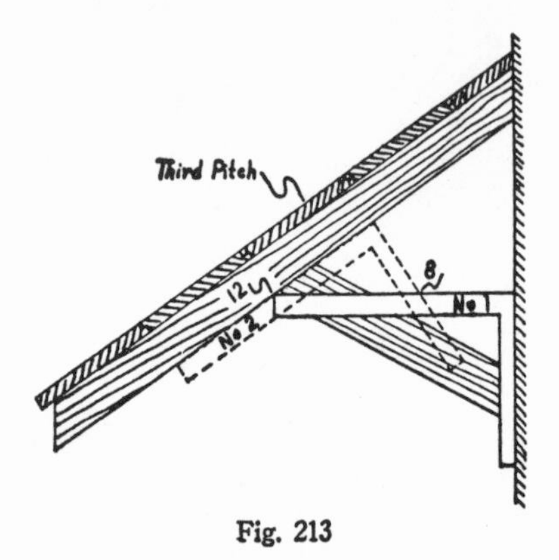

Fig. 213

shows the same brace and roof. Square No. *1* is in the position of the square shown by dotted lines in Fig. 212. To get the bevel for the cut where the brace joins the rafter, apply square No. *2* to square No. *1* as shown, using 12 and 8, the figures that will give the slope of the roof and also the cuts for the rafter. Another method is shown by Fig. 214. Here the line marked third pitch represents the slope of the rafter, while the line marked sixth pitch represents the slope of the brace. Applying the square to these lines as shown will give the points to use for marking the bevel of the brace where it fits the rafter, or 12 on the tongue of the square, and point *X* on the body—the tongue giving the bevel.

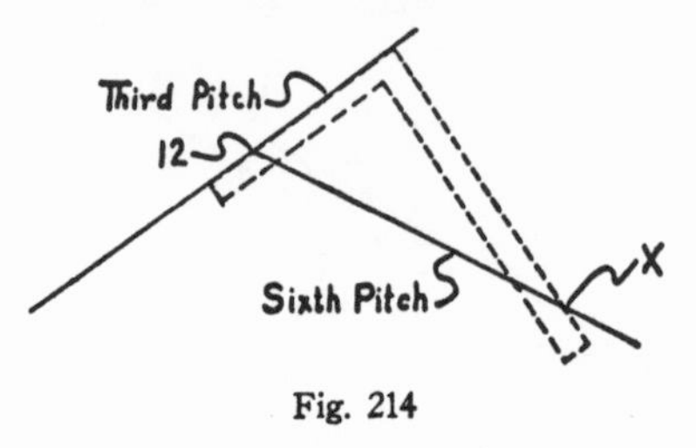

Fig. 214

The student will find the one-line diagrams a little harder to understand than the details, but when the principle is clear in his mind, he will find the diagram method simple, because all unnecessary lines have been eliminated. The important thing is to make the slope of the lines, respectively, the same as the slope of the roofs, or braces, as the case may be. Study the drawings, and practice finding different cuts, until the principle is fixed in your mind.

Chapter 24

FRAMING CURVED-EDGE RAFTERS

In this chapter framing rafters that have curved edges will be treated, especially the framing of hips and jacks. While only the ogee curve is used here, the principle will apply to problems involving any other irregular shape, and while only the regular hip roof is used in the illustrations, the principle will also apply to irregular pitch and irregular plan roofs.

Obtaining Curvature of Hips.—Fig. 215 is a diagram of a corner of a hip roof, whose rafters have an ogee curvature. To the upper right is shown a common rafter as if it were lying on the side. The base of this rafter is on line *c-a,* while line *c-b* gives the base of the hip rafter. To obtain the curvature of the hip rafter, divide the base of the common rafter into a certain number of equal spaces (unequal spaces will also work). The more spaces, the more accurate will be the results. In this case there are 12 spaces. After raising the 12 perpendicular lines, as shown, from the base of the common rafter to the curved upper edge of the rafter, drop these lines, as shown by dotted lines, from the common base to the base of the hip rafter. Where these lines intersect the hip base, draw 12 right-angle lines, as shown. Then make line *1* of the hip equal in length to line *1* of the common rafter, as indicated by the dotted part-circle to the left. With a compass transfer the respective lengths of the other lines to the respective lines of the hip rafter, and mark them as shown. This done, draw a curved line through the points marked, which should give you the curvature of the top of the hip rafter as shown by Fig. 216 at the bottom, in an upside down position.

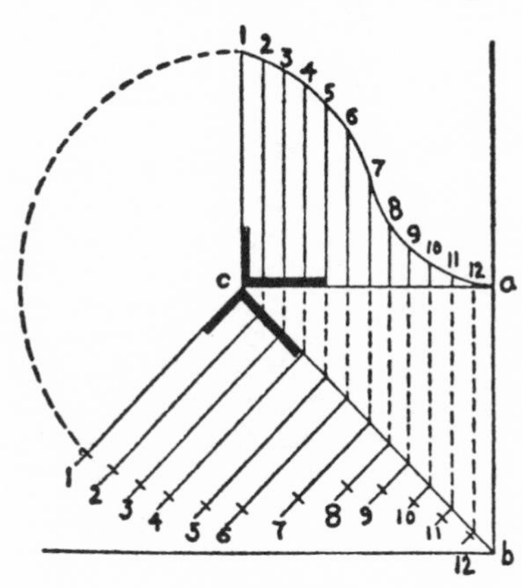

Fig. 215

Rafters Developed.—Fig. 216 shows the common rafter and the hip rafter developed in full. The two dotted part-circles to the left show how the top cut of the common rafter has been transferred to the hip. With the bottom edge of the hip drawn, the curvature of the upper edge of the hip rafter can be found by making lines *1, 2, 3,* etc., of the hip rafter respectively the same in lengths as lines *1, 2, 3,* etc., of the common rafters, measuring from the bottom edge of the rafters in both cases, and then drawing in the curved line as explained in the other case. A little study will show that the two methods are practically the same in principle.

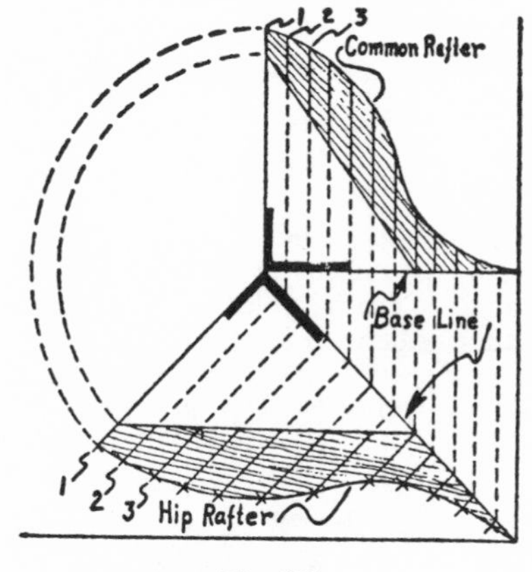

Fig. 216

Different Lengths of Jacks.— Fig. 217 shows how to get the different lengths of the jack rafters. The common rafter is shown on the side, and a one-line plan shown below with two hip rafters. The jack rafters are numbered from *1* to *6*. Now the points

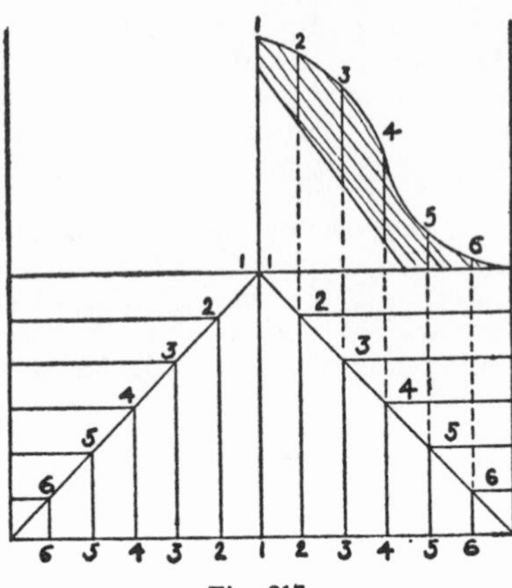

Fig. 217

where the jacks intersect the hip to the right, as shown by dotted lines, are raised to the base line of the common rafter and on up to the curved top edge. The top cut of the common rafter is shown numbered *1,* which gives the

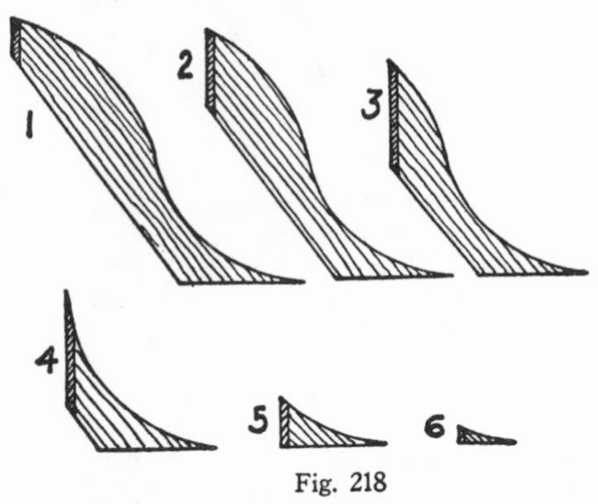

Fig. 218

length of the longest jack. Number *2* gives the next longest jack, and number *3* the next, and so on down to number *6*. Side views of the six jack rafters are shown by Fig. 218. The plan shown by Fig. 217 gives two hip rafters, so two pairs of each jack must be made, excepting the longest jack for which only one pair is needed, together with a single jack having a double bevel where it joins the two hips. This is numbered *1* at the bottom section of Fig. 217.

Stepping Off Common Rafter.—Fig. 219 shows two views of the common

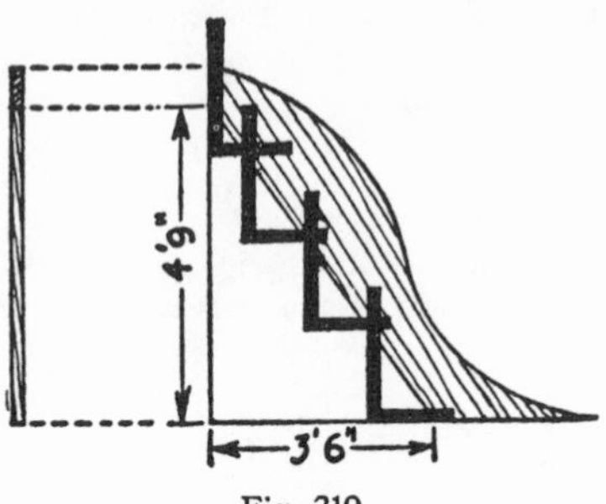

Fig. 219

rafter in a little larger scale. To the left is shown an end view of the top cut and the bottom edge of the common rafter. To the right is a side view, with the square applied for three full steps, and one half step, because there are 3 feet 6 inches in the run, measuring the run from the heel of the seat cut, rather than from the toe. This

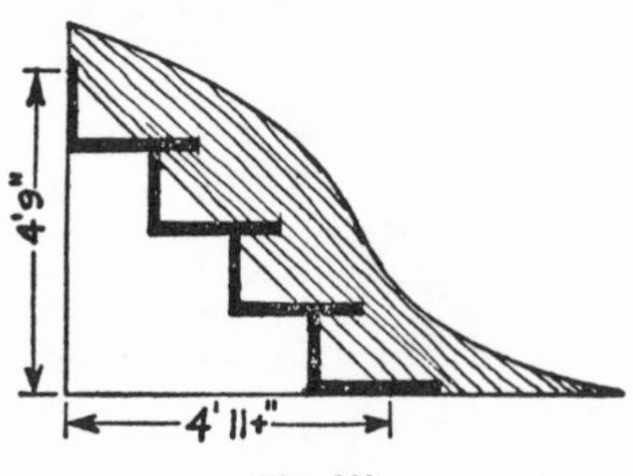

Fig. 220

should be kept in mind, for the stepping off and marking of such rafters must be done on the bottom edge.

Stepping Off Hip.—Fig. 220 shows, also in a larger scale, the hip corresponding to the common rafter shown in Fig. 219. The hip run, as shown, is 4 feet 11 inches, plus, and the rise is the same as the rise cf the common rafter, 4 feet 9 inches. While the stepping off here could be done on the basis of 4 full steps and one fraction of a step to take care of the 11 inches, plus; the method shown is a modification of the 12-step method, treated in an earlier chapter. The way this is done, multiply both the run and the rise by, say, 3, letting inches on the square represent feet, and take 4 steps. Why take only 4 steps? Well, because 4 is one third of 12. Multiplying the two figures by three increased the run and

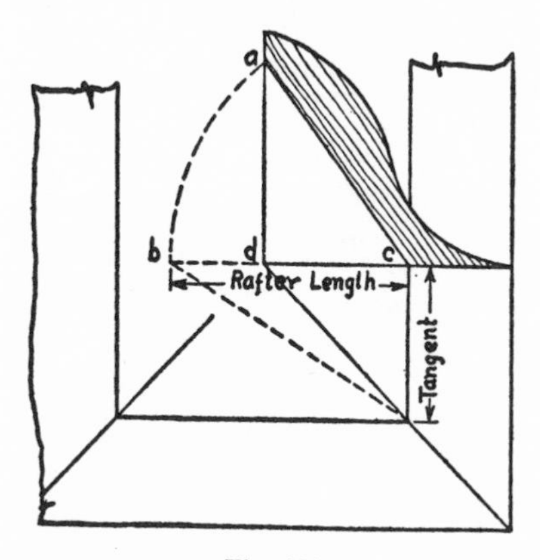

Fig. 221

the rise taken on the square, and at the same time reduced the number of steps. Had the multiplying been done by 2, then 6 steps would have been necessary. Study this until you understand it.

Edge Bevel for Jacks.—Fig. 221 shows how to get the points to use on the square for marking the edge bevel for the jack rafters. The drawing shows the common rafter on the side, where *c-d* represents the run, *d-a* the rise, and *a-c* the rafter length. Now set the compass at point *c*, and transfer the rafter length from *c-a* to *c-b*. The rafter length and the tangent will give the bevel. Mark on the rafter length.

Edge Bevel of Hips.—Fig. 222 shows how to get the points to use on the square for the edge bevel of the hip rafter. Here the hip is shown on the side, but in an upside down position. The run is represented by *c-d*, the rise by *d-a*, and the rafter length by *a-c*. Now set the compass at point *c* and transfer the rafter length from *c-a* to *c-b*. By taking the tangent on the tongue of the square and the rafter

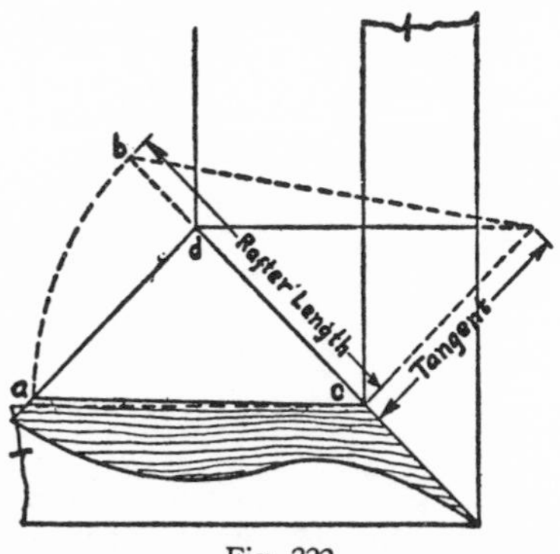

Fig. 222

length on the blade, you can mark the edge bevel along the blade of the square. It should be remembered that in all cases where the top edge of the rafter is curved or irregular, the marking is done on the bottom edge of the rafter. It should also be remembered that the run used in such operations is measured from the heel of the seat cut, rather than from the toe.

Principle the Same.—The student is advised to study the various problems given here, in keeping with similar problems that have been treated before. For in principle there is little difference, excepting the shape of the upper edges of the rafters.

Chapter 25

HAYFORK HOODS

Hay barns often have hayfork tracks installed directly under the comb of the roof. These tracks extend beyond the main roof on the ends that receive the hay. The hayfork hood is added to the barn roof to shelter this extension of track. The farming of such hoods is the subject of this chapter.

Skeleton of Hayfork Hood. — After studying Fig. 223, turn to Fig. 224 where is shown a diagram of the same

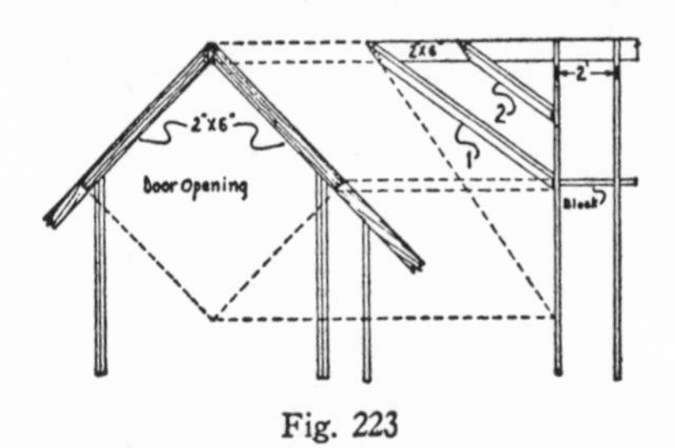

Fig. 223

hood. Locate points *A, D, B, C* of Fig. 224 on Fig. 223. Now carry point *D* to *k*, as shown by dotted line, and then on to *l*. Also carry point *B* to *j* and on to *i*. The triangle *A, D, B* in Fig. 224 gives the end view of the hood,

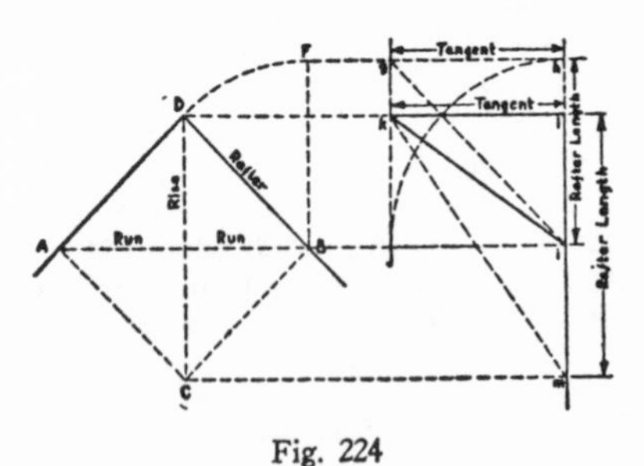

Fig. 224

while *k, l, i*, to the right, gives the side view. The line *l, i, m*, represents the last rafter of the main roof. Now to develop the diagram, set the compass at point *B* and with it strike the dotted part-circle from *D* to *F*. From *F* strike the dotted line through *g* and on to *h*. This done, set the compass at *i* and carry point *h* to *j*, as shown by the quarter circle. Carry *C* to the right to *m*, and join *k* with *m*, as shown by dotted line; also join *g* with *i*. To get the edge bevel of the hood rafters, use the triangle *i-g-h*. The rafter length will give the bevel. (In this case the tangent and the rafter length are the same in length, so either will give the bevel, but that is not always true.) To get the side bevel of the rafters where they join the ridge board, use the triangle *m-k-l*. The rafter length will give the bevel. Study Figs. 223 and 224 in connection with the explanations.

Skeleton Hood.—Fig. 225 shows the same skeleton hood framed a little differently. To the left is a front view and

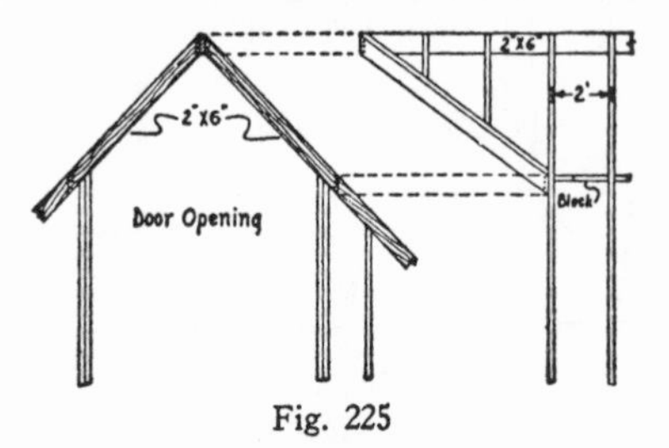

Fig. 225

to the right a side view, showing the spacing and the reinforcing block. The dotted lines give the relationship of the two drawings. These drawings should be studied in connection with the diagram shown by Fig. 226, where the diagram to the left represents a plan of the hood joined to the main roof. What the different lines represent is given in the diagram. To get the right conception of the triangle *a-b-c* in mind, you will have to imagine that line *a-c*, called "run," is on a level with the ridge, and that point *b*, when the rafter is in position will be directly under point *c*. In other words, imagine that the run is hinged on a level with the ridge board and in the position now shown. Then when point *b* swings down, it will be directly under point *c*.

This being true, what is now called "rise" would in reality be fall. This should be remembered to keep the proper setting of the diagram. The shaded bevel gives the bevel for the side of the rafter, both where it joins the rafter and where it joins the ridge-board. To get the points for the edge bevel, set the compass at point *a* and transfer the rafter length, *a-b*, to *a-d*. Then draw in the other lines as shown. Now the rafter length will give the edge bevel where the rafter joins the ridgeboard, and the tangent will give the bevel where it joins the rafter of the main roof. Since the main hood rafters are set in a plumb position, so far as the sides are concerned, they are framed exactly like a valley rafter.

Bevel for Backing.—The diagram shown to the right in Fig. 226, illus-

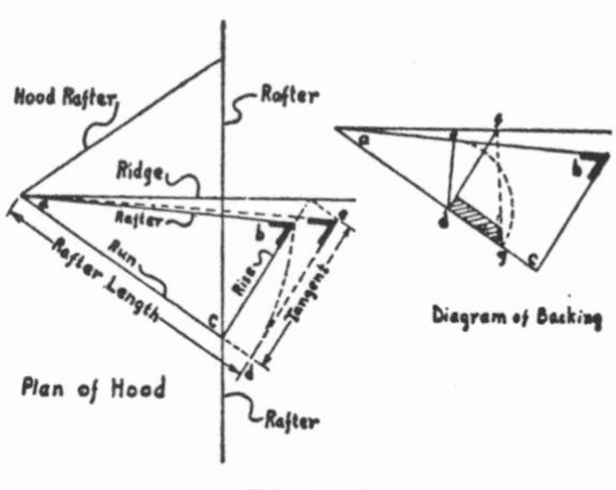

Fig. 226

trates how to get the bevel for backing, in case the rafters are backed. The triangle *a-b-c,* is the same as the one shown in the left diagram. At any convenient point draw line *e-d* at a right angle to the rafter line, *a-b*. With a compass set at *d,* transfer *d-e,* as shown by the dotted part-circle, to *d-g*. At a right angle to *a-c,* draw line *d-f*. Now join *f* with *g,* as shown by dotted line. The bevel at *g* is the bevel for the backing.

Practical Method for Framing Hoods.—Fig. 227, to the left, is a diagram of the drawing shown to the left in Fig. 223. The diagram to the right gives the side of the hood, looking straight at it, which makes it different from what is shown to the right in Fig. 223. In this case the distance *1-2* is transferred with a compass to *1-3*. Point *3* is then carried to the right, establishing point *4*. Distance *c* is made equal to distance *d* and then line *b* is drawn in. With these two simple diagrams you can obtain the cuts for the hood rafters. The side bevel where the rafters join the ridge is gotten by tak-

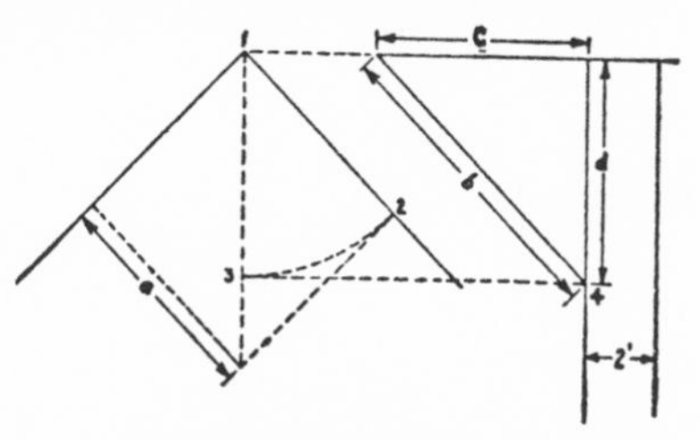

Fig. 227

ing distance *a* on one arm of the square, and distance *b* on the other. Distance *b* gives the bevel. The side cut where the rafter joins the main roof is square across. To get the edge bevel, take distance *c* on one arm of the square, and distance *d* on the other. In this case the distances are the same, making a true miter cut. In cases where the two distances are different, the former gives the edge bevel where the rafter

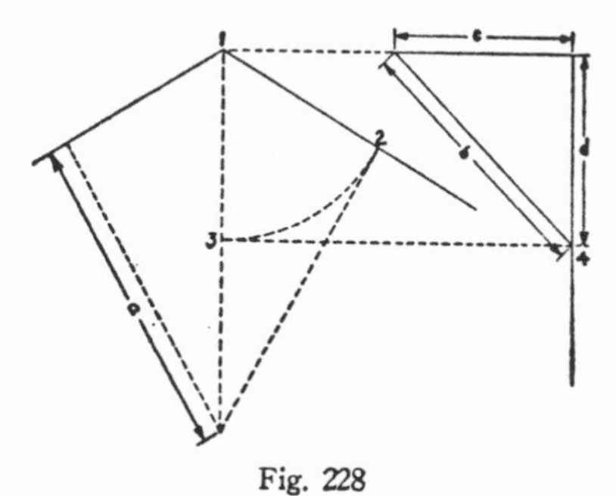

Fig. 228

joins the ridge and the latter gives the bevel where it joins the rafter of the main roof. The explanation just given, covering Fig. 227, will apply to Fig. 228, which shows a much lower pitch. The same practical method will apply to all other pitches, for the principle is the same.

Hood Cuts for Steep Roof.—Fig. 229, to the left, is a diagram of a hood on a much steeper roof. The triangle,

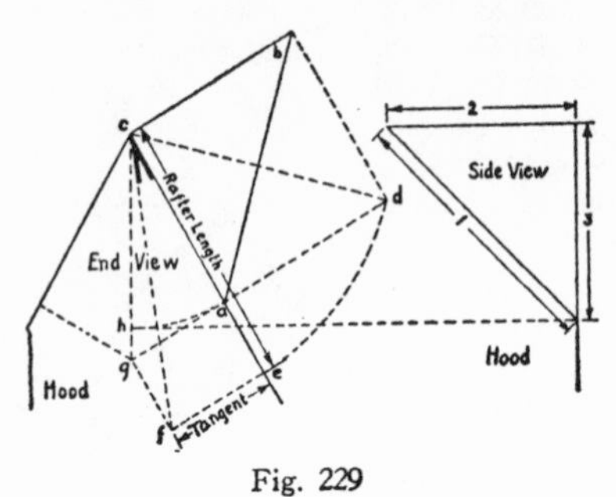

Fig. 229

a-b-c, shows a side of the hood turned up as if it were hinged on line *a-c*. Line *c-b* joins the ridge when down in position, while *b-a* gives the length of the rafter. For convenience, this triangle has been transposed in such a manner that *c-d* represents the rafter length, just as *a-b* does. Now then, set the compass at point *c* and transfer the rafter length, *c-d,* to *c-e.* Then the rafter length taken on one arm of the square and the tangent on the other will give the side bevel where the rafter joins the ridge. Mark on the rafter length. The diagram to the right shows the side of the hood, looking straight at it, as explained in Fig. 227. Distance *2* is made equal to distance *3,* which is the same as distance *c-a* in the diagram to the left. Distance *2* and distance *3* taken on the square will give the edge bevels for their respective positions. In this case the two edge bevels are the same, because the two distances are the same. When the distances are different, the respective bevels will also be different.

Chapter 26

SIMPLE APPLICATIONS OF SQUARE

Most of the applications of the steel square are rather simple—even those that are considered difficult by the ordinary carpenter. What the carpenter, particularly the apprentice, should strive for is the skill in handling the square, that will eliminate false motions to the extent that they will be kept at a minimum. In order to do this, it will be necessary to study new movements and make them deliberately until they become automatic. This achievement will cost mental effort and actual practice on the part of the student, but it is one of the best investments he can make, if he wants to be a master in the use of the steel square.

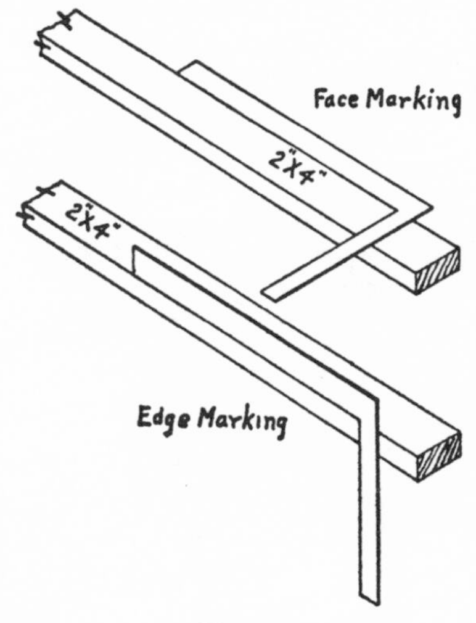

Fig. 230

The Square Cut.—Fig. 230 shows by the upper drawing the square applied to a 2x4 for marking a square cut across the face. The bottom drawing shows the square applied for marking a square cut on the edge of a 2x4. These two applications of the square are perhaps the simplest of them all, and are made on all kinds of sawed timbers.

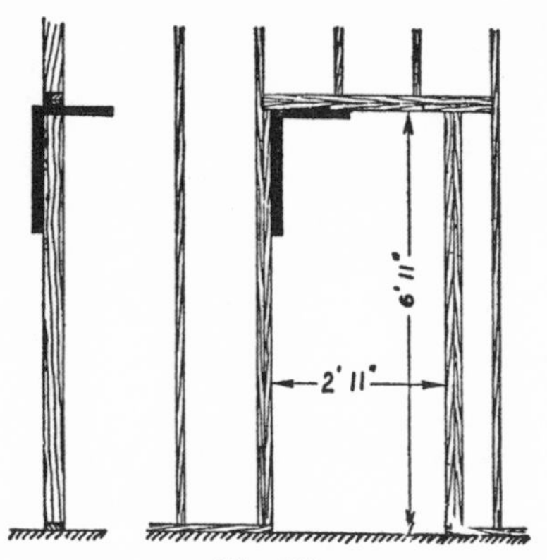

Fig. 231

Squaring Openings.—Fig. 231 shows to the right a rough door opening, framed for a 2′ 8″ by 6′ 8″ door. In this drawing the square is shown

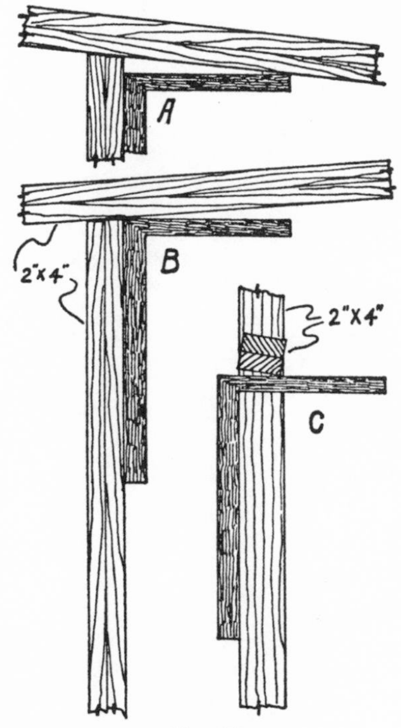

Fig. 232

placed in the left corner to test the squareness of the opening. The cross section of the opening, to the left, shows the square in position for testing the header as to whether or not it joins the studding square across. Fig. 232 shows three details, somewhat exaggerated, of out-of-square opening joints. At *A* the heel of the square does not contact the corner. At *B* the fault is in reverse, while at *C* the test shows a cross section of a poorly joined header.

Sixteen-Inch Spacing. — The upper drawing of Fig. 233 shows an outside 2x4 plate in part, with the built-up corner and three wall studding in place.

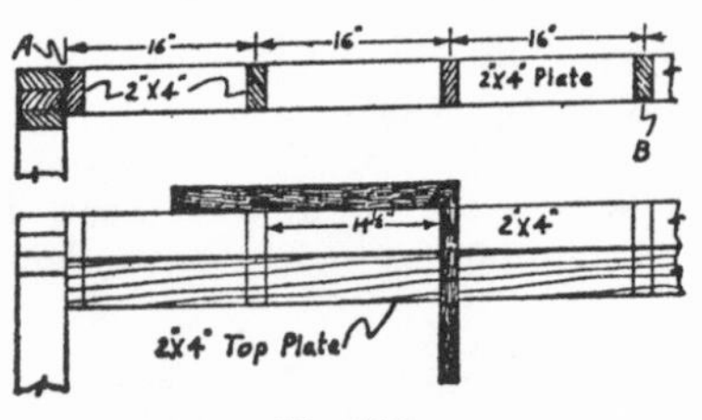

Fig. 233

It should be noted, that the first space is measured from point *A* to the center of the first wall studding, and following that the measuring is done from center of studding to center of studding. The reason for starting at point *A* is that when the lather butts the lath against the corner studding, they will center the wall studding at *B*. The bottom drawing shows the same plate marked for the built-up corner and for the three wall studding. The square is shown applied for marking the second space. Notice that the distance between the spaces is shown as 14½ inches. The distance between the studding when they are in place is only 14⅜ inches. The reason for this difference lies in the fact that the studding material is 1⅝ inches thick. For measuring the spaces for 1⅝-inch studding, use the 14½-inch point on the inside edge of the square, which will measure the distance between the spaces, to which the tongue of the square will add 1½ inches, making a total of 16 inches.

The top plate is shown shaded, placed in such a way that the two can be marked with one operation.

Two-Foot Spacing.—The upper drawing of Fig. 234 shows a plate in part, with the corner studding in place and two of the wall studding. Here,

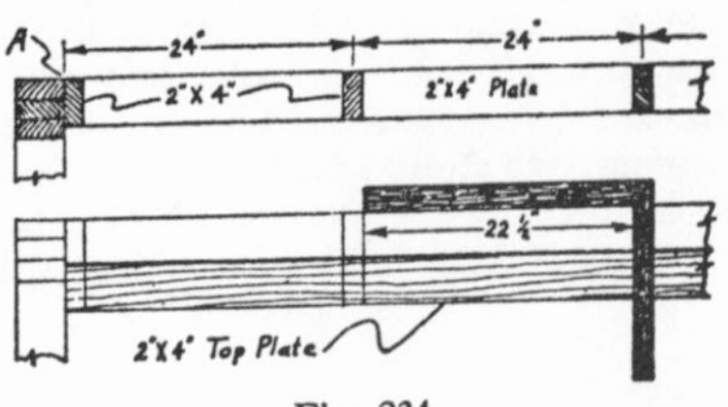

Fig. 234

as in the sixteen-inch spacing, the first space is measured from point *A* to the center of the first wall studding. The bottom drawing shows the same plate marked for the studding and the square in position for marking the second space. The top plate is shown shaded and placed in such a position

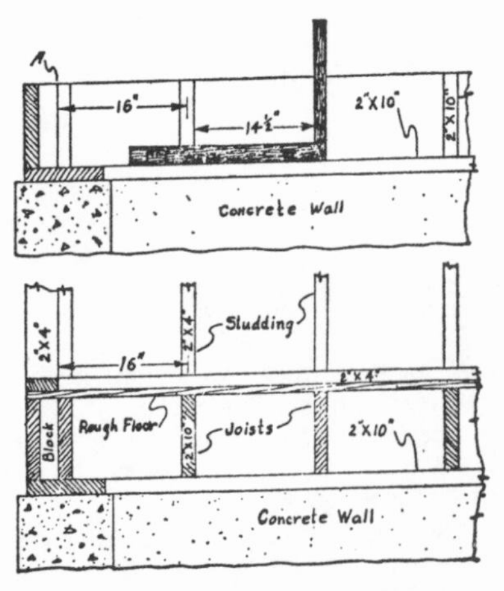

Fig. 235

that the two plates can be marked at the same time. The distance between the marked spaces in two-foot spacing is 22½ inches, as shown on the drawing.

Marking for Joists. — Fig. 235, the upper drawing, shows the square in position for marking the second space for the joists. It should be noted that the measuring for the first space is done from point *A,* just as it was in the spacing for the studding. The reason for this is to bring the studding directly over the joists. This is shown by the bottom drawing, where the rough floor is in place—also the foot plate and the studding, which are shown in part. Compare and study the two drawings.

Squaring up Joists.—Fig. 236 shows two ways to apply the square for squaring up joists. The upper drawing

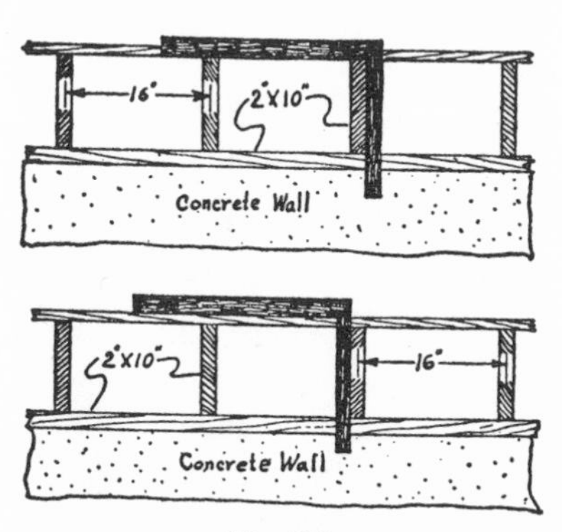

Fig. 236

shows the square, as it were, hooked over the joist and the blade resting on the edges of two joists. The inside edge of the tongue, in this case, contacts the joist and shows whether or not it is square. The bottom drawing shows the square resting on the spacing board and butting against the joist. Here the outside edge of the tongue contacts the joist, showing the squareness of the joist, or the lack of it.

Spacing for Ladders. — Fig. 237 shows two views of the uprights of a 12-foot ladder. The upper drawing shows to the left the square applied for marking the third space, and to the right the square is shown in position for marking the last step. It should be noticed that the steps are spaced 12 inches from top to top, which is the standard spacing for ladders. The bottom drawing shows how to square up

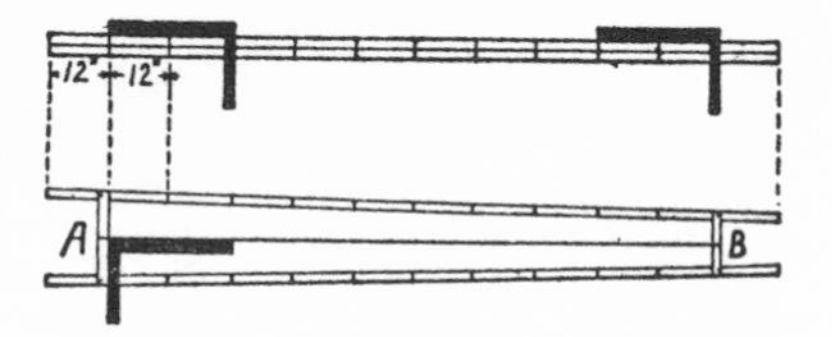

Fig. 237

a ladder when it is wider at the bottom than at the top. Nail the first and the last steps on with one nail to the bearing. Then mark the center of the two steps, as at points *A* and *B,* and stretch a line from one to the other. Apply the square to the first step as shown. When

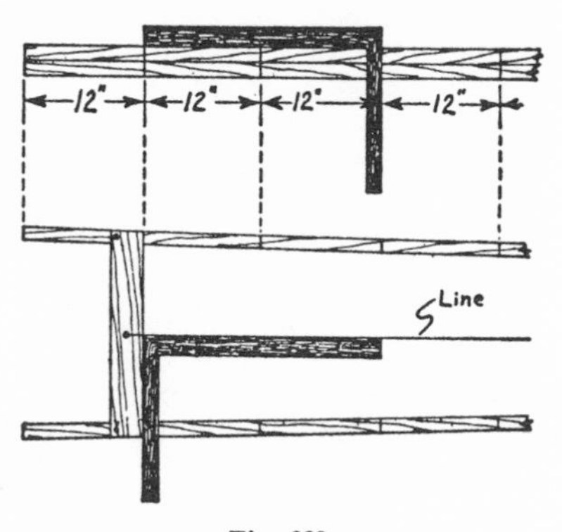

Fig. 238

the blade is perfectly parallel with the line, the ladder is square. Ordinarily this is not necessary, but in cases where accuracy is important, this is a good way to square up a ladder. Fig. 238 shows details of the left parts of the drawings shown in Fig. 237.

Chapter 27

DETERMINING SIZE OF STEP

Among the problems treated in the last chapter was that of making a ladder. This was done to connect the chapter to what is to follow, for this chapter is the beginning of a series, dealing with the steel square and its practical uses in framing porch steps and building different kinds of stairs.

Determining Rise and Run. — Fig. 239 shows a method of determining the rise and run of the step in building either steps or stairs. In this case 24 inches on the blade and 11½ inches on the tongue are taken as points, which are connected by the diagonal straight line, as shown. Now if the run of the steps were limited to 10 inches (or any other practical figure) square up from the established point on the blade, and the distance between the edge of the blade and where this squaring contacts the diagonal line would be the rise, or as shown by the drawing, 6¾ inches. Multiply this figure by the number of risers in the flight of steps, and you have the total rise of the steps. On the other hand, if the rise of the steps were limited to 6¾ inches (or any other rise) then square from that point on the tongue to the diagonal line, and the distance between the edge of the tongue and where this squaring contacts the diagonal line would be the run. However, it should be remembered that this method is basic; that is to say, the stair builder should not take the results as hard and fast. When the rise or the run, whichever the case may be, has been found, the workman should make such adjustments as in his judgment are necessary to make the finished job fulfill its purpose satisfactorily. The method is based on the rule, that the sum of the rise and the run in inches should be around 16½ inches; roughly, between 16 and 17 inches, going below or above the 16½ inches only in cases of necessity. To the right of the drawing is shown how the spacing of steps on a ladder can be determined by this diagram, although the standard spacing for ladders is 12 inches.

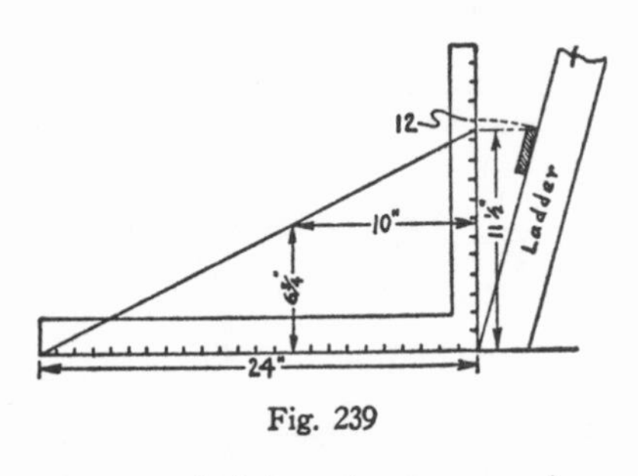

Fig. 239

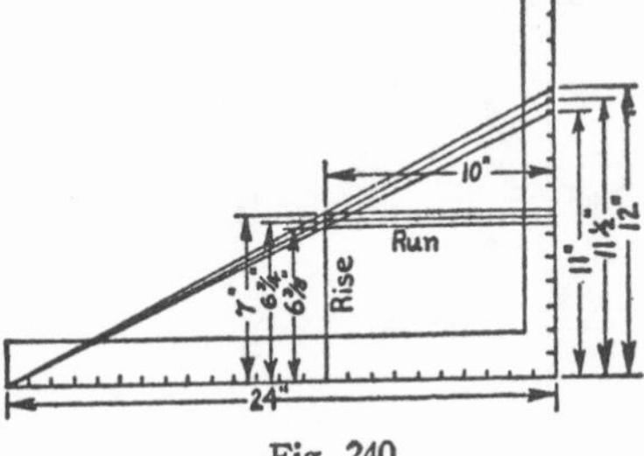

Fig. 240

Fig. 240 shows a similar diagram, giving three different points to be used on the tongue, 11, 11½, and 12 inches. Here the figure used on the blade, 24 inches, remains the same, which leaves the run of the step in the three cases the same, or as shown, 10 inches. But the figures taken on the tongue of the square being different, the figures that give the rise will also be different, as shown on the drawing. The sum of the rise and run of the largest step would be 17 inches, the second largest, 16¾ inches, and the smallest would be 16⅜ inches. This gives the stair builder enough leeway to make necessary adjustments in both the rise and the run of the steps. Study Figs. 239 and 240.

Determining the Rise. — Fig. 241 shows to the right that the total rise for a flight of porch steps is 27 inches. Now to find the number of steps and the rise for the individual steps, let's divide 27 by one of the rises shown in Fig. 240, say, 7 inches. Seven will go into 27 not quite 4 times, which determines the number of risers, but does not give the rise per step. Let's try 6⅜ inches: Four times 6⅜ equals 25½, which is too small, but four times

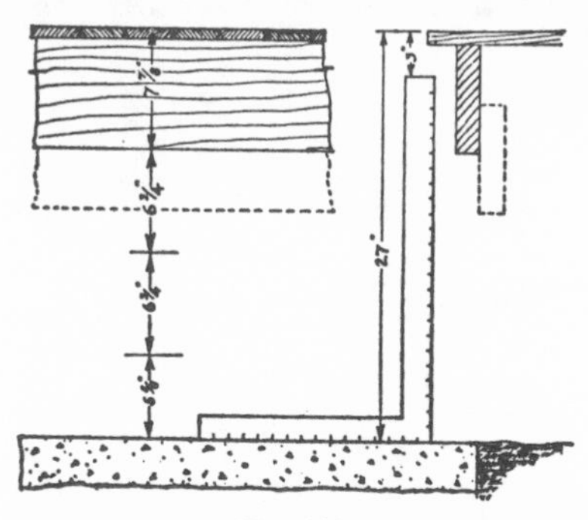

Fig. 241

6¾ equals exactly 27, and determines the rise per step as 6¾ inches. The last figures were purposely made to come out even, in practice adjustments usually have to be made. In using diagrams to determine the rise and run of steps, the workman should use as a basis, a rise or a run that will fit in

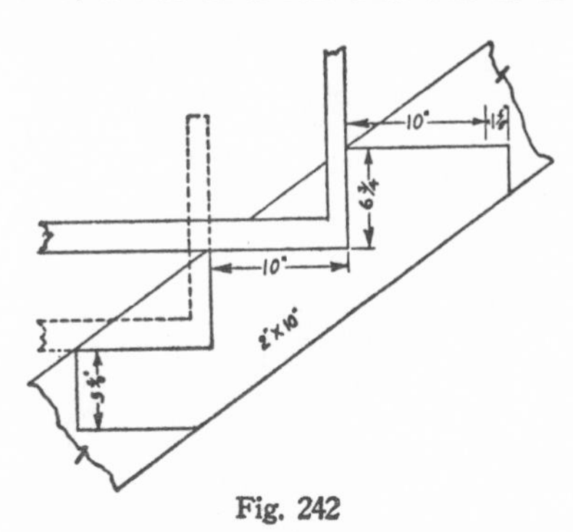

Fig. 242

with the circumstances under which he has to work. To the upper left of the drawing is shown in part the outside porch joist and the ends of the porch flooring. The figures shown to the left, give the risers for the rough horse, which will be more fully explained in following illustrations.

Stepping Off Horses.—Fig. 242 shows in part, a 2x10 from which a horse for the steps is to be cut. The two squares shown in part are in place for marking the rise and the run for two steps. The rough horse has only a 5⅝-inch rise for the first step. The other two risers are 6¾ inches. The

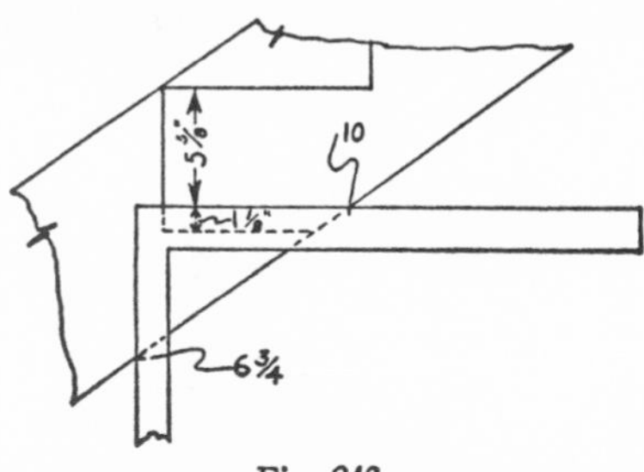

Fig. 243

fourth riser, as shown in Fig. 241, will be 7⅞ inches, which is not cut on the rough horse. Fig. 243 is a detail of the first rough step, with the square applied for marking the base cut. The dotted lines at the bottom of the horse show why the rough step has only a 5⅝-inch rise—enough is cut off the

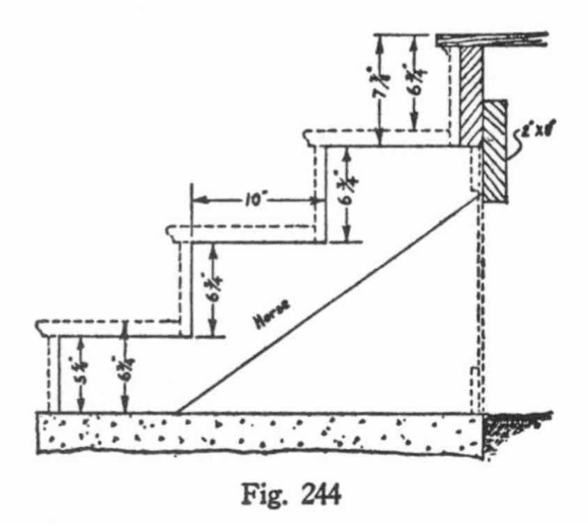

Fig. 244

bottom of the rough horse to take care of the thickness of the finished tread. By doing this, the thickness of the tread is added to the last rise of the flight of steps, which is as it should be. This is shown to the upper right in Fig. 244, where the treads are indicated by dotted lines. Study the figures given on this drawing, particularly those of the first and the last risers.

Finish Risers and Stringer. — Fig. 245 shows a top view of the steps without the treads, in part. Two rough horses are pointed out, three risers, one stringer, and the 2x8 porch joist—also the 2x8 to which the rough horses are fastened. The steel square is shown applied for marking the miter cut for the risers where they join the risers of the stringer. This, it will be seen,

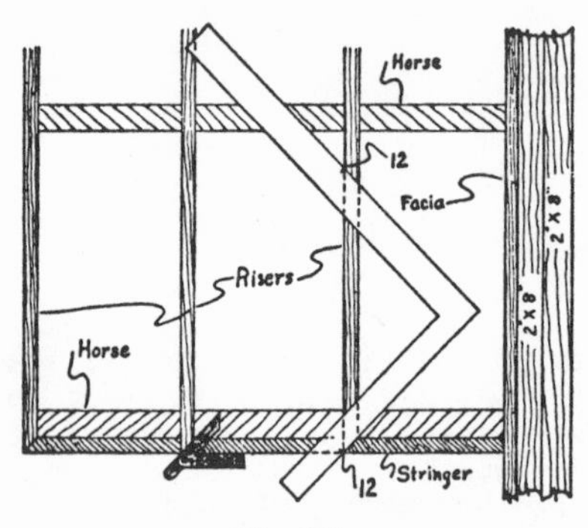

Fig. 245

would be rather inconvenient, or clumsy. To overcome this handicap, the workman establishes the bevel with the steel square, and sets his bevel square to it. Then with the bevel square he marks both the risers and the risers of the stringer. A bevel square in position for marking the bevel on a riser of the stringer is shown applied to the second riser at the bottom of the drawing.

The Finished Steps.—To the upper left of Fig. 246 are shown three different joints that can be used in joining the risers to the risers of the stringer. The one numbered *1,* is a true miter joint, such as shown in Fig. 245. This

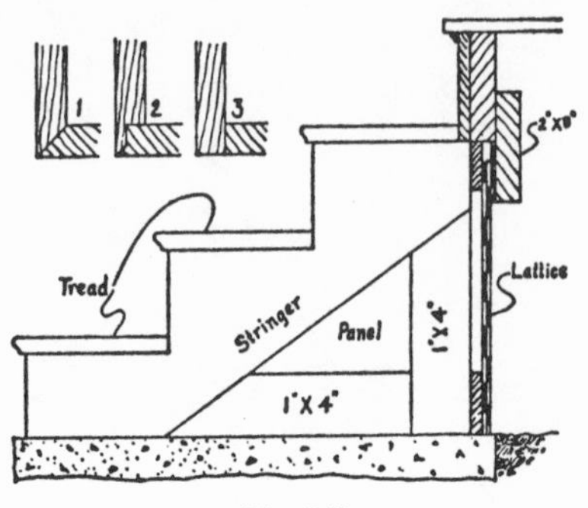

Fig. 246

joint is commonly used. A shoulder-miter joint is shown at number *2,* while number *3* shows a simple joint that is often used. This joint is stronger than the other two, gives better service and drains better, consequently it will last longer, but it doesn't look quite as well as the other two. The main drawing in Fig. 246 gives a side view of the finished steps. The illustrations should be studied as a whole, giving special attention to what is in between the lines.

Chapter 28

PLANK STAIRS

Plank steps and plank stairs are built primarily for service. Stairs to basements in a great many instances are made of plank, both the stringers and the treads. Steps to loading docks, porches, stands, and so forth, are also often made of plank. The rules for building plank stairs are practically the same as the rules for building any other kind of stairs, excepting that fewer rules are necessary. Otherwise the specifications for such stairs or steps can be written in two words, substantial construction.

Old Fashioned Fence.—Fig. 247 shows a square with an old fashioned wooden fence fastened to it. The figures to which the fence is set are 8 on the tongue and 8¾ on the blade. An edge view of the fence is shown at the bottom of the drawing. This fence gives good service in all cases where the material is perfectly straight. If the material that the square is used on is crooked, then it is unreliable.

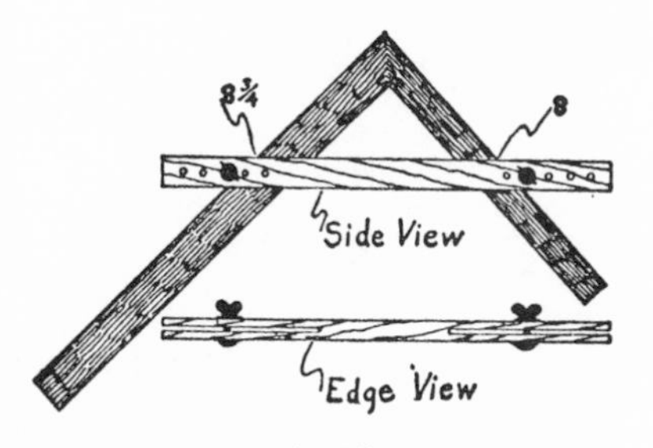

Fig. 247

Metal Guides.—Fig. 248 shows by the upper drawing a square with a pair of metal guides on it. One contacts the figure 8 on the tongue, and the other is set at the 8¾ point on the blade. These figures are just about right for the run and rise of a plank stair step. Two views of a metal guide are shown directly below the square. At *A* is shown a side view, while *B* shows the inside edge of the guide.

How to Use Fence.—Fig. 249 shows a square with a fence on it, in position for marking a plank stair horse. The first three steps are shown marked.

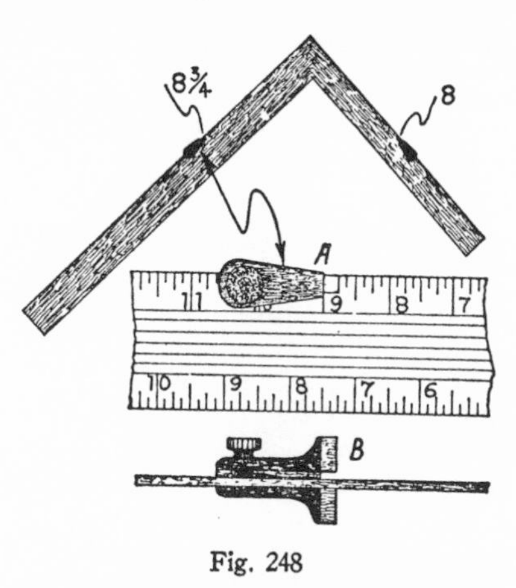

Fig. 248

The word "out" on the triangles with the slanting marks, indicates the parts that are cut out when the horse is framed. Notice the figures given for

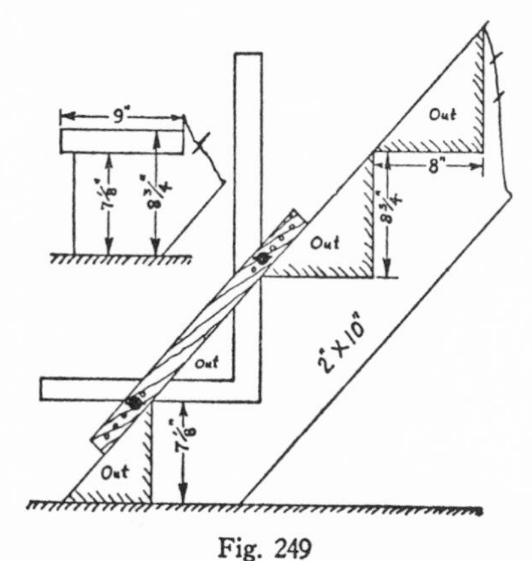

Fig. 249

the rise and run of the upper step shown. The rise and run of all the other steps are the same, excepting that the first rise of the horse is 1⅝ inches lower than the others to take care of the thickness of the tread. To the upper left is shown an end view of the first step, giving in figures the rise for the horse and also the rise for the completed step.

How to Use Guides.—Fig. 250 shows a plank stringer in part. The dotted-line square shows the first position of the square in laying out the housing for the treads on the stringer. In this position the bottom and the right end of the housing are marked. Then the square is moved up to the shaded position and the top of the housing is marked. The part that is to be housed for the tread is shown shaded. The tread, as the figures will show, will be 1⅝ inches by 8 inches. Two other housings for treads are shown marked. The guides on the square are again fastened at 8 and 8¾, as indicated on the drawing.

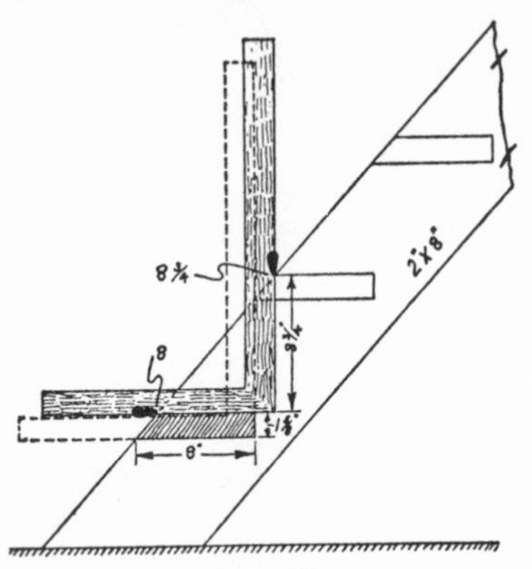

Fig. 250

A Perspective View.—Fig. 251 shows a perspective view of the first step of a housed plank stair. The stringers are made of 2x8's, while the tread is 1⅝ by 8 inches, full. The run of a step of this stair is 8 inches, while the rise is 8¾ inches, just as shown in the other illustrations of this chapter. Where the corners of the tread should be dubbed off, is shown shaded at *a* and *a*.

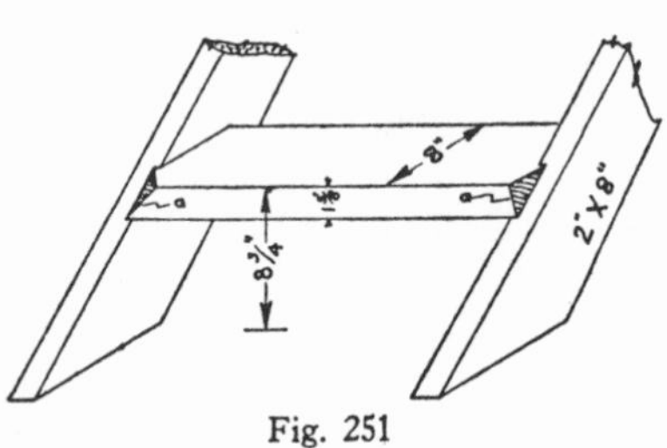

Fig. 251

Notched and Housed Stringer.—Fig. 252 shows to the left a side view of a stringer in part, gained out for the tread and notched for a lug that is to be cut on the tread. To the right is shown an outside view of the other stringer, showing the end of the lug, and by dotted lines the housing for the tread.

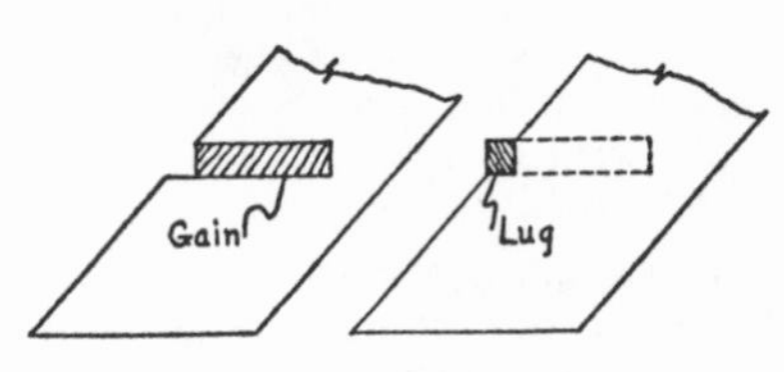

Fig. 252

Tread with Lugs.—Fig. 253 shows by the top drawing the elevation, or front view, of the first step of a housed plank stair. The bottom drawing shows a plan of the step, showing the lugs on each of the ends. The depth of the housing is shown by dotted lines. The two stringers are shown only in part.

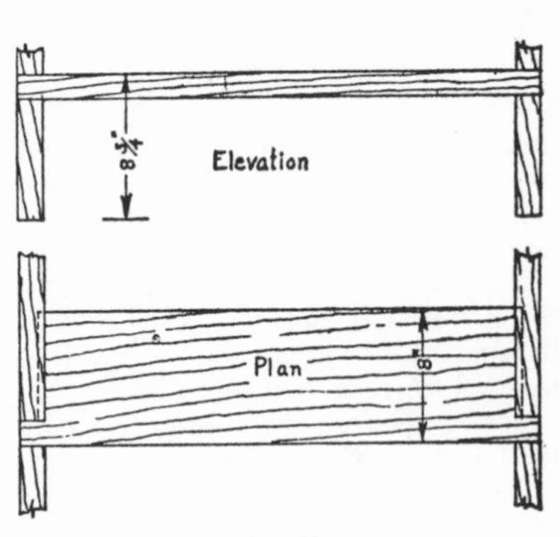

Fig. 253

Plank Cellar Stairs.—Fig. 254 shows a stringer in place of a plank stair, which has a full run of 96 inches, and a full rise of 105 inches. It should be noted that there are 12 runs and 12 rises shown on this stringer. The bottom and top black squares are applied to the stringer for marking the foot and top cuts, respectively. The dotted-line squares, show how the stepping off is done. The blades of the dotted-line squares are in position for marking the top of the housing for the treads. This is also true of the top shaded square. The blade of the bottom shaded square, as stated before, is in position for marking the foot cut.

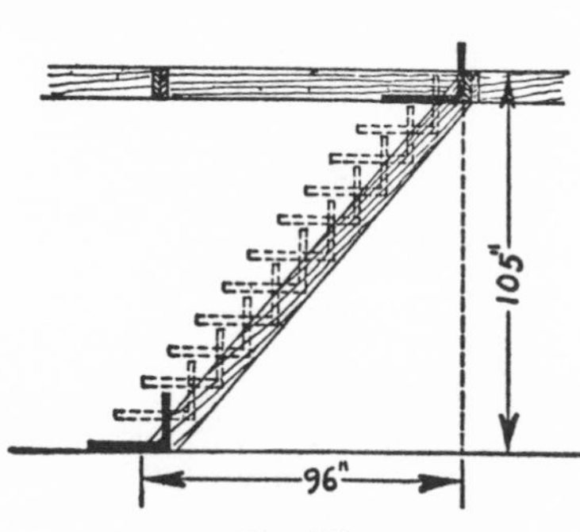

Fig. 254

Headroom.—The same stair is shown by Fig. 255, where the steps are pointed out. How to obtain the headroom is shown to the left. A distance of 7 feet from the headroom header down to the nosing line, as shown, will give ample room for going up and down on this stair without bumping the head.

Other Ways of Building Plank Stairs.—Nothing has been said about plank stairs in which the unhoused stringers are nailed to the treads, which is a cheap construction and should be used with caution. Another cheap construction is to nail cleats to the stringers for supporting the treads. Such stairs, will give good service when they are well built, and are always dry. If they are used in wet places for only short periods of time they are all right. Otherwise there is danger of the nails rusting and the wood rotting. The marking of the stringers, in both instances, is the same as for any other stair stringers.

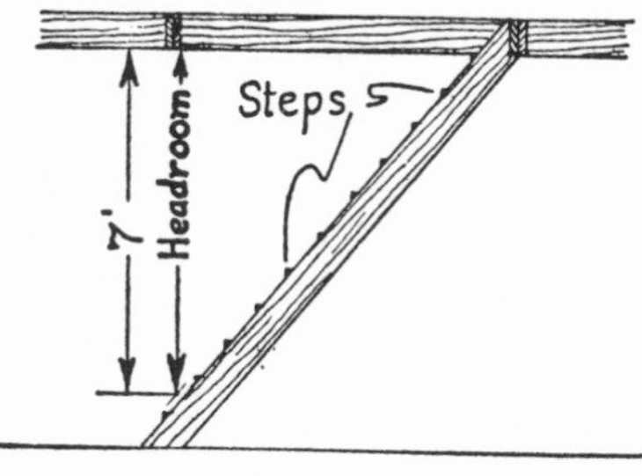

Fig. 255

Chapter 29

SERVICE STAIRS

Stairways have at least two definite purposes for which they are built. The first one is to give service. That is to say, the stairway is built to make it possible for passing from one floor to another, or from one elevation to another by means of steps. In order to fulfill the requirements necessary to give this service, the steps must be designed in such a way that they will be easy to ascend and descend. If the steps for stairways are designed too large or too small, persons who are not used to them will have difficulties in passing over them. The second purpose is to add decorative value to the room in which the stairway is located, giving it besides utility value, the value of beauty.

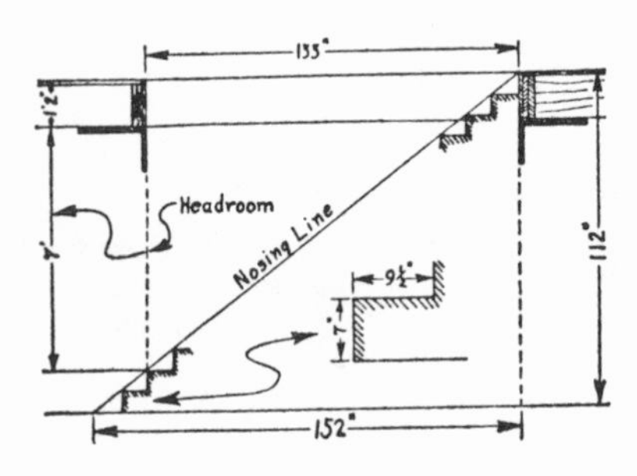

Fig. 256

Figuring Size of Well Hole.—Fig. 256 is a diagram showing how to find the length of a well hole, so as to give it the proper amount of headroom. The first thing to do is to find the full rise of the stair. In this case it is 112 inches, as shown to the right. A step with a run of 9½ inches is about right, which is based on the rule that the sum of the rise and run of a step in inches, should be around 16½ inches, not over 17 inches. Now subtract 9½ from 16½, and you will have 7, or the rise per step in inches. The next thing to find is the number of 7-inch risers there will be in the stair. This is done by dividing 112 by 7, which gives 16. By multiplying 9½ by 16, you will have the full run in inches, or 152. Now draw the nosing line, as shown by the diagram, using a 152-inch run and a 112-inch rise. To locate the headroom header, use the square as

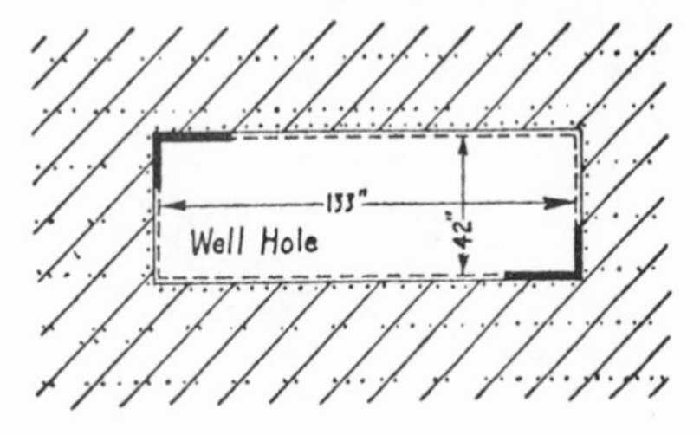

Fig. 257

shown to the left, and when you have 7 feet headroom, mark the joist for the header, as indicated on the drawing. Another way to get this is to add the thickness of the floor, in this case, 1 foot 2 inches, to the 7 feet headroom, which makes 8 feet 2 inches, or 98

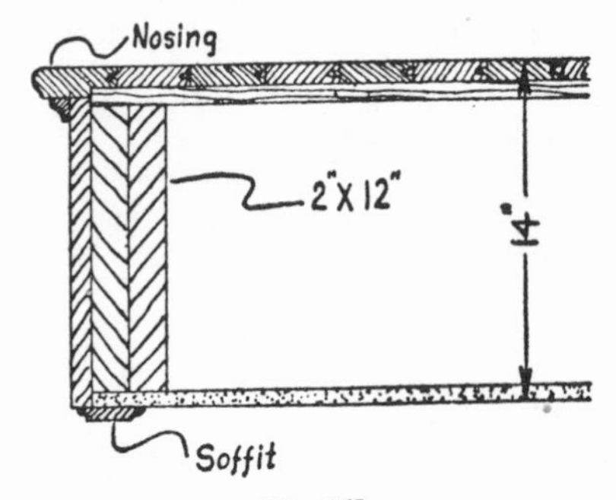

Fig. 258

inches. By dividing this by 7 you will have the number of risers in the stair to the point where the well hole stops, or 14. Now multiply 14 by 9½, the run of a step, and you will have the length of the well hole in inches, or as shown, 133. In framing the well hole, care must be taken to allow for the finish around the well. The well hole cut out and the rough flooring in place is shown by Fig. 257. The dotted lines indicate the nosing line of the well. The width in this case is 42 inches. Fig. 258 shows a detail of both the rough construction and the finish around the well hole.

Unhoused Stair.—Fig. 259 shows by the upper drawing a full-width skirt board nailed to the wall—also a 1⅝ by 2 spreader that holds the rough horse away from the wall. Cross sections of the skirt and the spreader are shown at *a*. At *b*, by dotted lines, the location

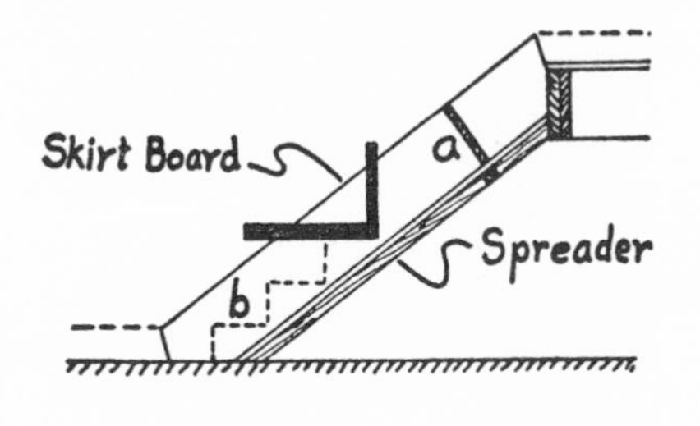

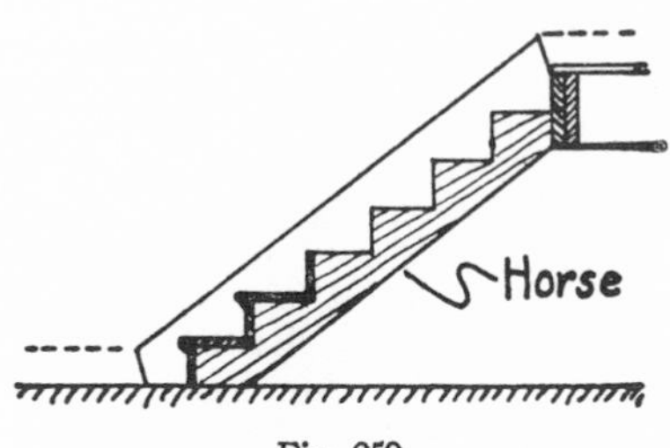

Fig. 259

of three steps are indicated. The bottom drawing of Fig. 259 shows the rough horse in place, with two treads and three risers also in place. The risers and the treads are cut to a length that will fit between the two skirt boards. The skirt boards should be nailed tight to the wall, so that they will not give when the risers and treads are put in place. The risers and the treads must be cut carefully—not too long or too short—so that they will not spring the walls or show open joints.

Scribing Skirt Boards. — Fig. 260 shows a good way to build an unhoused service stair. Here the rough horses are put in place and the finish risers are nailed to them. Then the skirt boards are cut, as shown by the upper drawing, and placed against the risers for scribing, which is shown by the

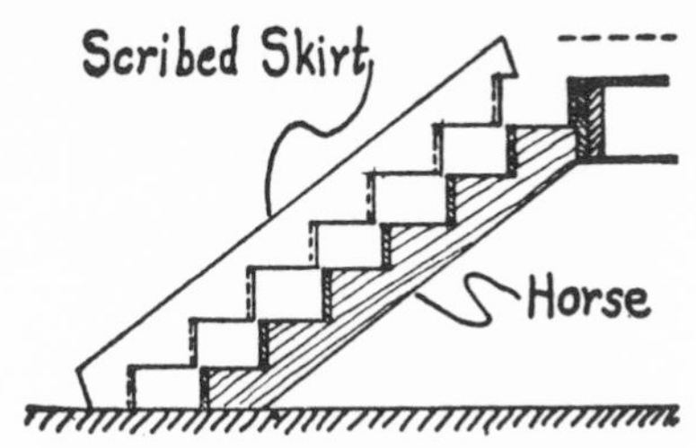

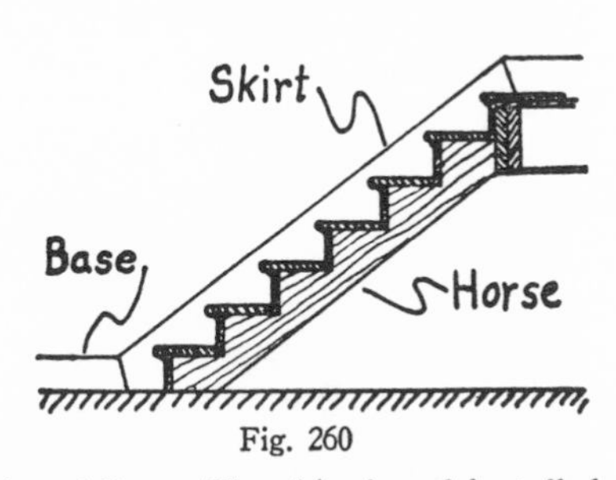

Fig. 260

dotted lines. The skirt board is pulled back, showing it separated from the risers and horse. The risers of the skirt board are then cut to the scribing, and nailed into place. This done,

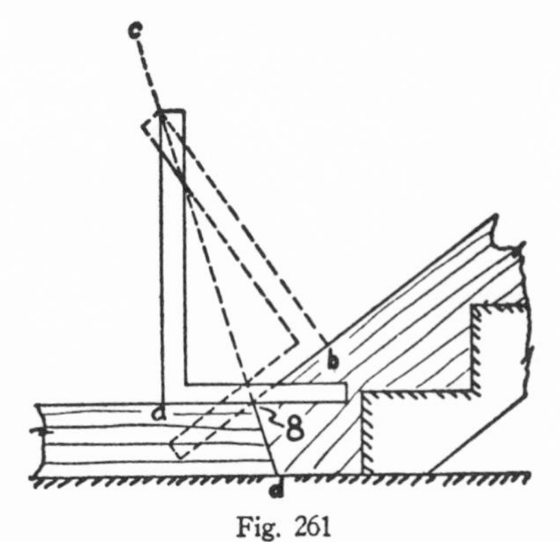

Fig. 261

the treads are cut and nailed in. A cross section of this stair completed, is shown by the bottom drawing.

Joining Base to Skirt Board.—Fig. 261 shows a diagram illustrating how to get the cut with the square for the joint between the base and skirt board. Place the square, first as shown by dotted lines, and then as shown by the full lines, using a convenient figure on the tongue, in this case 8, which is held to the angle between the base and

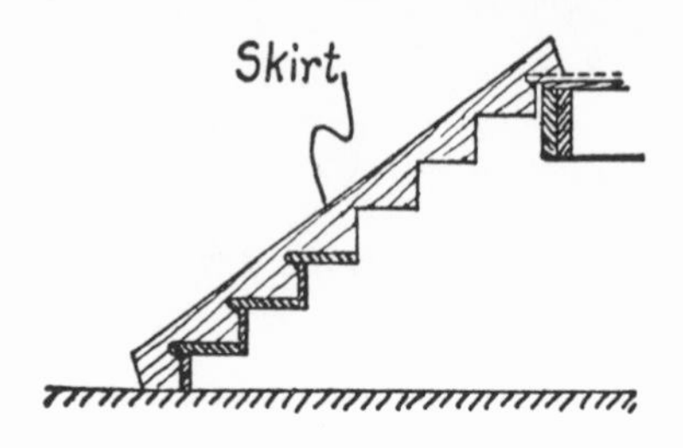

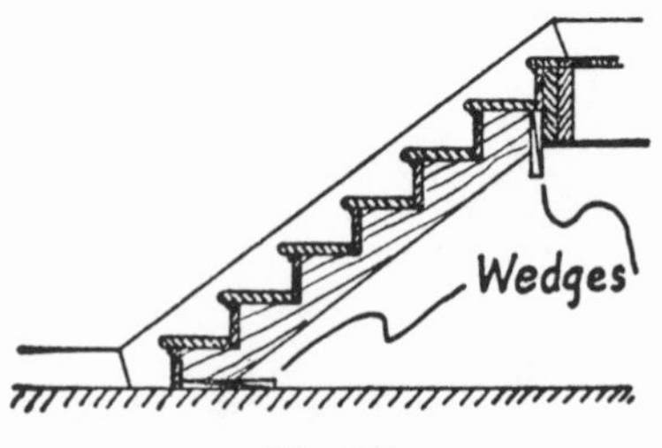

Fig. 262

the skirt. The distances from the angle to both point *a* and point *b* must be the same. Now strike the line *c-d,* making it cross the point where the blades of the squares cross, or come to a point, and also where the tongues of the squares cross, which is the angle between the two boards. The line from *d* to the angle, is the cut for both the base and the skirt board.

Install Horses Last.—**Fig.** 262 shows another way to build a service stair that will insure good joints. The top drawing shows the skirt board cut out for the steps, nailed to the wall. Three treads and three risers have been nailed to the skirt boards just enough to hold them in place. After all of the treads and risers have been nailed in, put the rough horses in place, as shown by the

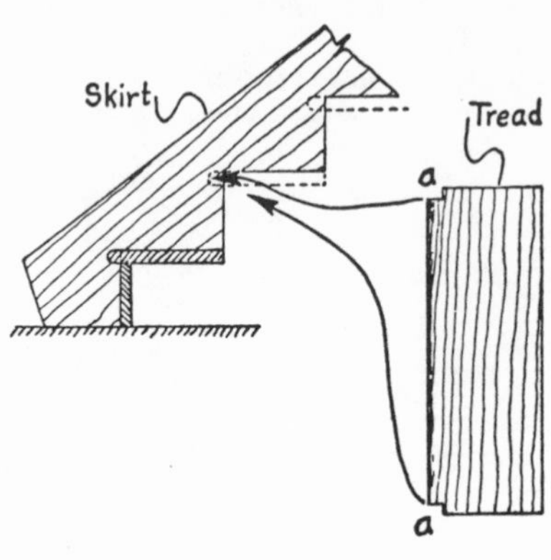

Fig. 263

bottom drawing, wedging them as indicated. These wedges should be glued and nailed so they will not get out of place. Fig. 263 shows a detail of the skirt in part, and a full face view, to the right, of a tread. Notice the notches at *a* and *a,* which are cut as deep as the nosing. The arrows indicate how the tread is brought up to a level position and put in place. The tread is nailed in place first, and then the riser that supports it is nailed in also. This is repeated until all the treads and risers are in place.

Chapter 30
HOUSED STAIRS

The owner, when he builds his permanent home, always gives particular attention to the main stairway. This part of the home must be the show place. It is to the stairway that he wants to point with pride, and when admirers call, the stairway must bring them to the climax. Consequently the work on such a stairway must be first class, and if it is first class, it will be a housed stairway.

Adjustable Templet. — The housing of the skirt boards for housed stairways can be laid out with the steel square, or it can be laid out by means of a templet. Such a templet is shown

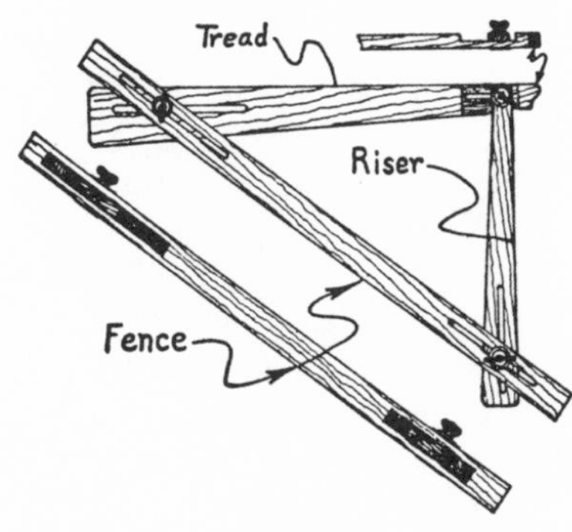

Fig. 264

by Fig. 264. There are three parts to this templet, as shown on the drawing. The upper horizontal part is used for marking the housing of the treads, the perpendicular part is the part by which the risers are marked, while the sloping part is the guide or fence, as it is commonly called. In the drawing two views of the fence are shown, an edge view and a side view. The three parts are fastened together by means of three thumb bolts, which run through the slots, as shown, and clamp the parts together. Fig. 265 shows a steel square with a fence fastened to it, and by dotted lines it shows two parts of the templet shown in Fig. 264, the part for the tread and the part for the riser.

Housing the Skirt Board.—Fig. 266 shows to the bottom left, the bottom edge view of the skirt board shown in the main drawing. This skirt board was laid out with a templet similar to

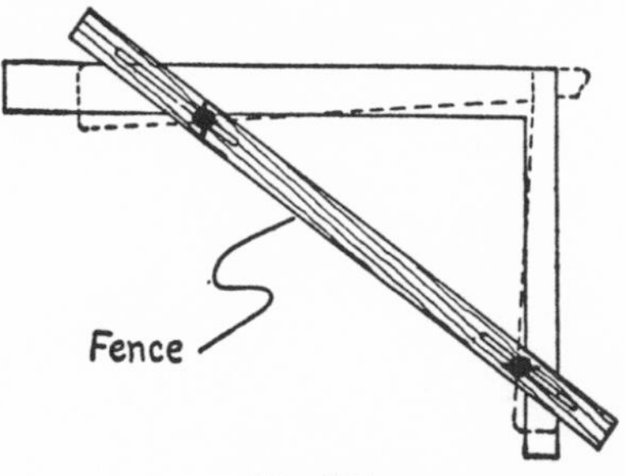

Fig. 265

the one shown in Fig. 264—or it might have been laid out with the steel square and fence, shown in Fig. 265. The same skirt board is shown in perspective view at the bottom in Fig. 267. The upper drawing of this figure shows

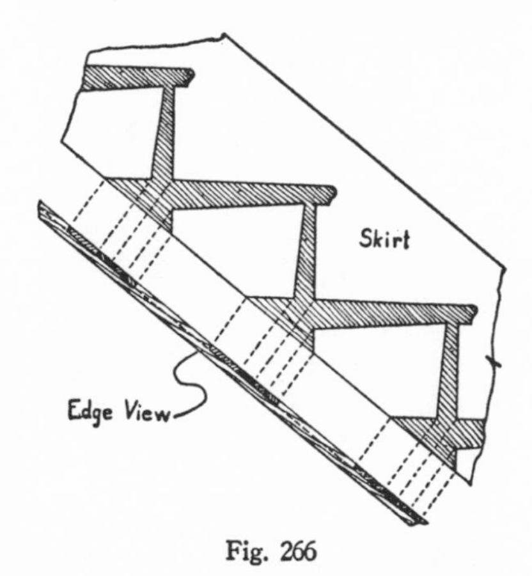

Fig. 266

in part, a tread and a riser fastened together, ready to be put into the housing of the skirt board, as the downward pointing arrows indicate. As shown, the tread and the riser are put together before they are glued and wedged into the housing of the skirt board that joins the wall. The skirt board on the open side of the stair is joined to the risers of the steps by means of some suitable joint, usually a miter joint.

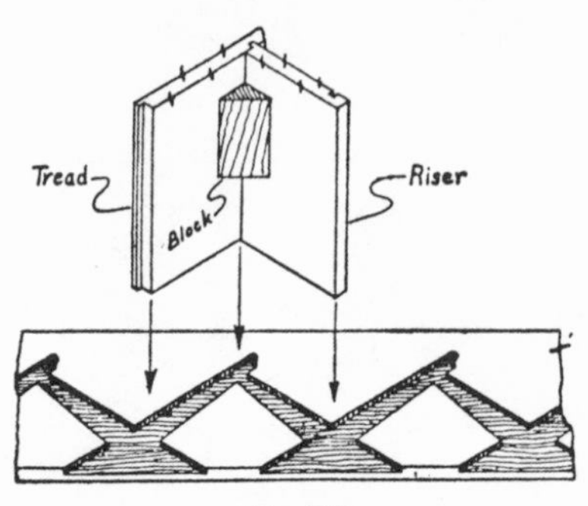

Fig. 267

The block pointed out on the upper drawing, is put in place after the flight of stairs has been put together, but before it is put in place. These blocks are glued into the angles of the under side of the steps in order to reinforce the joint and prevent squeaking.

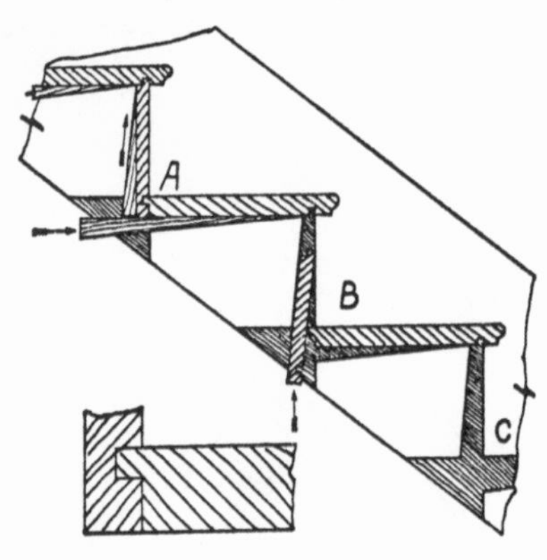

Fig. 268

Putting Treads and Risers in Place.—Fig. 268 shows a skirt board for a housed stairway that is built between two walls. When such a stairway is installed, the housed skirt boards are nailed to the walls. After that the treads and risers are put in place and wedged, as shown by the arrows back of the tread and riser at *A*. At *B* the tread is shown in place and the riser is in the process of going into place. This is indicated by the arrow at the bottom. At *C* neither the tread nor the riser have been put in. The detail to the bottom left shows the joint at the back of the tread and the bottom of the riser. In the other joint, the riser tongues into the tread just back of the nosing, as shown. This joint is practically the same as the one shown in the detail.

Open End of Stair.—An open end of three steps is shown in Fig. 269

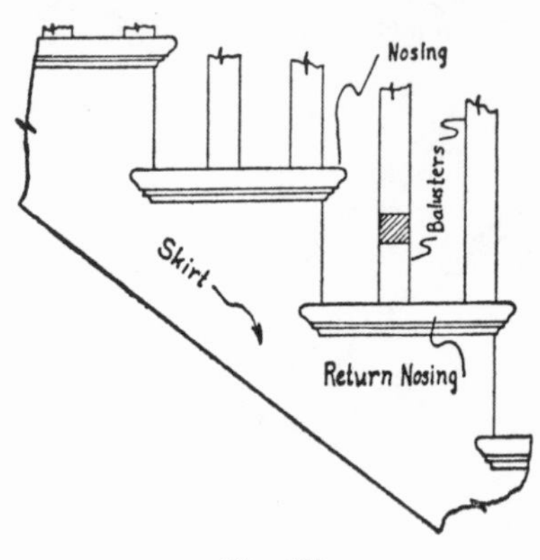

Fig. 269

completed. Pointed out are nosing, return nosing, balusters, and skirt. The top view of the same three steps is shown by Fig. 270. The first step has the tread in place, with the return nosing pointed out. The balusters are indicated by dotted lines. The miniature squares show the application of the square for obtaining the miter cuts for the nosings and for the risers. In practice this squaring is usually done with a bevel square or a miter square. The rough horses, the risers, and the skirt boards are also pointed out. The dotted line at the top of the drawing shows the depth of the housing in the skirt board.

Detail of Return Nosing.—Fig. 271 gives a detail of the open end of a tread, showing the return nosing ready to be put in place, as indicated by the dotted lines. The mitered return to the left on the return nosing, shown shaded, sometimes is not mitered, but the straight nosing has a return cut onto it, which is perhaps just a little more substantial,

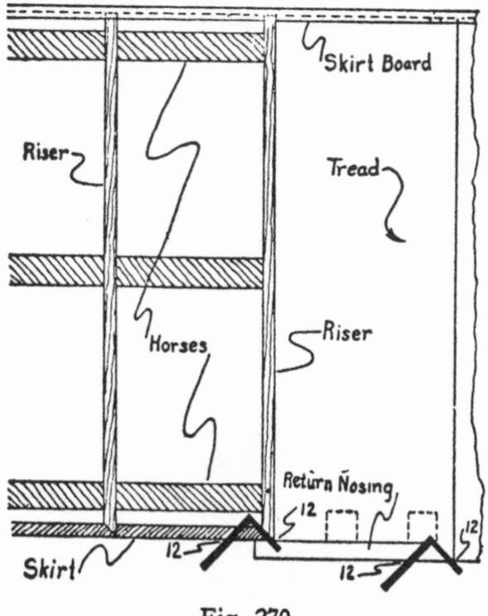

Fig. 270

but not quite as classy as what is shown. The balusters of this step are pointed out, also the riser and the rough horses, in part, of the next step.

Mill Work.—In these days of machine efficiency, most of the material for housed stairways is gotten out by the mills. In many cases the material for service stairs is prepared in the same way. This leaves the field carpenter stripped of everything but installing the stairway. And that in many instances is taken over by specialists, sent out by the mill that furnishes the stair material. But notwithstanding this, every carpenter should know how to build any kind of stairway, including the housed stairway. For if he knows

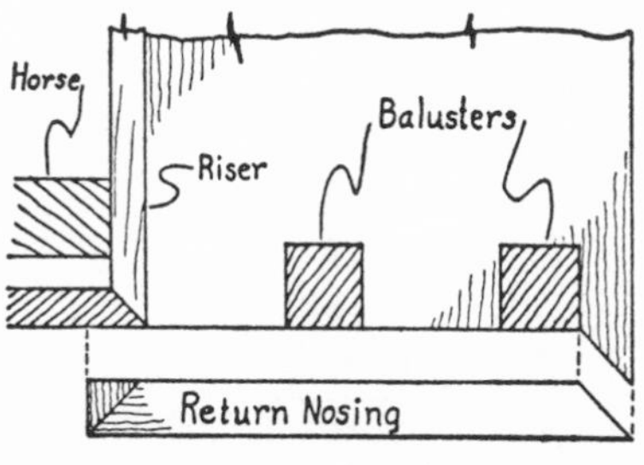

Fig. 271

how to do this work, should he be called on to do it, he can go right ahead. It will also help him in installing the stairs that are mill-made.

Glue and Wedges.—These two items should be mentioned, even though they have little connection with the steel square. Only a good quality of glue should be used, and the wedges should be made so that they will fill exactly the housing space left for them.

Chapter 31

ROUGH HORSES

The commonly used rough stair horse has been touched upon in the chapters dealing with stairs, that have gone before. In this chapter different kinds of stair horses are treated. The chapter begins with the old fashioned pitch board, which is still frequently used by stair builders. It should be remembered, however, that in this chapter, as in all other chapters of this series, the subject covers the steel square. Even when the drawings do not show the square, the principle of this important tool is still present. The student should keep this in mind, and when the square is omitted, he should figure out the relationship of the steel square to the drawing.

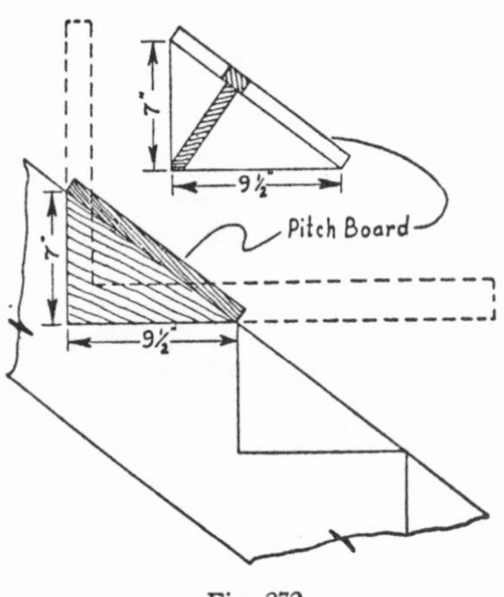

Fig. 272

Pitch Board.—Fig. 272 shows by the main drawing a pitch board in place for marking the run and the rise of a step. The board is shown shaded, and the connection to the steel square is shown by the dotted lines representing a square. At the top, center, the pitch board is again shown, with a cross section inset.

Pitch Board for Marking Housing.—Fig. 273 shows a pitch board applied to a skirt board for marking the run and rise of a step. The board is shown shaded, as in the other case, but it will be observed that the board is larger than the one shown in Fig. 272. The reason for this is shown by the dotted line on the main drawing, which touches the angles between the run and the rise of the steps, indicating that the skirt board is to be marked for housing. The pitch board shown in part at the bottom left, shows how by using a wedge, the board can be used for marking the back lines of the housing. At *a* the end of a wedge is shown

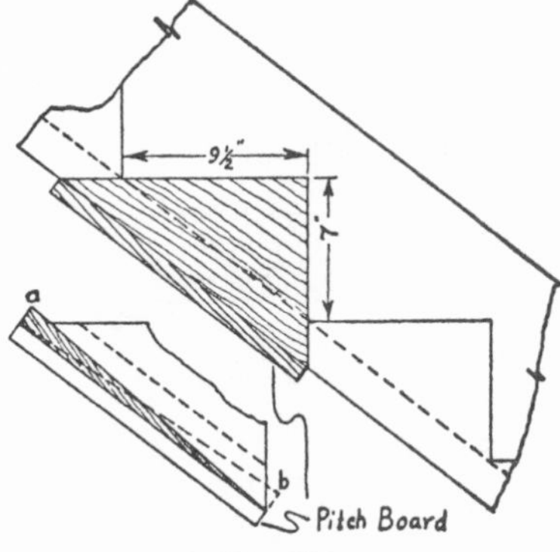

Fig. 273

in place for marking the back line for the tread, while *b* by dotted lines, shows the application of the wedge to obtain the back line for the riser. Fig. 274 shows the pitch board in place for marking the back line of the tread. The end of the wedge is pointed out to the left. To mark the back line for the riser, reverse the wedge, bringing it into the position shown at *b*, Fig. 273, and the riser of the pitch board will give the proper line for the back of the riser housing.

Nosing Templet.—At *A*, Fig. 274, a templet for marking the nosing of the treads is shown in place. This is done after the housing for the treads and the risers have been completed. Two views of the templet are shown at the bottom left. At *a* a side view is shown, and at *b* is an edge view.

Built-Up Horses. — Two ways of building up horses are shown by Fig. 275. The one shown at the top, will have practically no shrinkage up and down, but it will shrink horizontally. The bottom one is built in reverse order, so far as the shrinking is concerned. To overcome most of the shrinkage difficulties, use well seasoned material. The boards that form the rough steps are nailed to 2x6's, alternately, one on one side and the next on the other side. This balances the horse well. But in cases where this is not practical, the boards that form the rough steps can all be nailed to one side.

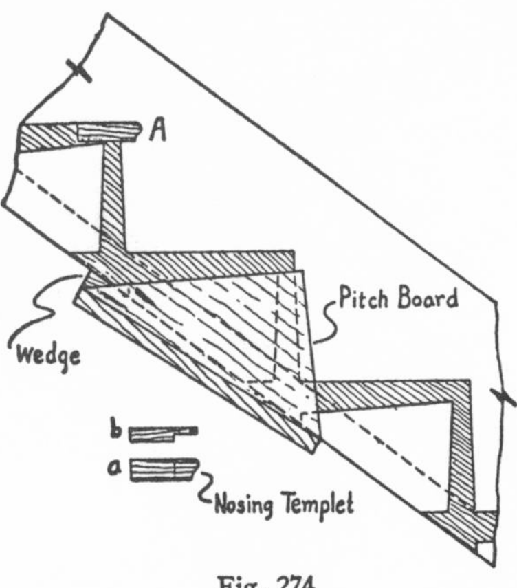

Fig. 274

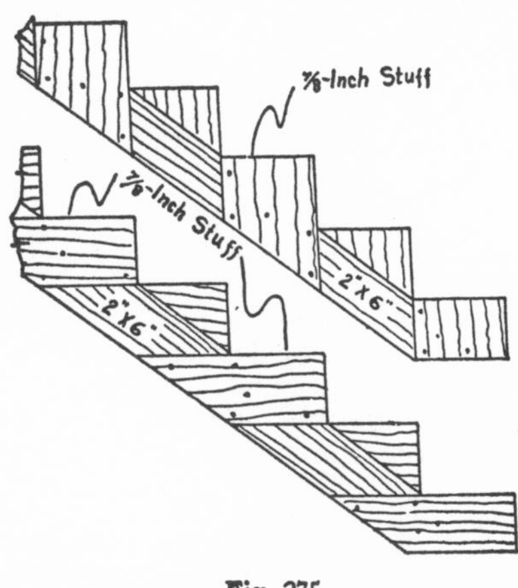

Fig. 275

Regular and Built-Up Horses.—Fig. 276 shows how to use the blocks sawed out in making a regular stair horse, to build up a stair horse by nailing them to a 2x6, as shown by the drawing. The arrows to the left, with tail feathers, show how blocks, numbered *1* and *2,* have been transferred from the main timber to the 2x6 timber. The arrows to the right indicate how two more blocks, numbered *3* and *4,* are to be transferred in the same way.

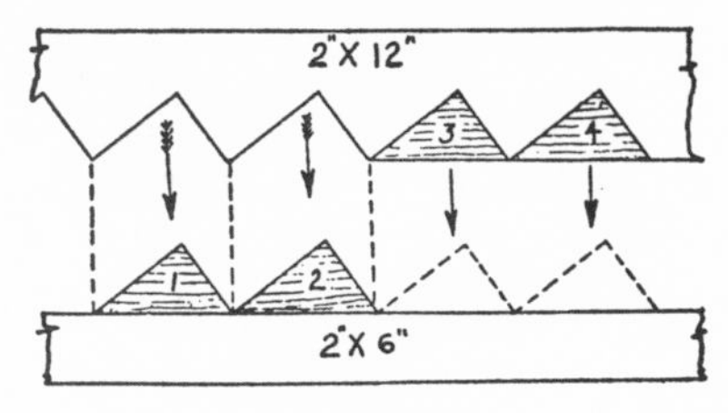

Fig. 276

Shrinkage.—Fig. 277 shows at *A* the bottom step shown to the right, top, in Fig. 275. Here the amount of shrinkage is indicated by dotted lines, which is somewhat exaggerated. At *B* is shown by the dotted lines how the rough step of a regular stair horse shrinks. This is also shown exaggerated. The remedy for holding shrinkage to a minimum, as stated before, is to be sure to use well seasoned material.

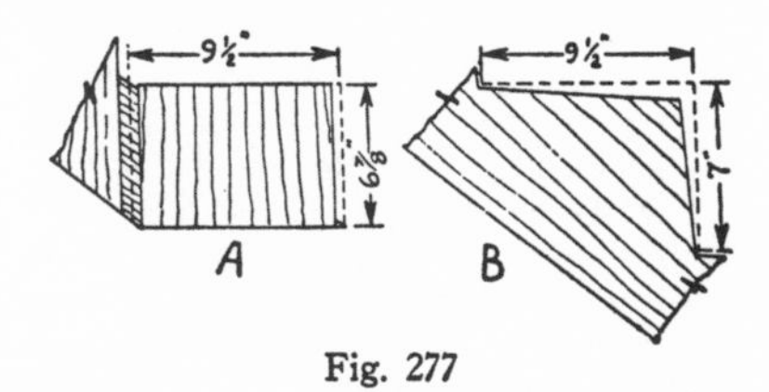

Fig. 277

Gaining Headroom.—Fig. 278 shows a 2x10 that has been framed into the studding of one of the walls of a staircase partition, in such a way that the face of the 2x10 comes flush with the surface of the plastering. Then the rough steps, in the form of blocks, are nailed to the 2x10, as shown by the drawing, where blocks numbered *1, 2,* and *3* have been nailed in place, and *4, 5,* and so on, shown by dotted lines, are yet to be nailed on. The blocks numbered *1* and *2* have been sawed out in the regular way, while the block numbered *3* was sawed out in such a

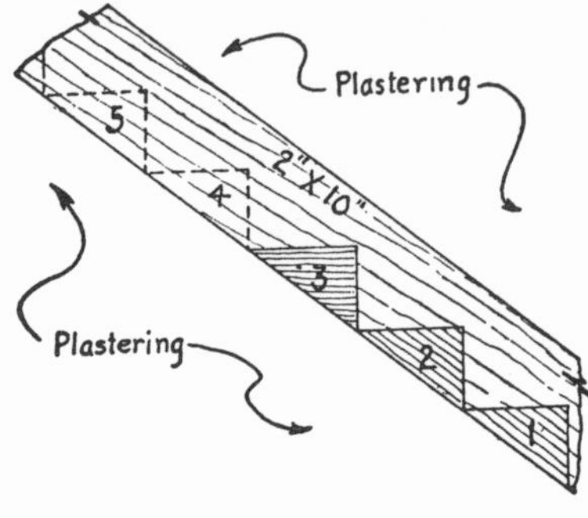

Fig. 278

manner that the grain will run horizontally. This method of building rough horses is justifiable only in cases where it is necessary to increase the headroom for a stair immediately below the one in question. The lath for the plastering is fastened directly to the under corner of the steps. In this way 2 or 3 inches in headroom can be gained.

Grandstand Steps.—Fig. 279 shows four grandstand steps, or rather seats. In this case the rise is 14 inches and the run 28, which makes a rather comfortable seat for spectators. The steps can be made smaller or larger, just as the circumstances might dictate. If the grandstand is made for children, a smaller step might be better, but if it is made primarily for grown-ups, the

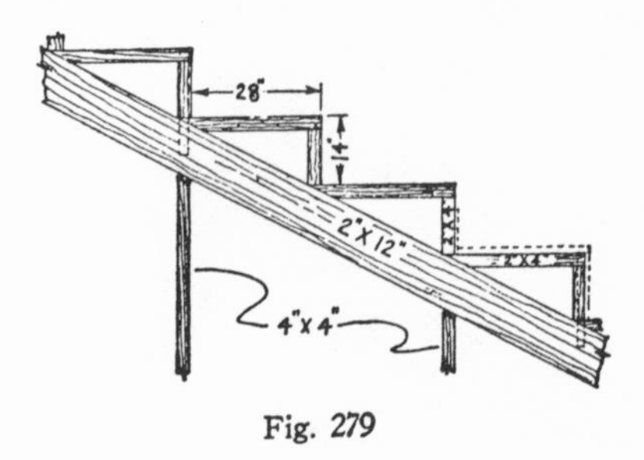

Fig. 279

steps could even be made larger. The floor of a step and the boarded riser are shown by dotted lines to the right. Cleats for holding the rough steps together are indicated to the right and

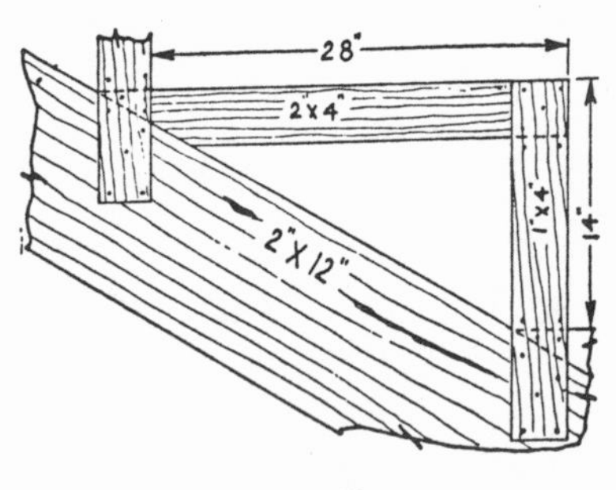

Fig. 280

left by dotted lines. A detail of a rough step is shown by Fig. 280. Here the 1x4 cleats are shown nailed to the side of the rough riser. The ledgers for the floor are made of 2x4's.

Chapter 32
NEWEL POSTS

The Newel.—Just as the main stairway is the show place of the room in which it is located, so the main newel of that stairway is the center of the artistic attraction. The wood used in such stairways is carefully selected and the workmanship on them must be of the very best available. In fact, the men who build such elaborate stairways are artists, even though they may not be recognized as such.

A Newel in Place.—The drawing in Fig. 281 shows a part of an open stairway. The drawing is as simple as it can be made, because in that way all lines that might confuse the student,

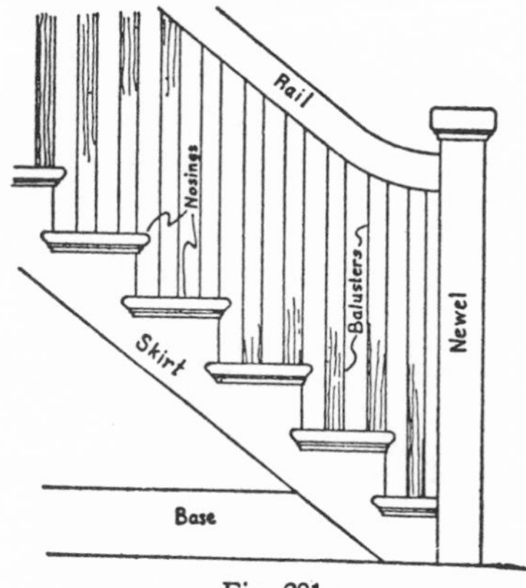

Fig. 281

can be omitted. Pointed out are the base, the rail, the newel, nosings, and balusters. The curved part of the rail where it joins the newel is called an easement. Quite frequently, however, the rail joins the newel without an easement at all.

Installing Newel.—Fig. 282 shows a detail of the first two steps shown in Fig. 281. Here a part of the newel is shown to the upper right, ready to be put in place. The dotted lines show how the inside corner of the bottom of the newel has to be cut out in order to let it slip over the corner of the step. The downward pointing arrows show how the newel is placed. One of the balusters is shown also ready to be put in place. The heavily shaded housing for the dovetail of the baluster is shown

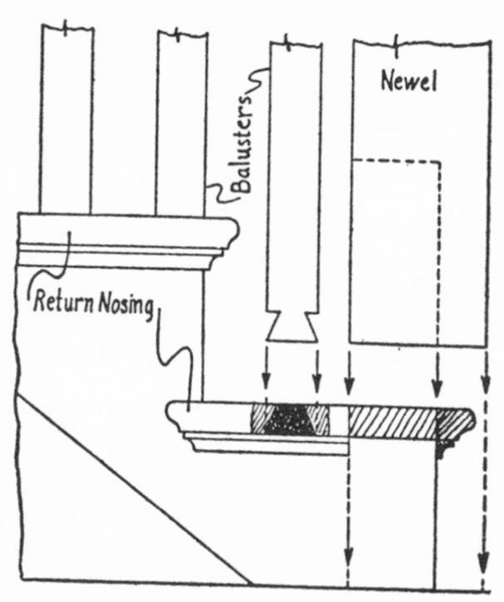

Fig. 282

directly below it, where the return nosing has been cut out enough to show the housing. The arrows indicate how the baluster is to be put in place. The

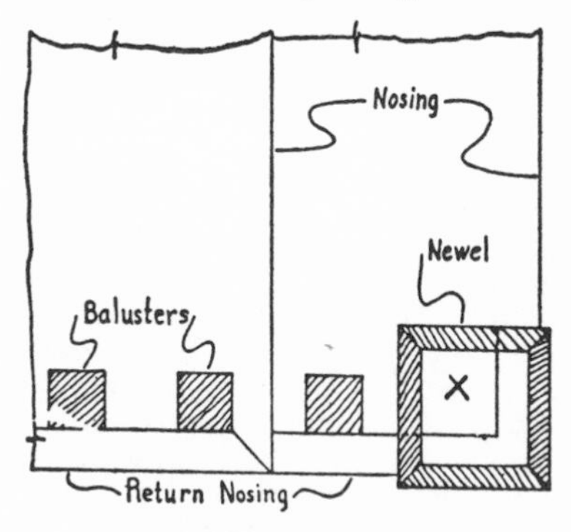

Fig. 283

return nosings are pointed out to the left. The second step shows the balusters installed. A sort of plan of what is shown in Fig. 282 is shown by Fig. 283. The part marked *X* is the corner over which the newel is set. To accomplish this the inside corner of the bottom of the newel must be cut out, as indicated by dotted lines in the two figures. The nosing and the return nosing of this step must be cut out in such a way that they will die into the newel.

Platform Newel.—Fig. 284 shows a detail, in part, of a platform newel in place, but not fastened. Before the

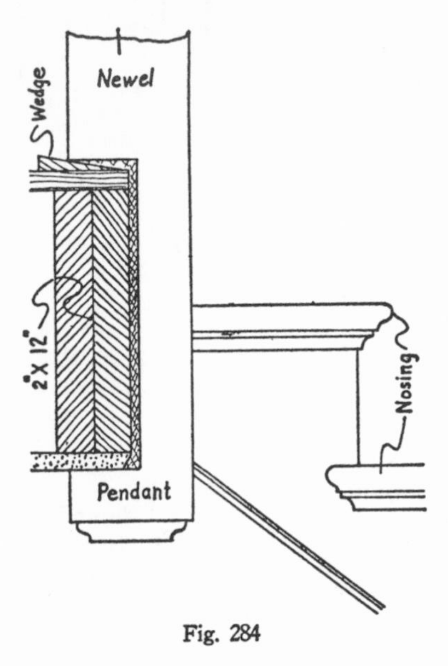

Fig. 284

flooring of the platform is laid, the newel should be set. The cutting-out that is necessary must be carefully marked with the square, and just as carefully cut out. The nosings also must be carefully cut to let the newel in, although, the return nosings are usually left off until the newel is permanently fastened. In the drawing the return nosing is shown in place. It will be noticed that the pendant fits the ceiling plastering of the platform. Above can be seen the wedge used for holding the newel up. When the newel is in perfect alignment and perfectly plumb, then it is fastened permanently. The open space where the wedge is shown should be filled with good blocking, fastened so that there will be no

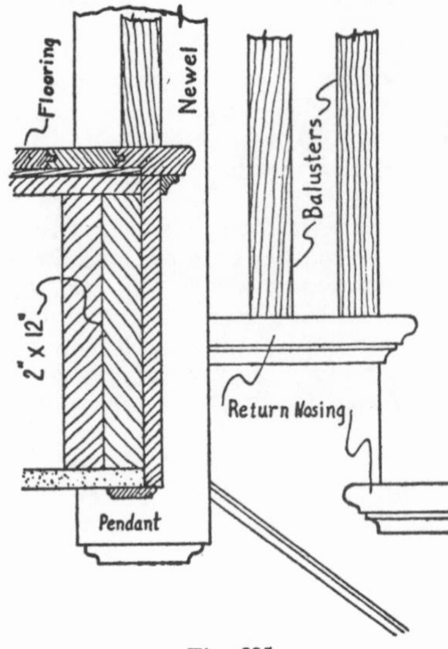

Fig. 285

danger of it working out of place. Fig. 285 shows the same newel permanently installed. Here the balusters are also in place, which are shown shaded. The flooring is in place, and in this case it is laid on strips, which is sometimes done. The 2x12's could in practice be 2x10's, or even 2x8's, depending on

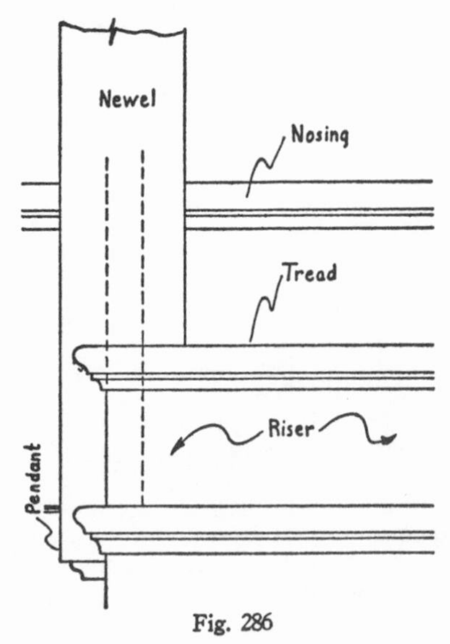

Fig. 286

what is necessary to meet the requirements of the particular situation. This drawing should be compared and studied with Fig. 284. What is shown in these drawings can be taken as repre-

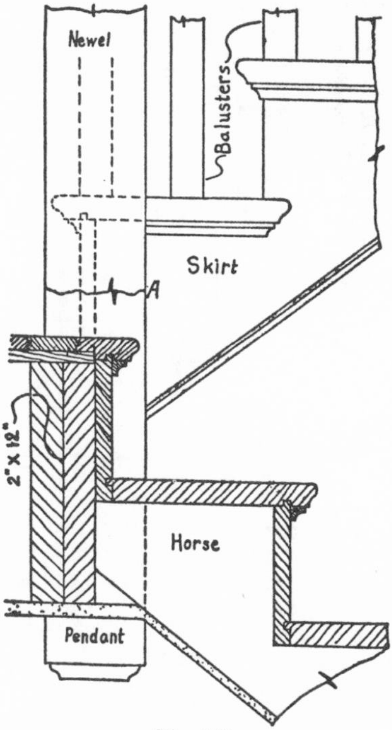

Fig. 287

senting a flight of stairs joining a platform landing or the top landing of a stairway. Fig. 286 shows a finished front view of the same newel layout. The newel in part, nosing, tread, riser, and pendant are pointed out.

Return Flight.—How a newel is set for a return flight on a landing is shown by Figs. 287 and 288. Fig. 287 shows the newel in part, joining the upper flight of stairs, where the balusters are pointed out and the work has been completed. The details of the steps at the bottom are shown in cross section, giving the relationship of the

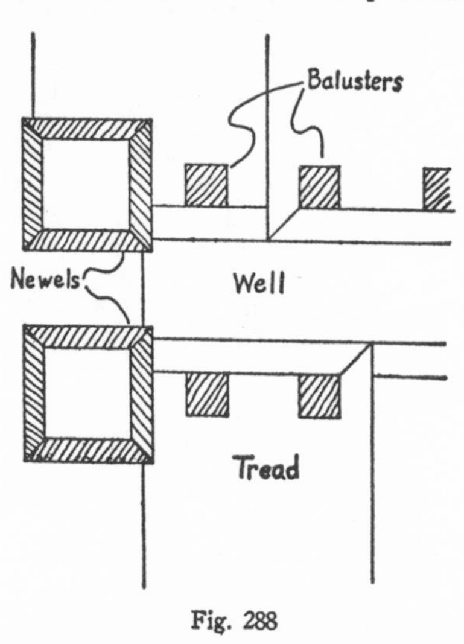

Fig. 288

steps to a second newel, shown cut off at *A*, but running down to the bottom, where the pendant is shown. Fig. 288 shows a sort of plan of what is shown in Fig. 287. The newels and balusters are shown in section, while the treads in part are shown in plan.

The Finished Stair.—The man who installs a stairway, especially the main stairway, must be a man who knows the value of careful workmanship. He must know the importance of accuracy in everything that he does, uppermost of which is accuracy in measurements. Another important thing that he must remember is that no mechanic can do good work with poorly sharpened tools.

Chapter 33

HANDRAILING

While it might seem far-fetched to take up the matter of handrailing in dealing with the principles of the steel square, it nevertheless is a part of the subject. Even though the handrail cuts are made in a miter box, the principles of the square are present, because when the miter box was made it was laid out with the square. If the newels are set just a little out of square, it will show up in the fitting of the handrails, but if the newels are set perfectly, and the handrails are not cut in keeping with the square, then the mechanic again will have trouble in making good joints. Accuracy is important in all stair work,

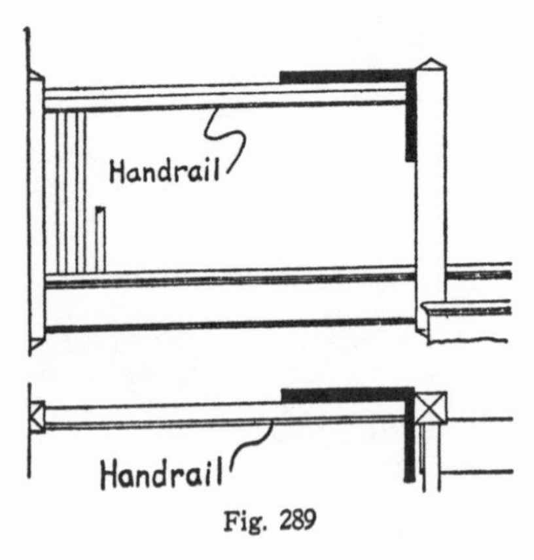

Fig. 289

but it is especially important in fitting handrails.

Straight Handrail.— Fig. 289, the upper drawing, shows a straight handrail between a half newel and an angle newel with some of the balusters shown in place to the left. The square is applied to the right, showing how the rail and the newel must have a square joint. The bottom drawing gives a top view. The square to the right again shows how the principles of the square must be applied in making the joint.

Sloping Handrail.—Fig. 290 shows the open side of a flight of an open stair. Here also a straight handrail is used. The application of the two squares shown on the railing is the same as the application of the square shown on the steps, which is to say. the figures used for marking the steps will also give the cuts for the handrail.

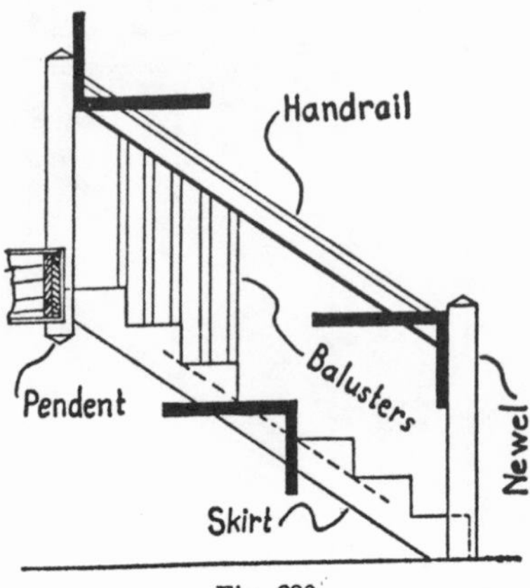

Fig. 290

For instance, if you have a rise of 7 inches and a run of 9½ inches, then 9½ and 7 taken on the square will give the cuts for the handrail.

Joining Handrail to Newel.— Fig. 291 shows two details of handrails

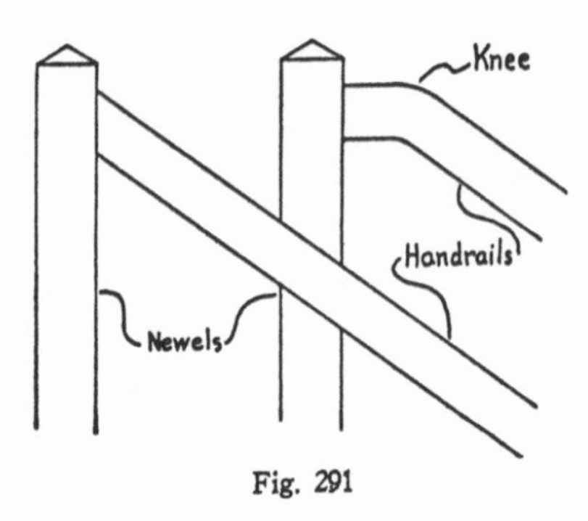

Fig. 291

joining the upper newel. The drawing to the left shows a joint between a straight handrail and a newel, while the detail to the right shows a handrail with a knee on the upper end, which joins the newel with a square cut. Fig.

292, at the bottom left, shows a straight handrail joining the bottom newel. The center drawing shows a handrail with an easement at the bottom, which joins

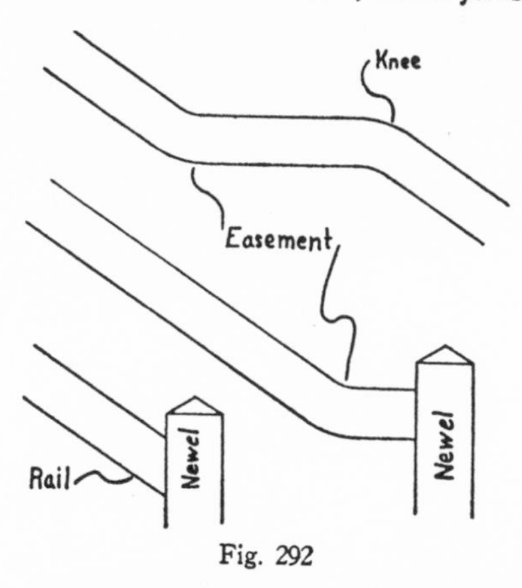

Fig. 292

the newel with a square cut. The upper drawing shows an easement to the left and a knee to the right.

Goosenecks. — Fig. 293 shows two kinds of goosenecks. The one to the left has a right-angle turn at the top, and joins the newel with a square joint,

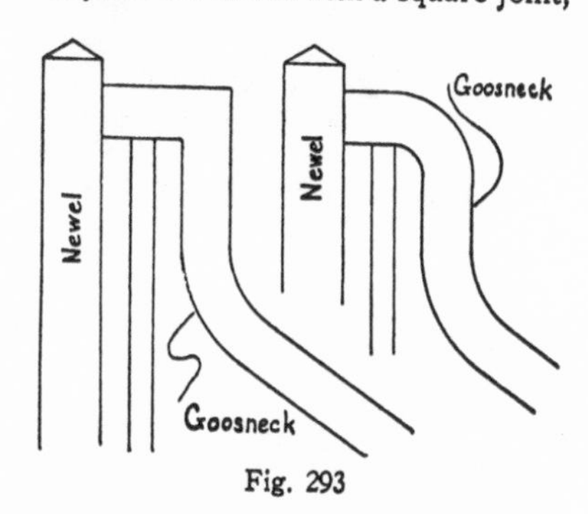

Fig. 293

while the gooseneck shown to the right has a curved turn at the top, which also joins the newel with a square joint. Each of these details shows a baluster in place. It is important that the turned part be made in such a way that the newel, baluster, and railing will balance properly.

Railing Joints.—Fig. 294 shows four railing joints. The one shown at *A* is the best. It is a square joint and made with a stair bolt. The railing is shown cut out so as to show the nuts of the bolt. The bolt is shown by dotted lines. A cross section of the railing is shown inset at *E*, which will also answer for

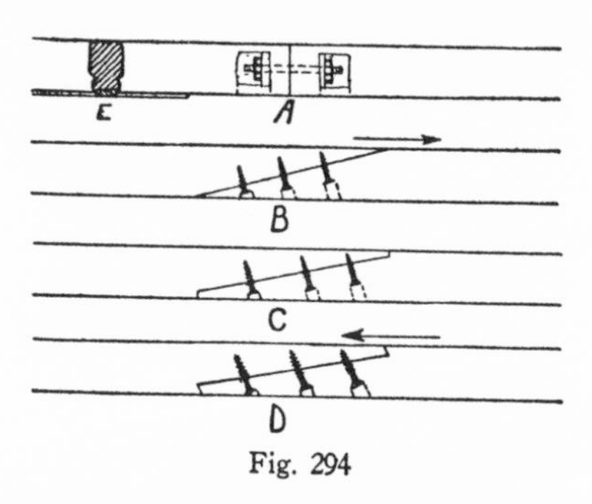

Fig. 294

the other three rails shown. The joint shown at *B* is the least satisfactory of those shown here, however, if it is made perfectly, and a good quality of glue is used, it gives good service. When this joint is used, the downward slope of the handrail should be in the direction of the arrow. The joint shown at *C* is a much better joint, but it also depends on good workmanship and good glue. The joint shown at *D* is better than either the one shown at *B*

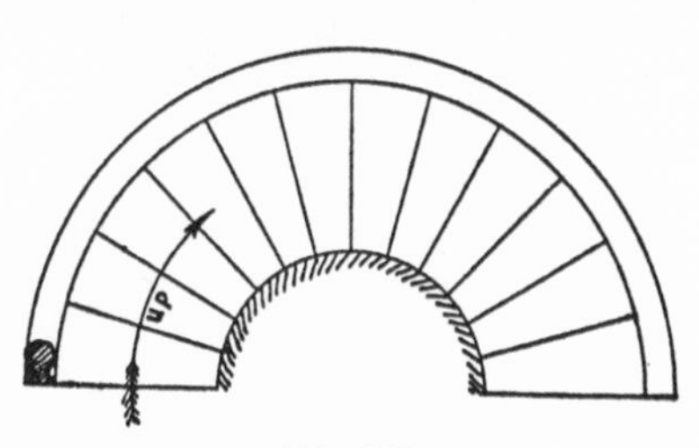

Fig. 295

or the one shown at *C*, provided, of course, that it is well made and well glued. When this joint is used, the downward slope of the rail should be in the direction of the arrow.

Circular Handrail. — Fig. 295 is a plan of a circular stairway with twelve treads and thirteen risers. It should be mentioned before going any further, that circular stairs should never be used, excepting in cases where a circular stair is the only solution to the problem. They are not only difficult to pass over, but they are dangerous, because the treads change in width, which

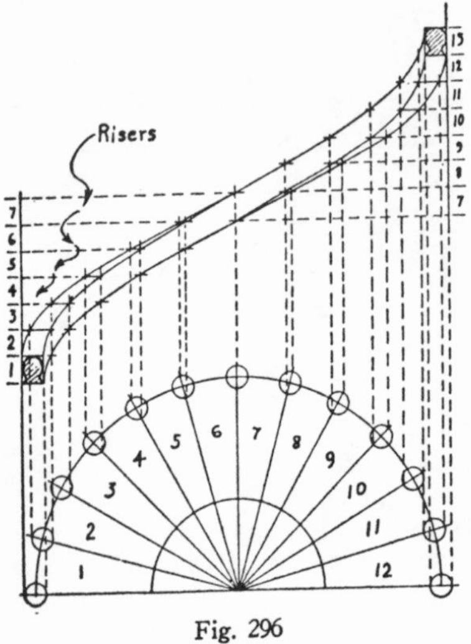

Fig. 296

is likely to cause persons to be injured by falling. A narrow circular stair is perhaps the most practical of circular stairs, but even then it should not be used, excepting to solve a problem. The two outside circular lines in the diagram, represent a circular handrail, which is somewhat exaggerated, as the inset to the left indicates. Another diagram of the same circular stairway is shown by Fig. 296. Here the little circles that are centered at the intersection of the risers of the steps and the center line of the railing, are used in describing the handrail, as shown above. The perpendicular dotted lines are raised from where the riser lines cross the little circles up to where the handrail is to be described. Each little circle from right to center and from left to center brings the perpendicular dotted lines closer together, until you come to the center riser, where only one dotted line is shown. By these lines, the varying widths of the handrail drawing are obtained, as a study of the diagram will show. The risers of the stairway are indicated to the left and to the right on the upper part of the diagram, numbered *1, 2, 3, 4* and so on up to *7*, to the left. To the right the risers are shown beginning with riser *7*, and continuing with *8, 9, 10* and so on up to *13*. As stated in Fig. 295, the handrail shown is somewhat exaggerated, in order to make the diagram easier to understand. To simplify the drawing, the handrail is shown square, while cross sections of the finished rail are shown on the two ends. Study Figs. 295 and 296 carefully.

Chapter 34
BALUSTERS

No attempt has been made to give anything but the simplest of baluster designs in this chapter. The best reason that can be offered for doing this, is that the simpler the design the clearer will be the drawing, as a rule. With reference to the base, balusters can be placed in two classifications, the square

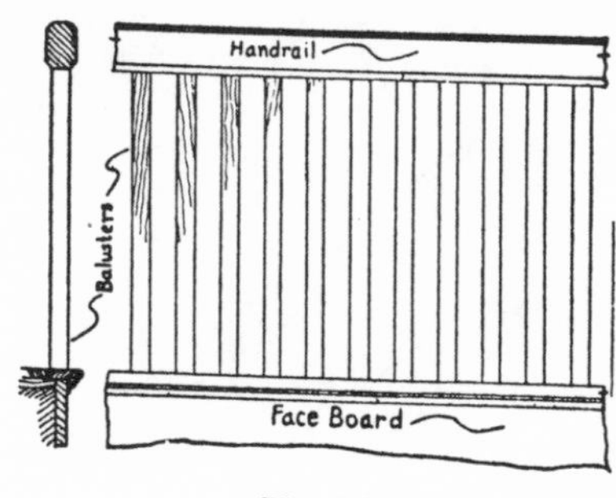

Fig. 297

and the round. The round ones that taper toward the top are often called spindles. Bannister is another name for baluster.

Cheap Baluster Installation. — Fig. 297 shows widely used square balus-

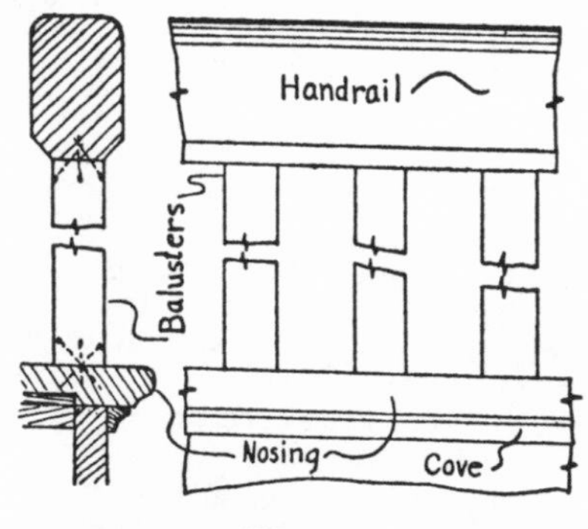

Fig. 298

ters, which are about 1½ inches square. The method of installing such balusters, as shown by the details in Fig. 298, is the cheapest that can be used—toenailing. This method should not be used on railings for first class stairways, although, on cheap work it is often justifiable. And when such work is well done, it gives good service. The details shown in Fig. 299, show the same kind of installation used on a stairway with a closed stringer. The handrail design used here shows little development, indicating that economy and service are the two outstanding considerations.

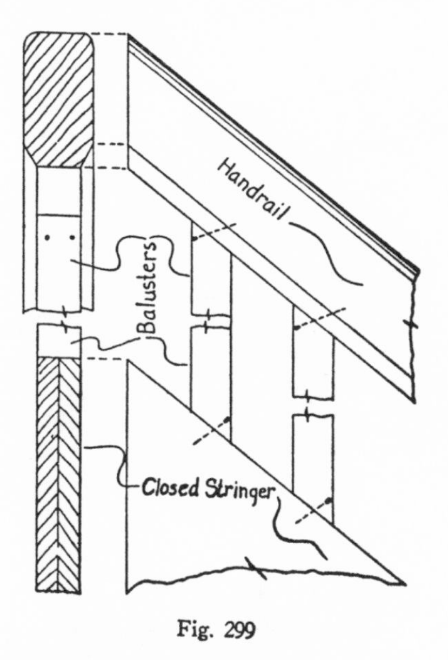

Fig. 299

Better Construction.—Fig. 300 gives details of a little better developed handrail, and the construction, so far as installing the balusters is concerned, is much better than what is shown in Figs. 297, 298, and 299. Here the balusters have tenons worked on the ends, as pointed out on the drawing, which fit into mortises on both the handrail and on the closed stringer. It will be noticed that the stringer is made of one piece, whereas in the other case, two pieces were used, joining each other flatwise.

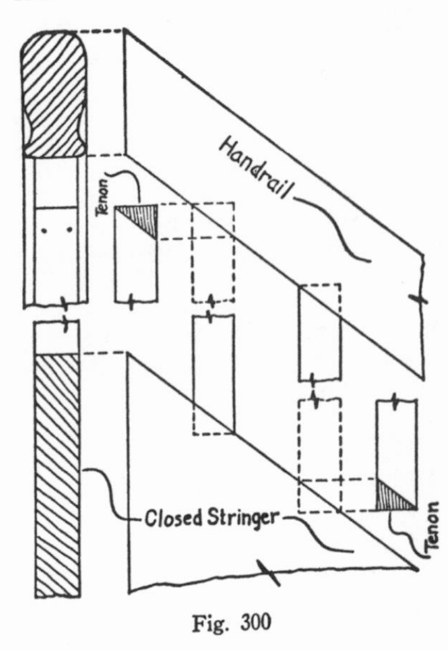

Fig. 300

Good Construction.—Fig. 301 shows a well developed handrail, which has a wide groove at the bottom to receive a fillet, as shown. There is also a fillet used on the closed stringer. These fillets are spaced and nailed to the balusters. Then this part is put in place and nailed to the top of the stringer, as shown by the details. The handrail is then placed, and the upper fillet is nailed into the groove of the handrail, as shown by the cross section to the left. This construction makes a pretty good job, if it is carefully done. Study the drawings.

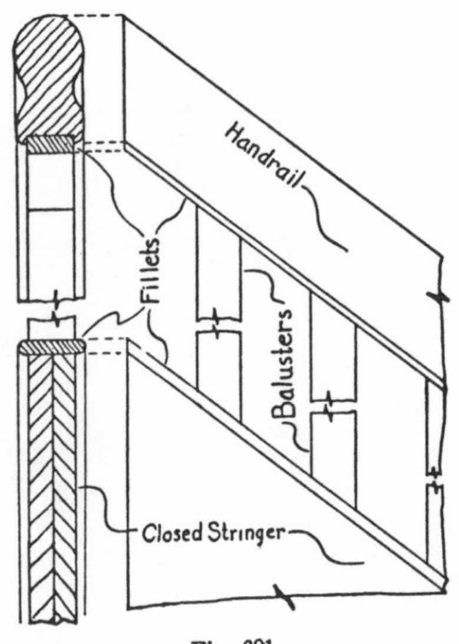

Fig. 301

Approved Method.—Fig. 302 gives details of the approved method of installing balusters. Here an astragal is nailed to the stringer, as shown. Then the handrail is put in place. After this

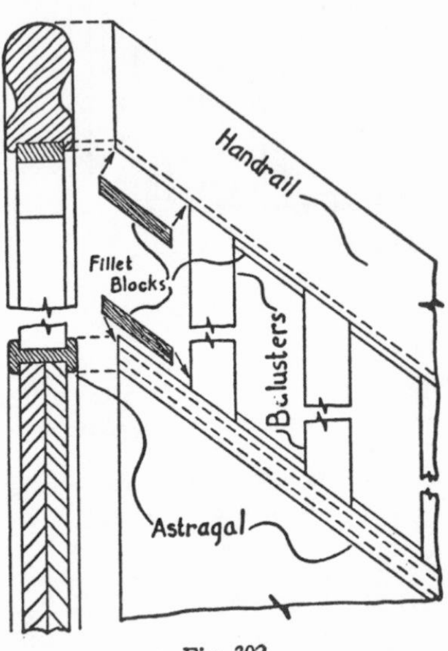

Fig. 302

the balusters are cut to the proper length and put in place, using fillet blocks cut to the right length for spacing the balusters. Two of the fillet blocks are shown shaded, ready to go into place. This is the best way to install balusters on an open stairway that has a closed stringer.

Open Stringer.—Fig. 303, the upper drawing, shows in part, two steps of a stair with an open stringer, in which four different baluster fastenings are shown. To the extreme left is shown a two-nail fastening. This is often used on cheap work, as mentioned before. The second fastening from the left is made by means of a dowel worked on the base of the baluster. This dowel is slipped into a hole bored for it in the tread. To the extreme right is shown the dovetail fastening, while the second from the right shows the half-dovetail fastening. The last two named are commonly used on first class stairways. The bottom drawing shows a plan in part, of what is shown by the upper drawing. At *A,* by dotted lines, is shown the location of the toenailed baluster shown in the upper drawing. At *B,* also by dotted lines, the location of the baluster with the dowel fastening is shown. The hole for the dowel is indicated by the heavy shading. The

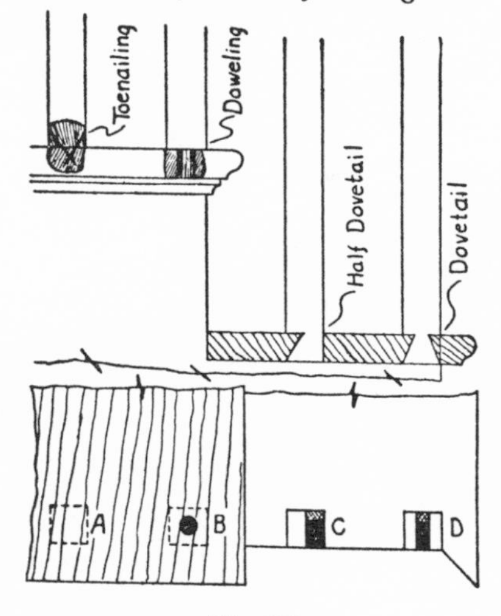

Fig. 303

tread shows the grain running to the end, indicating that the return nosing is worked onto the end of the tread, which is suitable for the two baluster fastenings shown. At *C* is shown the half-dovetail fastening. In some instances the half dovetail is cut the full width of the baluster, as shown by the two shadings. There is one objection to this, which is that if the baluster shrinks very much, it will leave an open crack on the side where the baluster is dovetailed into the tread. To overcome this, the half dovetail is cut out on the inside of the baluster, about as indicated by the light shading, with the housing done accordingly. The shrinkage in the direction of the width of the tread will not cause an open crack, because the tread will also shrink enough to offset the shrinkage of the baluster. At *D* is shown the plan of the dovetail

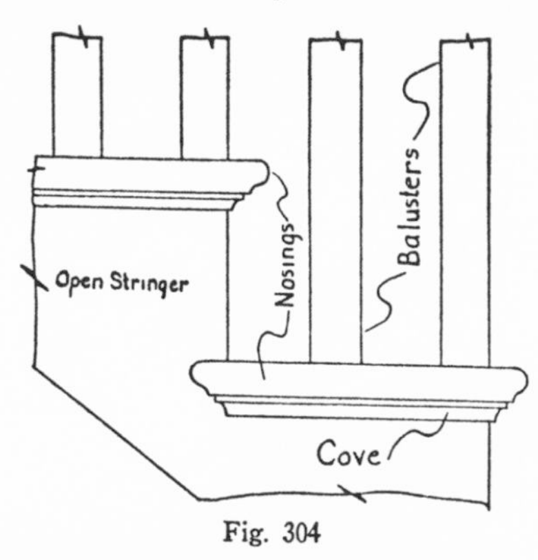

Fig. 304

fastening. Here also provision for shrinkage can be made by reducing the dovetail and housing, as suggested by the lighter shading. The return nosing is not shown on this tread. A com-

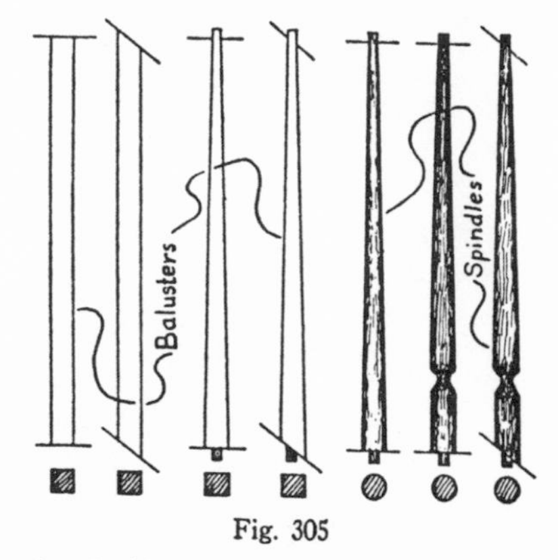

Fig. 305

pleted side view of what is shown in Fig. 303, is shown by Fig. 304, where the different parts are pointed out.

Balusters and Spindles.—Fig. 305 shows to the left two square balusters. One is shown as it would stand in a horizontal railing, and the other as it would be in a sloping railing. At the center are shown two square tapering balusters, one for a horizontal railing and the other for a sloping railing. To the right, shaded, are shown three spindles. One is a plain tapering spindle, and the other two have a suggestion of ornamentation worked on them. The two first ones are shown as they would stand in a horizontal railing, while the last one is shown in position for a sloping railing.

To obtain the bevel for the baluster cuts of a sloping railing, take the rise and the run of a step of the stairway on the square—the rise will give the bevel. For the baluster cuts of a horizontal railing mark square across. The dovetail cuts must be governed by the requirements of the situation. A bevel running between 45 and 60 degrees will give good service.

Chapter 35

WINDING AND CIRCULAR STAIRS

The stairways that are treated in this chapter should never be used, if it can be avoided. The only reason these stairways are discussed here is to give the reader the knowhow, so that in cases of emergency he can build them. There was a time when it seems winding stairs were used as a sort of novelty, rather than because there wasn't any way to prevent using them. A middle-aged man, who was a good carpenter and later took up architecture, when he built his own home, used winders in the stairway. This writer has seen the stairway several times, and he can find no reason in the world why winders were necessary, unless it was the novelty of the thing. It was during that period when winders were used extensively without justification. There are, however, geometrical stairways that are built for art's sake, rather than for comfort, safety, and service.

Framing Winders.—Fig. 306 shows a plan of a stairway with six winders in it and two flights of straight steps. There are fourteen steps, as shown in figures on the drawing, and fifteen risers. The rough horses are shown as if the three walls had been taken apart at the corners and laid back, as one would lay back the covers of a folder. To the left are shown the horse for the straight flight with four steps, and a horse for a winding step and a half, numbered *5* and *6*. The top drawing shows a horse for two full winders, *7* and *8*, and two half winders, *6* and *9*.

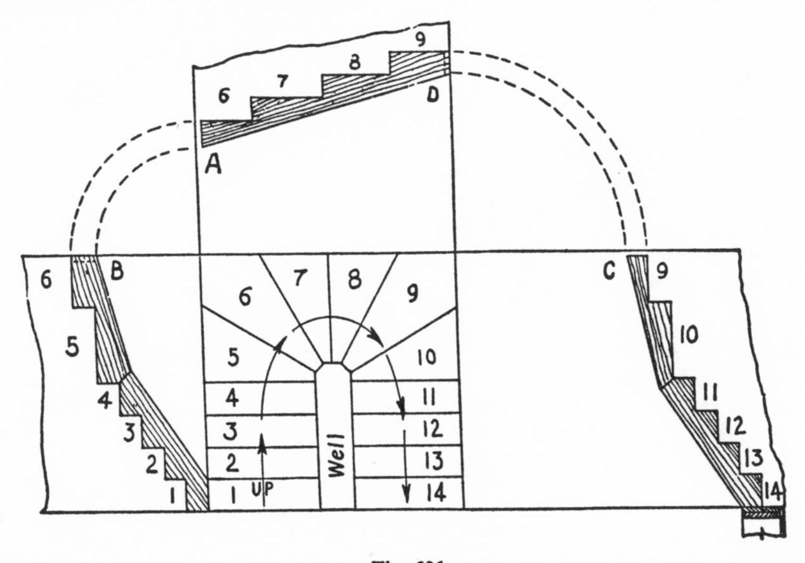

Fig. 306

The end of this horse at *A*, when the walls are in position, joins the horse shown to the left at *B*—that is, *A* joins *B*, as the dotted lines indicate. In the same way the end of the winder horse to the right, marked *C*, joins the winder horse shown at the top, as indicated by dotted lines, at *D*. The winder horse to the right supports a half winder and a full width winder, numbered *9* and *10*. Also to the right is shown a horse for four straight steps, numbered *11*, *12*, *13*, and *14*. The numbers given with the steps of the horses correspond with the same numbers given on the plan.

Horses Around Well.—Fig. 307 shows the horses for the straight steps and the supports for the winders around the well, as if they were laid back flat on the floor. The relative elevations of the different horses and supports are

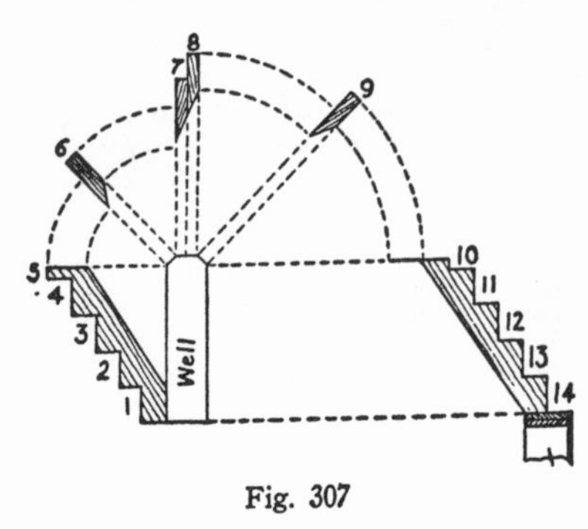

Fig. 307

shown by dotted lines. The horse to the left has four straight steps, as numbered on the drawing, and the fifth step is for the narrow end of the first winder, which is numbered *5*. At *6* is shown the support for the narrow end of the second winder. At *7* and *8* the supports for the narrow ends of the third and fourth winders are shown, while at

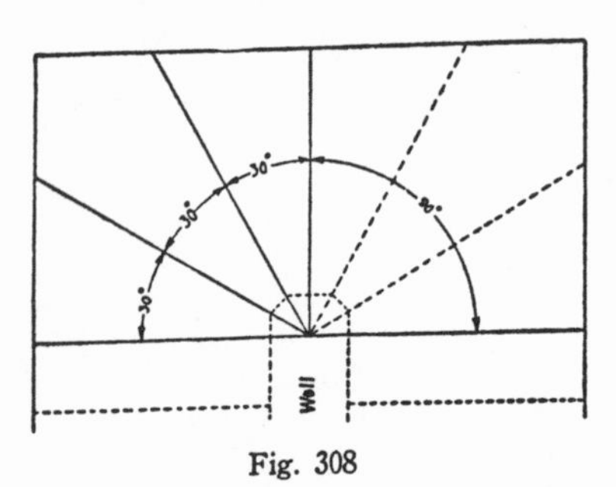

Fig. 308

9 the narrow-end support for the fifth winder is shown. The horse shown to the right shows at *10* the support for the narrow end of the sixth winder and four straight steps. Figs. 306 and 307 should be compared and studied.

Laying Out Winders.—Fig. 308 is a plan of the 6 winders shown in the previous figures. The three winders to the left are shown with straight lines radiating from a common center with 30-degree angles. To the right the 90-degree angle can be taken as a quarter-turn landing, or as a continuation of winders, as shown by the two dotted lines. How to apply the square to get the 30-degree angle is shown by Fig.

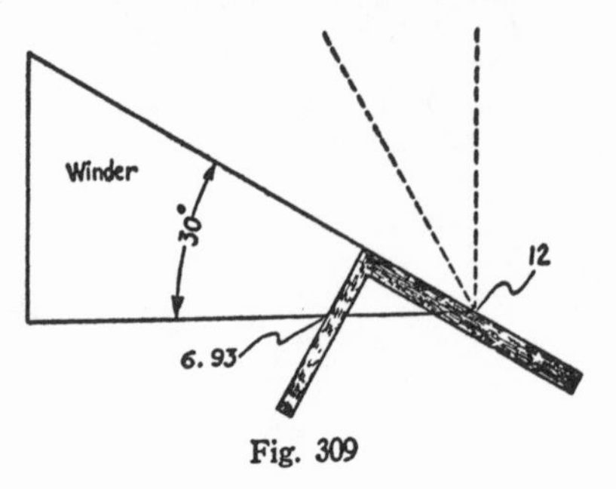

Fig. 309

309. Two additional winders are indicated in their relative position by dotted lines. The figures to use on the square are 12 and 6.93, as shown on the drawing.

Line of Traffic.—Fig. 310 is a plan of the same 6 winders shown in previous figures, with the line of traffic in-

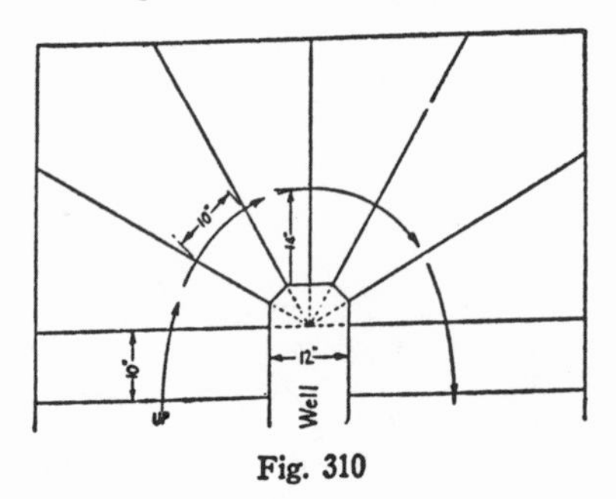

Fig. 310

dicated by the arrows. It should be noted that the line of traffic around the well is about 14 inches from the narrow ends of the winders. At this line the run of the winders should be the same as the run of the straight steps, or as shown, 10 inches.

Dancing Winders.—Fig. 311 shows in part, a winding stairway with dancing winders. These winders are laid out from what is called the dancing center, indicated at *D. C.* The common center is given by the heavy dot at *C*, from the center of which the circular end of the well is described. A quarter-

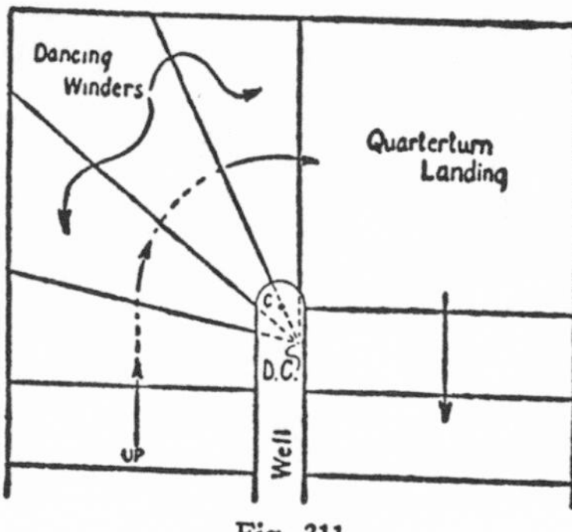

Fig. 311

turn landing is shown to the right. Fig. 312 shows six dancing winders that were laid out from the dancing center, as indicated. The common center is again shown at *C* by a heavy dot, the

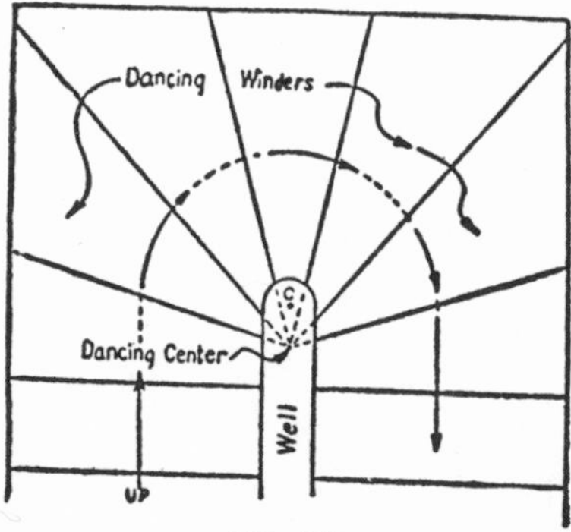

Fig. 312

center of which is used for describing the circular end of the well.

Swelled Steps. — Fig. 313 shows a circular stair with swelled steps. The arrow indicates the line of traffic, at which line the steps should have a run and rise per step that will equal between 16½ and 17 inches. A templet should be used for marking the swell of the steps, and when the plan is laid out on the floor the templet should be fastened to a common center in such a manner that the steps will radiate from it properly. The dotted lines that join the small circle show how the

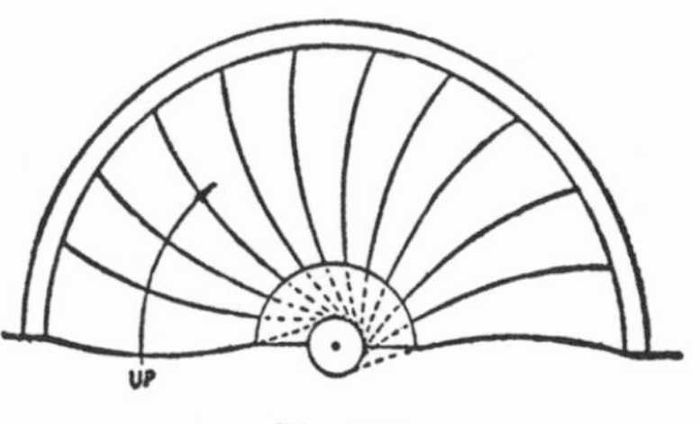

Fig. 313

curvatures of the steps miss the common center.

Horses for Circular Stairs. — Fig. 314 shows the left end of the swelled steps shown in Fig. 313. To the left is shown how to lay out the horse that supports the wide ends of the steps. The dotted half-circles and dotted per-

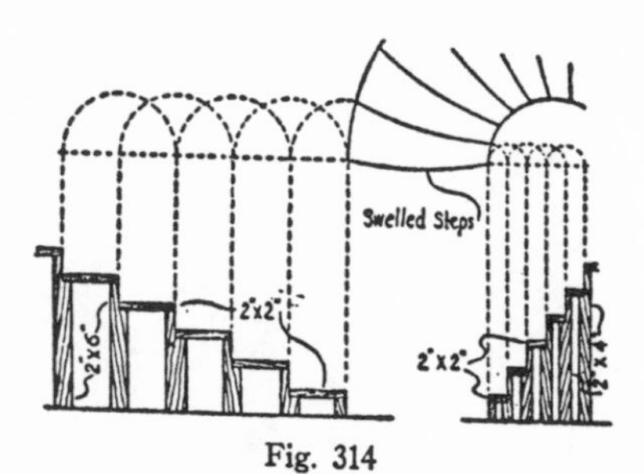

Fig. 314

pendicular lines show the relationship of the wide ends of the treads to the steps of the horse. To the right is shown the same development of the horse that supports the narrow ends of the steps. The dotted half-circles and dotted perpendicular lines again show the relationship of the treads to the steps of the horse. Study the drawing.

Chapter 36

CONCRETE STEPS

Forms for Steps.— Building forms for concrete steps still belongs to the field carpenter, and in framing the form material the steel square plays a major part. Concrete steps are widely used when the steps are exposed to the weather. The principal reason for this is that concrete steps will not rot, while wooden steps will. If the forms for such steps are built right, the steps will not only look well, but they will be properly proportioned, making them easy to pass over. There are a number of important features about building forms for steps that will be pointed out in this chapter, which every form builder should know.

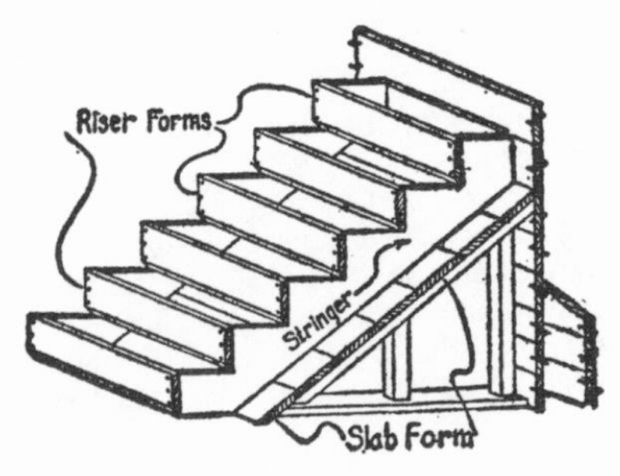

Fig. 315

Self-Supporting Steps.— Fig. 315 shows the form for a flight of self-supporting concrete steps, ready for the reinforcing and then the concrete. It should be noted that the bottom edges of the risers are beveled, making it possible to trowel the top of the steps up to the angle between the step and the riser. To get a first class job, the stringers and the risers should be dressed on the side where the concrete will come, and then treated with a paraffin preparation that will prevent the lumber from absorbing the moisture in the concrete. Another thing, the paraffin treatment will keep the concrete from sticking to the forms, leaving the concrete steps with smooth surfaces. Fig. 316 shows the completed concrete steps with the forms shown in Fig. 315 removed. As shown, these steps have a rise of 7 inches and a run of 9½ inches.

Forming for Nosings.— The risers for the steps shown in Figs. 315 and 316, are shown at a right angle with the steps; that is, the risers were set in a plumb position. This kind of concrete step is unsatisfactory, although it is quite frequently used. With but little, if any, extra expense, forms for concrete steps can be made so that the steps will have nosings, somewhat on the order of the nosings of wooden steps. A few practical designs are given in this chapter. Fig. 317, *A,* shows a cross section of a simple design for

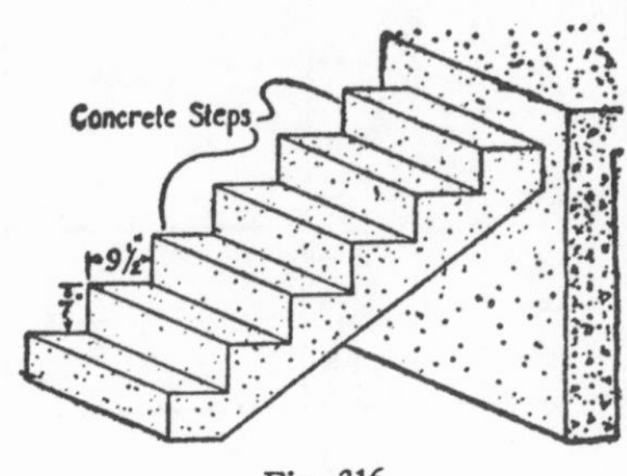

Fig. 316

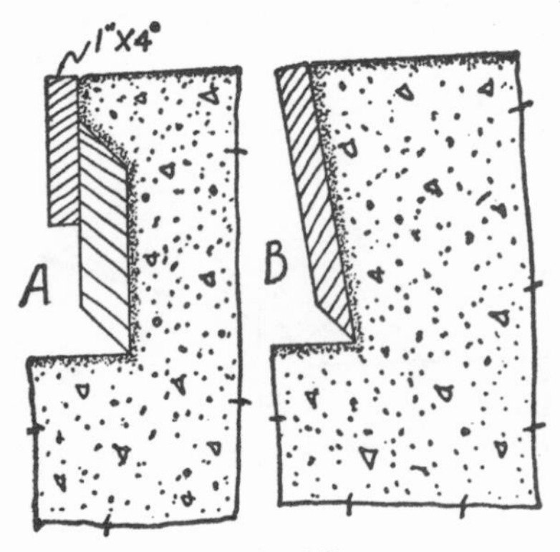

Fig. 317

forming a nosing. The piece with the beveled edges should be as thick as the nosing is to project, usually 1¼ inches. The upper piece can be 1x3 or 1x4, whichever is best suited for the place, or is available. At *B* is shown a cross section of a riser form that is commonly used, and gives quite satisfactory service. In this form the riser is set out of plumb enough to give the step a sort of nosing, which provides the step with toe and heel room. The bevel at the bottom of the riser form, makes it possible to trowel the step up to the angle. The material for such risers, as stated before, should be dressed and treated with a preparation of paraffin.

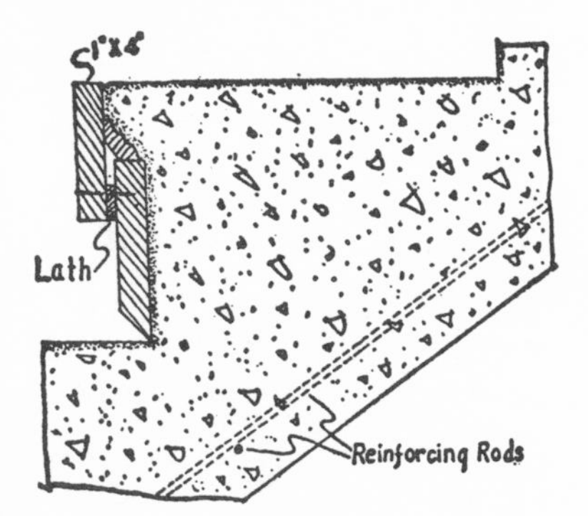

Fig. 318

The dotted shading shown near the surfaces of the concrete, indicates that the steps are to have monolithic surfaces; that is, a rich mixture of sand and cement is used for the surfaces, which is placed for the riser just before the concrete is poured, and on the steps right after it is poured.

Molded Nosing Forms.—Fig. 318 shows a cross section of a riser form that will give the step a nosing having a molding effect. The riser form is made with a beveled board of the proper width, to which a 1x4 with a lath between is nailed. Then a bed molding is placed in the angle, as shown. A little study will make the construction clear. A cross section of another riser form with a molding effect nosing is shown by Fig. 319. In this design the step has added a metal safety tread, indicating that this is a better step design, but more expensive. Notice the reinforcing and the thickness of the slab.

Clearance for Troweling.—Fig. 320 shows a cross section of a form with

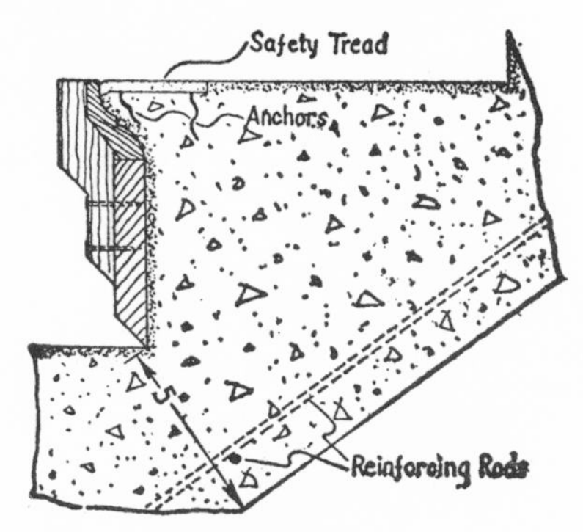

Fig. 319

the concrete steps already poured, that provides ample space for troweling the top of the steps. The risers in this case are the same as the riser shown in Fig. 317, *B*.

Fig. 321 shows five different ways

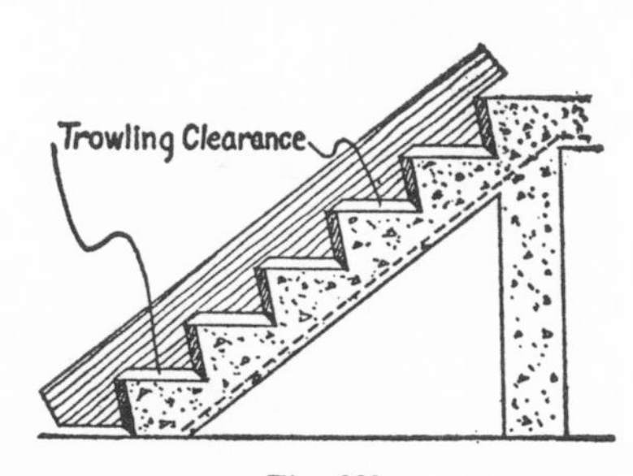

Fig. 320

to hold the riser forms for concrete steps, so as to provide ample room for troweling. At number *1* is shown a commonly used method of making such forms. Here a short piece of board is nailed to a 2x6 in such a way that it will hold the riser in place. At number *2* is shown an especially good method. At *3* and *4* are shown two modifica-

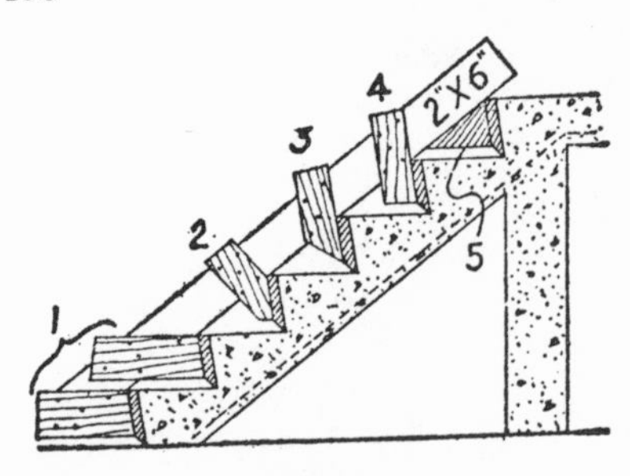

Fig. 321

tions of what is shown at *2*. At number *5* is shown an entirely different method. Here a block that was cut out of a plank in making a horse for step forms, is nailed to the edge of a 2x6, and to it the riser form is fastened. When this method is used, one of the horses is framed by cutting out the blocks that are then nailed to a 2x6 (or 2x4), which makes the second horse. The two are then used for holding the riser forms in place.

Steps with Curbs.—Fig. 322 shows a flight of concrete steps with the forms still in place. These steps have a curb on each side. The bottom edges of the risers are shown beveled, indicating that the steps are to have finished surfaces when the forms come off. This means that the form material must be surfaced and treated.

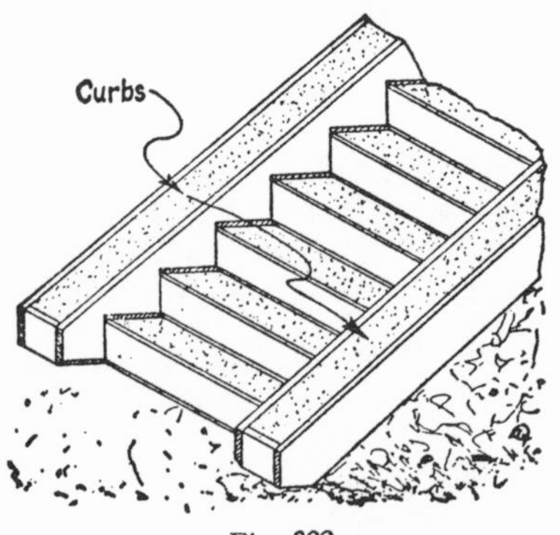

Fig. 322

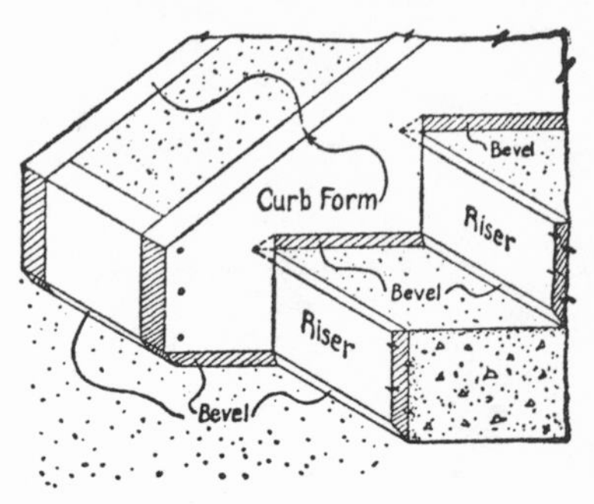

Fig. 323

A detail showing the beveling of the forms, so as to provide room for troweling, is shown by Fig. 323, where a part of a curb and two steps, also in part, are given. Compare and study Figs. 322 and 323.

Chapter 37

CIRCULAR CEMENT STAIRS

Perhaps the most practical of circular stairs is the cement circular stair. This is especially true when the stair is not wider than two feet. For then the traffic is kept almost constantly on the line of traffic, where the steps are of normal size. When the stair is made wider than two feet, the danger of accident increases. This is true of all circular stairs, whether they are made of cement or of some other material.

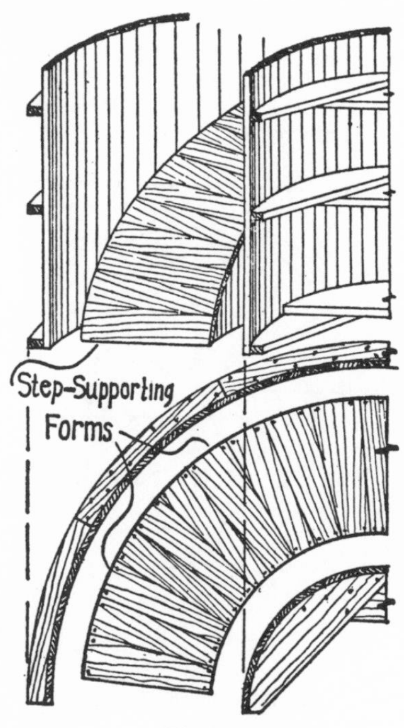

Fig. 324

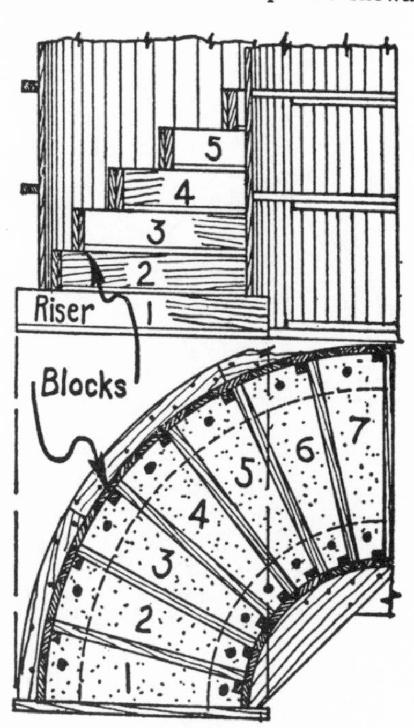

Fig. 325

Forms for Circular Stairs. — Fig. 324, at the bottom, shows a plan of the forms for a part of a circular stair. The upper drawing shows a sort of perspective view of the forms. The step-supporting form is pointed out on both drawings. It will be noticed that the circular forms have the boards running up and down, with the nailing girts cut to the proper radiuses. Study and compare the two drawings. Fig. 325 shows the same part of the stair with the concrete poured up to the finished steps. The dotted part-circles show the thickness of the circular walls. The heavy dots at the ends of the steps indicate perpendicular reinforcing rods. Seven steps are shown by the bottom drawing, where the riser forms are still in place. These riser forms are held in place by blocks that are nailed to the circular forms, as the indicators point out on both drawings. The top drawing shows the elevation. These two drawings should also be studied and compared. Fig. 326 shows the same part of the stair shown in the previous drawings, with the steps finished and the forms removed. The steps in both drawings are numbered

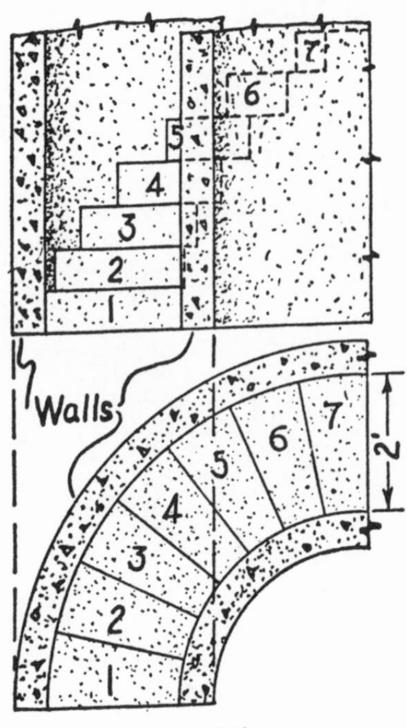

Fig. 326

from *1* to *7*. To the right of the bottom drawing the width of the stair is given, which is 2 feet. A detail of one of the steps, cut at the traffic line, is shown by Fig. 327. Here the reinforcing rods are pointed out—also a

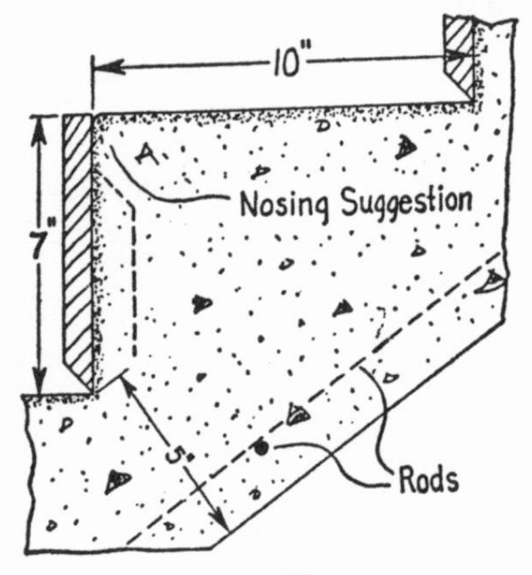

Fig. 327

suggestion for forming a nosing. Notice the bevel at the bottom edge of the riser form. The rise is 7 inches and the run is 10 inches, and as stated before, these figures are taken at the line of traffic.

Self-Supporting Circular Stairs.—The most practical of the circular stairs is the one that winds around a center column, and is well anchored to a firm base, from which the stairs receives its support. This stair can be built economically and substantially. Such a stair will not have the hazards that usually accompany circular or winding stairs. This is true because the narrowness of the stair keeps the traffic always on the traffic line, where

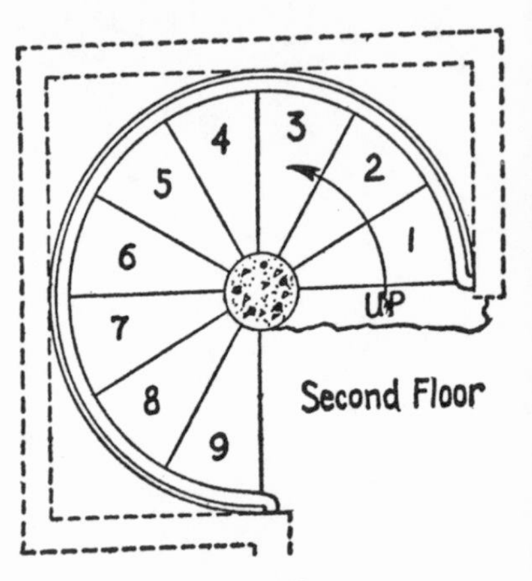

Fig. 328

the steps have the normal size. Fig. 328 shows a plan of a self-supporting circular stair with 9 steps. The arrow shown is on the traffic line and points up. The ninth step joins the second floor landing. The dotted lines show how the stair can be built within a staircase, or how it can have a staircase built around it after it has been constructed. An elevation of this stair is shown by a sort of diagram, in Fig. 329. The steps, handrail, and column should be studied and compared with Figs. 328 and 330. Fig. 330 shows the full second floor view of the same plan. Here the arrow points down, and the

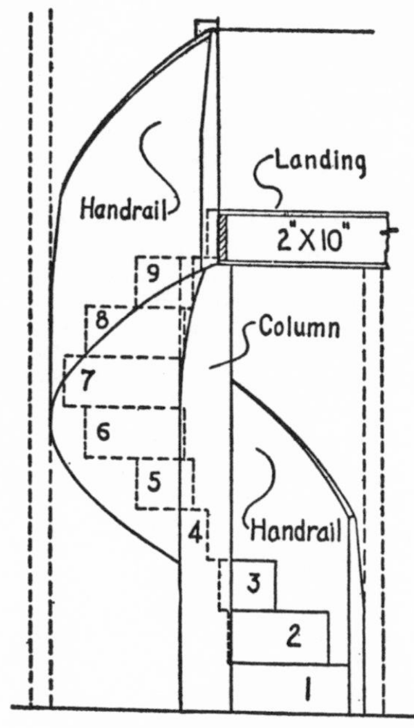

Fig. 329

wall shown by dotted lines, joins the reduced column that projects above the second floor.

Detail of Column and Step Reinforcing.—Fig. 331 shows a section of the column. This column is poured into a metal form after the reinforcing rods for the steps are placed, by running them through holes that are made in such a way that the rods will come right for the steps. When the rods are

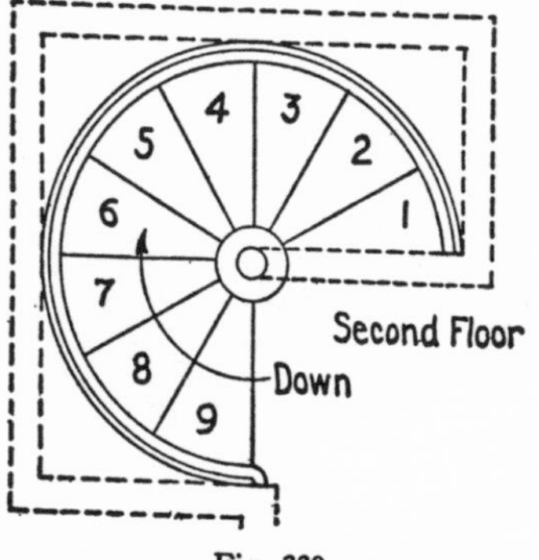

Fig. 330

in, the column form is filled with a rich mixture of concrete. Before the concrete has had time to set, the reinforcing rods for the steps should be adjusted so that they will radiate from the center of the column—they should also be kept on a level. The rods for steps number *1* and *2* are shown to the

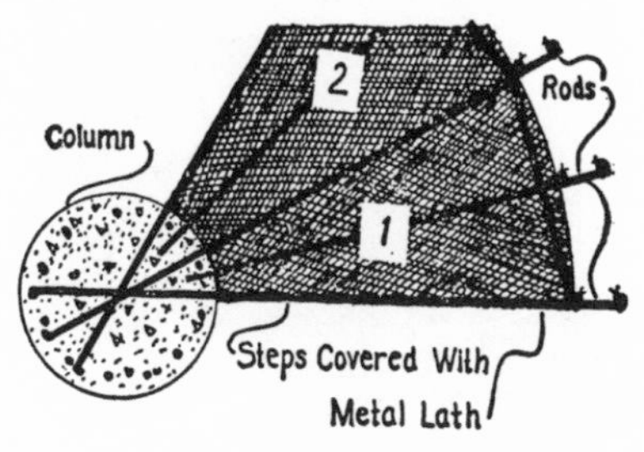

Fig. 331

right, covered with metal lath. How the rods for the risers and for the railing should be placed is shown by Fig. 332. Here a part of the column and one riser are shown in elevation—also the bottom and top of one railing rod. The inset drawing, at a much smaller scale, shows the same layout with the

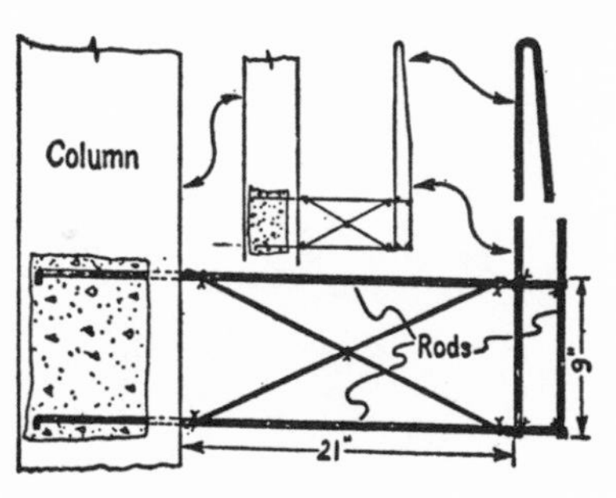

Fig. 332

railing rod in full. In both of these drawings the column is cut out, showing how the rods are anchored to the column. The horizontal rods for the risers are the rods that are shown in Fig. 331 for the steps, over which the metal lath is placed. The center rods of steps *1* and *2*, shown in the detail of these steps, do not run through the column. The heavy dots shown to the right in Fig. 331 indicate how the railing rods are to join the reinforcing rods of the steps. The rods are shown wired, but if they can be welded it will hold them much firmer until the cement mortar is in place and set. Fig.

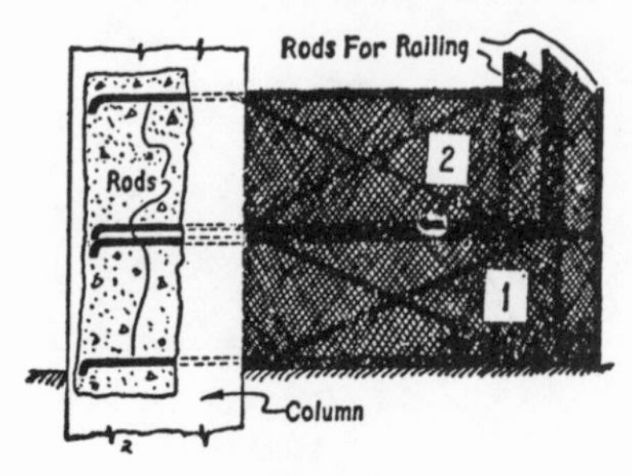

Fig. 333

333 shows a detail of the risers for steps *1* and *2*, and of the bottom part of the handrail, covered with metal lath. To the left the column is cut out, showing how the rods are bent in order to anchor them to the column. Study and compare Figs. 331, 332, and 333.

Finishing the Stairs.—When the concrete in the column is thoroughly set, then the metal form can be removed, either entirely or only where the steps will join the column, whichever is desirable. This done, the rods should be wired or welded and covered with metal lath, both for the steps and for the handrailing. Then a rich cement mortar should be used to cover the metal lath and reinforcing completely, leaving it rough. When this has set enough to thoroughly hold its own, it should be back plastered under the steps with the same kind of cement mortar. After that, metal lath should be fastened to the bottom angles of the steps with wires that were placed before the back plastering was done, and then plastered with the same kind of mortar. When the rough plastering has had time to set, the railing and the bottom of the steps should be finished first. Then the steps can be finished with cement. If this is all done in a workmanlike manner, you will have a serviceable, self-supporting winding stairs that will have a pleasing appearance.

Chapter 38
THE PROTRACTOR

Degrees.—Whether or not the mechanic who handles the steel square knows it, every operation with the square involves degrees. This is true because the circle and the square are closely related. Every right-angle triangle is related to both the square and to the circle. This is especially true with regards to roof framing. The base of the triangle represents both the run of a rafter and the radius of a circle. The altitude represents both the rise of a rafter and the tangent of a circle, while the hypothenuse represents the rafter and gives the angle between the rafter and the horizon in degrees. Of course, in these operations the degrees are not visible, but they are present just the same. The instrument that will give the degrees of any angle is a protractor, and the mechanic whose work involves the use of degrees, should have a good protractor in his kit of tools.

Protractor.—Fig. 334 is a drawing of a simple protractor, giving the number of degrees in a half circle, or 180 degrees: For there are 360 degrees to a circle; 90 degrees to a quarter circle, or right angle; 60 minutes to a degree, and 60 seconds to a minute—and that is bringing it down to rather fine points, if you are thinking in terms of roof framing.

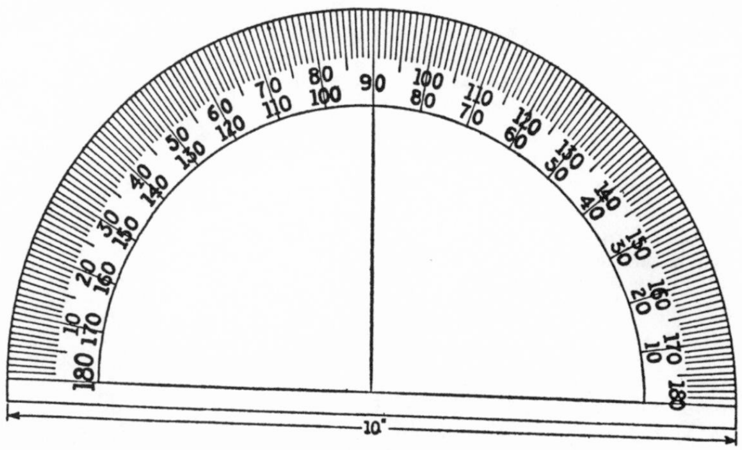

Fig. 334

Degrees and Pitches.—Fig. 335 shows a shaded square at a much smaller scale than the scale that was used in drawing the protractor shown on the same drawing. This was done to bring out the points more clearly. To the left of the drawing will be found a section of 10 degrees divided into degrees. The line running from 12 on the tongue of the square to the left, is drawn at an angle of 5 degrees, as shown. To the right, reading from the bottom up, are

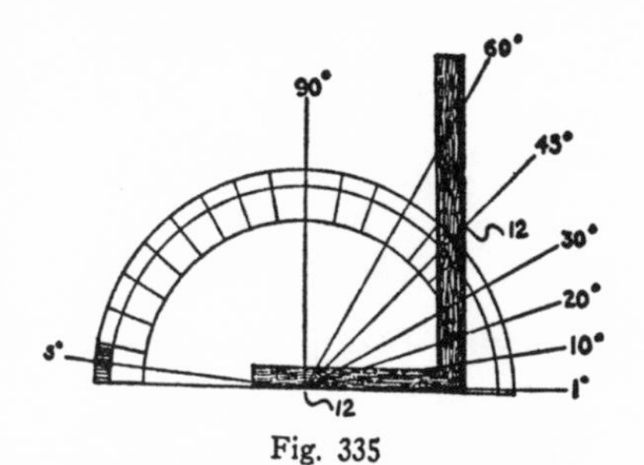

Fig. 335

shown lines radiating from 12 on the tongue of the square, at different angles: The first line is at an angle of 1 degree, the second is at an angle of 10 degrees, the third, at 20 degrees, the fourth at 30 degrees. The line drawn from 12 on the tongue through 12 on the blade is at a 45-degree angle, and gives the slope of a half pitch roof. The line drawn to 60 degrees, gives the angle of a triangle. The perpendicular line at the center is drawn at an angle of 90 degrees, or a right angle.

Polygons.—Fig. 336 shows, besides the right angle, the angles for the ten most important polygons, as follows: Triangle, square, pentagon, hexagon,

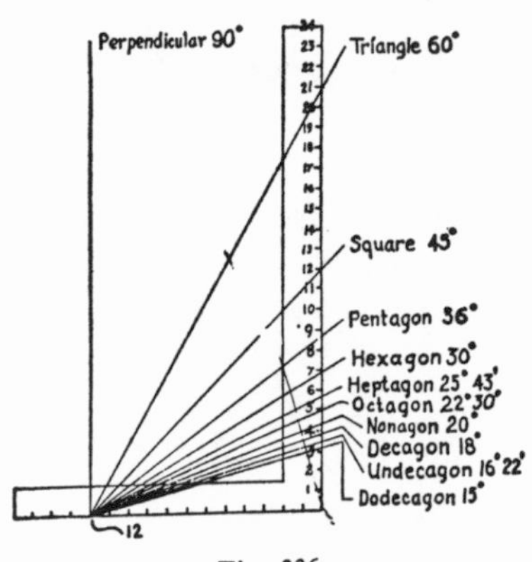

Fig. 336

heptagon, octagon, nonagon, decagon, undecagon, and dodecagon, giving the degrees of each of these angles. When those angles are made with the square, take 12 on the tongue of the square and the point where the line intersects the outside edge of the blade. Any other angle described in degrees, can be made with the square, using the protractor to find the angle, as shown in Fig. 335, and without showing the protractor, in Fig. 336.

Degrees of Twenty-Four Angles.—Fig. 337 shows a square with twenty-four lines running from 12 on the tongue to each of the twenty-four figures on the blade of the square, which represent the twenty-four inches. At the end of the lines will be found the

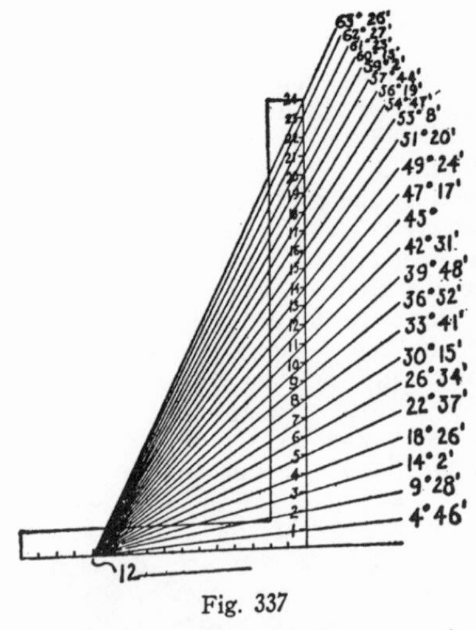

Fig. 337

degrees and minutes of the respective angles.

Common Roof Pitches.—Fig. 338 shows a square with five lines running from 12 on the tongue to the points on the blade representing five of the pitches commonly used in roof framing. To find the degrees of the angles, make a drawing of the run, the rise, and the rafter of the pitch in question, and apply the protractor to get the number of degrees in the angle. In the same way, the degrees of any other roof pitch can be found.

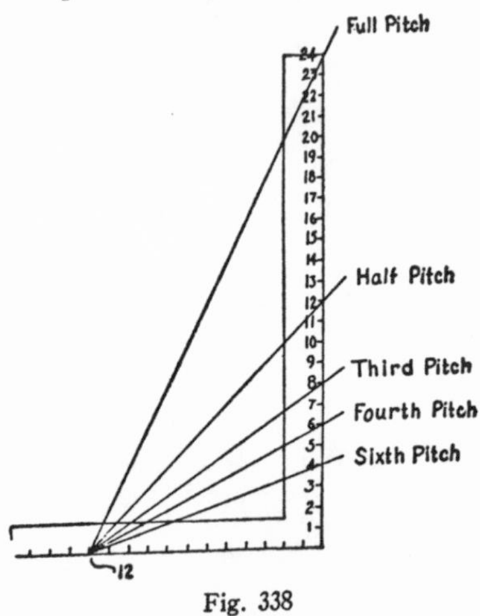

Fig. 338

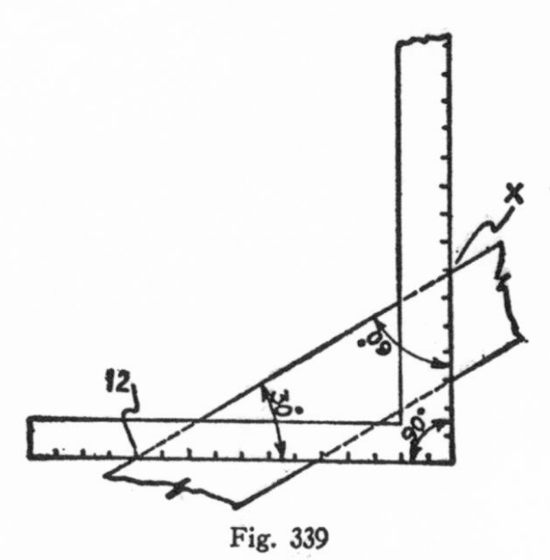

Fig. 339

Right-Angle Triangles.—Fig. 339 shows the square applied to a timber, shown in part, with an angle at the point of the level cut, of 30 degrees. The angle at the point of the plumb cut in this case is 60 degrees. The right angle where the run and rise join is, of course, 90 degrees. Now if you know the degrees of one of the two sharp angles of any right-angle triangle, you can get the other by subtracting the known angle, in degrees, from the number of degrees in a right angle, which is 90, and that will give you the unknown number of degrees of the other sharp angle. Take the illustration in Fig. 339, which shows the degrees in the two sharp angles: Subtract 30 from 90 and you have 60, or in reverse order, subtract 60 from 90 and you have 30. These figures all come out even, but the results will be just as accurate when the angle is expressed in degrees, minutes, and seconds. The points taken on the square are 12 on the tongue and X (6.93 to be exact).

Fig. 340 shows the square applied to a timber shown in part, for obtaining the level and plumb cuts. In this case the degrees in the two sharp angles are the same, or 45. Add the two and you have 90, which proves that they are correct, provided that you are sure of one of them. To be sure, use the protractor.

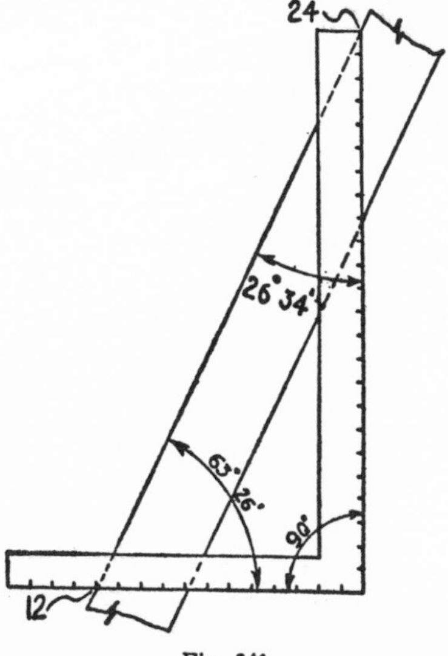

Fig. 341

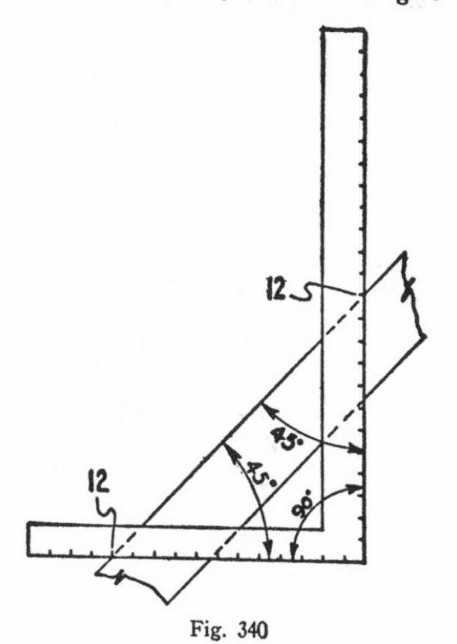

Fig. 340

Degrees and Minutes.—Fig. 341 shows the square applied to a timber shown in part, for obtaining the level and plumb cuts of a rafter for a full pitch roof. Here the two sharp angles are expressed in degrees and minutes. If you add the two together you will have 90 degrees.

Chapter 39

ROOFS AND ROOF PITCHES

Regular Roof Framing—A Short Course.—By regular roof framing, as it is used here, is meant common or simple roof framing as opposed to irregular plan or irregular pitch roof framing. This course, chapters 39, 40. and 41, covers the different kinds of roofs, stepping off common rafters and common rafter cuts, stepping off regular hip and valley rafters and their cuts, and regular jack rafters and their cuts. These are the most practical parts of roof framing, enabling the mechanic to frame most of the commonly used roofs.

Single-Pitch Roof.—Fig. 342 is a sort of diagram, showing a single-pitch roof. This roof is frequently used on temporary buildings, such as material sheds, stands and tool houses. When

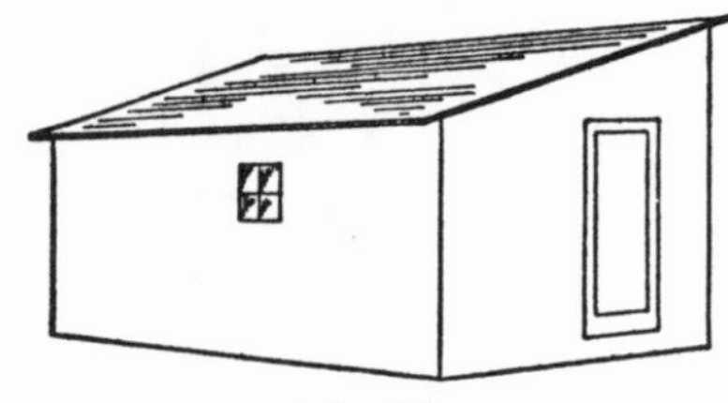

Fig. 342

it is used on permanent buildings it is done as a matter of economy or expediency. It is also called pent roof, shed roof, to-fall, and erroneously, lean-to.

Lean-To Roof.—Fig. 343 shows what is in reality a lean-to roof, which is to say, that the roof leans against some other building and derives some of its support from that building.

Saw-Tooth Roof.—The saw-tooth roof shown in Fig. 344, so far as the individual parts are concerned, is much on the order of a shed roof. This roof is mostly used on factory buildings, garages, and similar structures. The advantage of this roof is that it makes possible a great deal of window space for the admission of light and for ventilation.

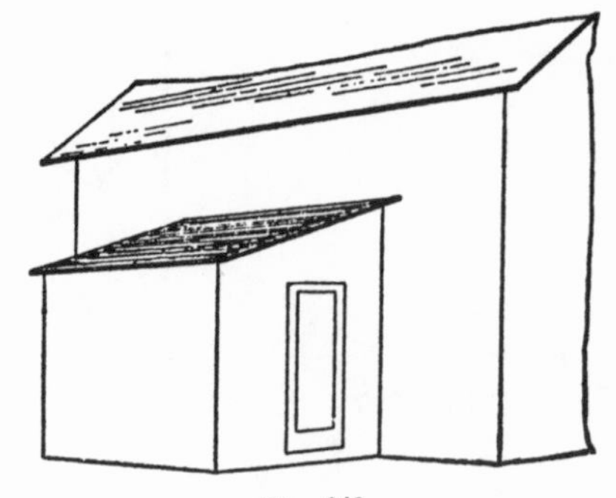

Fig. 343

Double-Pitch Roof.—A simple diagram of a building with a double-pitch roof is shown by Fig. 345. This roof is, perhaps, the basis for all other roofs

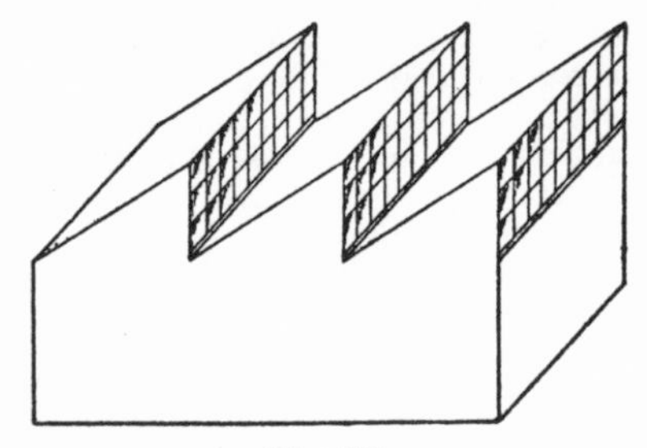

Fig. 344

that are classified as pitch roofs. It is also called gable roof and saddle roof.

M Roof.—Fig. 346 shows what is known as an M roof, which is made up of two double-pitch roofs, as shown by

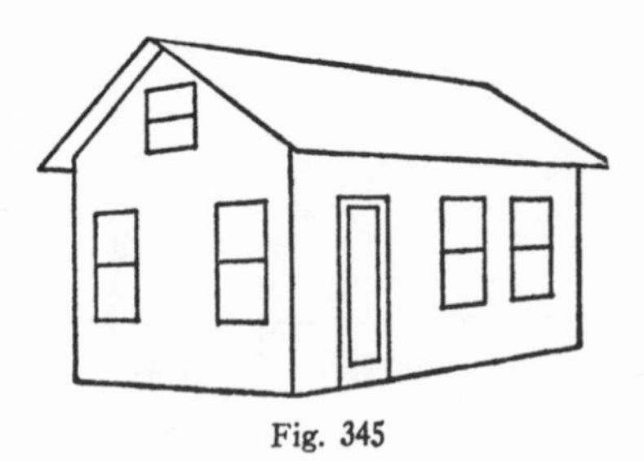

Fig. 345

the drawing. This roof has two advantages; first, it reduces the elevation of the building, and second, much shorter material for rafters can be used in its construction. These advantages will be clear when one takes into consideration

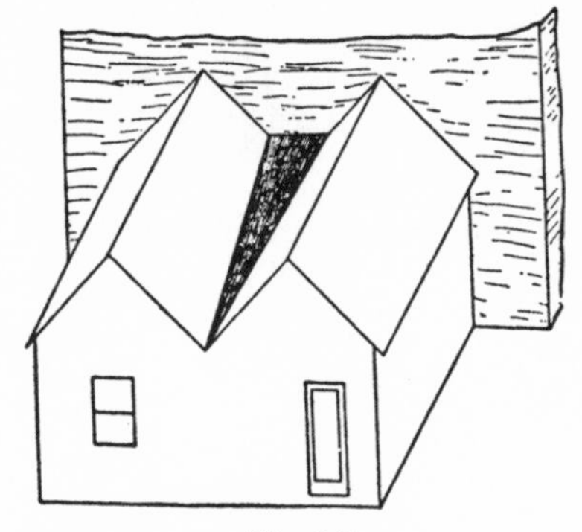

Fig. 346

that the diagram shown represents a rather small building.

Hip Roof.—Fig. 347 shows a building with a hip roof. This roof is perhaps the strongest of the pitch roofs. It is much more substantial than ordinary conditions require; however, in lo-

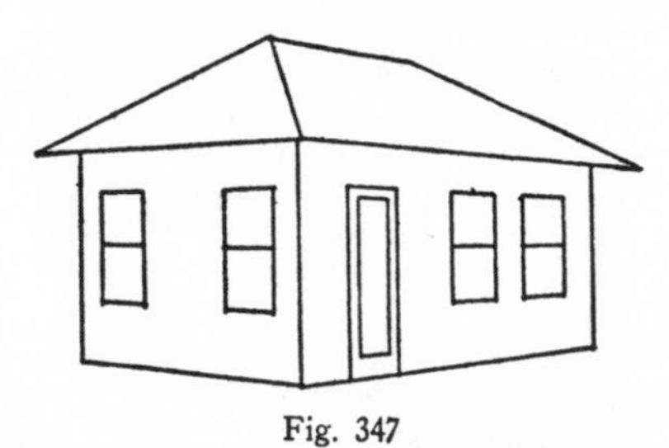

Fig. 347

calities where there is danger of wind damage, this roof is quite suitable, but in such cases it must be firmly anchored to the building supporting it, which in turn must be anchored to the foundation.

Gambrel Roof.—A gambrel roof is shown by Fig. 348. The advantage of this roof is that it increases the attic space, and when dormer windows are used it is almost equivalent to a second

Fig. 348

story. A gambrel roof has little in its favor from the standpoint of economy and appearance.

Mansard Roof.—Fig. 349 shows a mansard roof, which is a modification

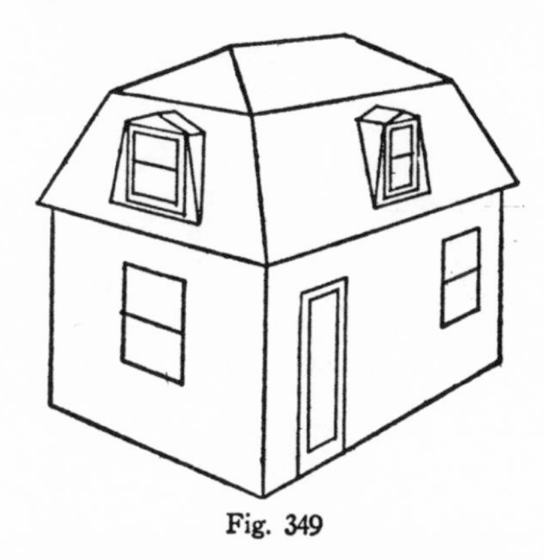

Fig. 349

of the gambrel roof, or rather the hip version of it. Its advantages lie in the space added to the attic and in the additional strength of the construction.

Semicircular Roof.—Fig. 350 shows a building with a semicircular roof.

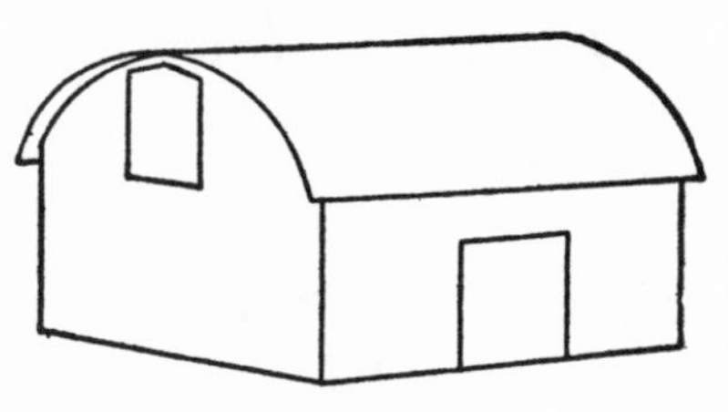

Fig. 350

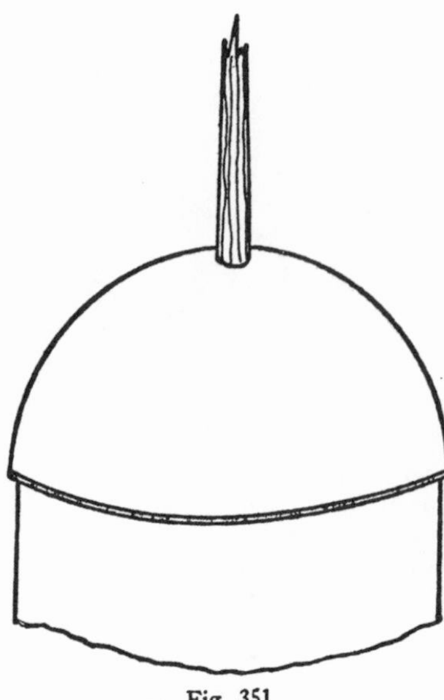

Fig. 351

This roof is often used on barns. Sometimes a ridge is added to this by reversing the curvature slightly about four or five feet from the center at the top, which gives a sort of English Gothic effect. The curvature at the eaves is also reversed enough to give them a little more drip.

Dome and Turret Roofs.—Fig. 351 shows a dome roof, while Fig. 352 shows to the left a square turret and to the right a cone-shaped turret.

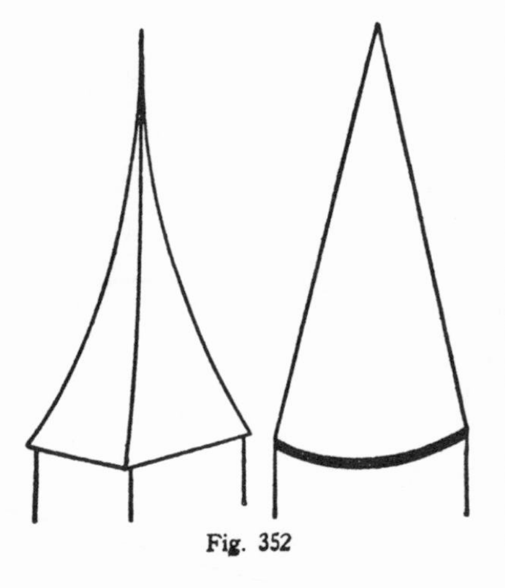

Fig. 352

Other Roofs. — Other roofs that might be added are bell roof, ogee roof, gable-and-valley roof, hip-and-valley roof, irregular-pitch, or uneven-pitch roof, deck roof, and flat roof. These all are modifications of the roofs shown by the illustrations.

Old-Fashioned. — This writer has been criticized for using the old-fashioned terms one-fourth pitch, one-third pitch, one-half pitch, full pitch, and so on. However, these terms are basic and give the student a practical conception of the thing under consideration. The same critics advocated the use of roof framing tables, instead of the steel-square method of roof framing. But imagine the carpenter who attempts to do roof framing without understanding anything about it, excepting what the roof framing tables give him. What would he do if he couldn't find his roof framing table when he is ready to start framing a complicated roof, with carpenters standing around waiting for orders? At such a time he would give anything to be able to frame any kind of roof by the simple application of the steel square. The steel-square method might be old-fashioned, but if thoroughly understood and carefully applied, it will work.

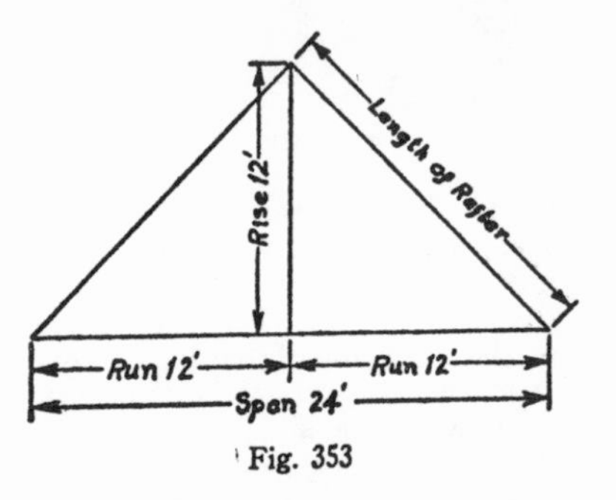

Fig. 353

Span, Run, and Rise.—The diagram shown in Fig. 353 gives the first requisites of roof framing; namely, the span, the run, the rise, and the length of the rafter. These are all basic principles and they remain the same whether you use the terms used here, or express them by some other means—they are fundamental.

Basic Pitches.—Fig. 354 is a diagram showing four basic pitches brought together for comparison; they are, read-

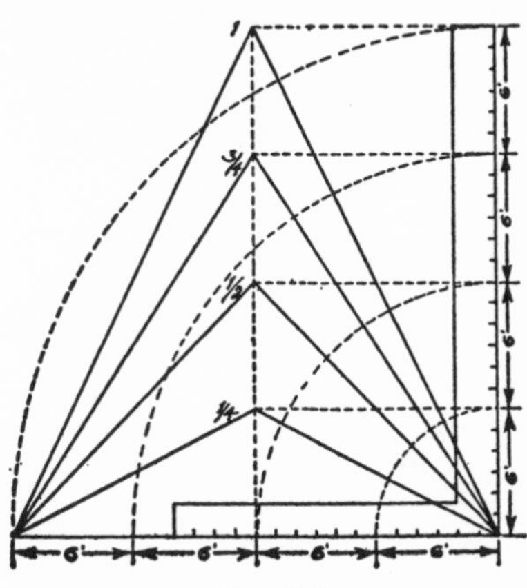

Fig. 354

ing from the bottom up, one-fourth pitch, one-half pitch, three-fourth pitch, and full pitch. These basic pitches are used here, because they are easily expressed and the figures to be used on the square do not involve fractions. It should be remembered that the different pitches that a roof can be framed to are unlimited. Every time there is a change made in the rise, no matter how small it might be, it gives a different pitch. The run, though, in roof framing is basic.

Full Pitch and Steeper.—The four basic pitches shown in Fig. 355 were used a great deal for steeples, up to a generation or so ago. These pitches are, reading from the bottom up, full pitch, double pitch, triple pitch, and quadruple pitch. The student can read-

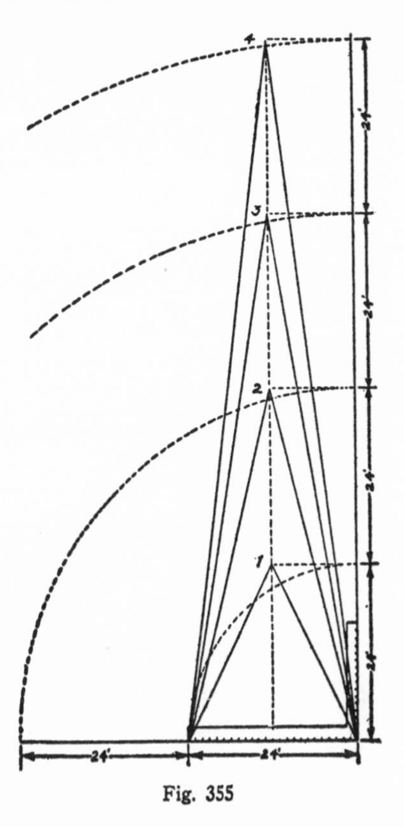

Fig. 355

ily see how this process can be carried on indefinitely, but steeper than the quadruple pitch could hardly be called practical, unless it would be in metal towers.

Chapter 40

THE STEEL SQUARE, GUIDES, AND STEPPING OFF

The Steel Square.—The steel square, which always holds first place in roof framing, is one of the most useful, if not the most useful tool a carpenter can own. It is so important in roof framing that it would almost be impossible to frame a roof without it. And when roof framing is done without the use of the steel square, the principles of the square are nevertheless used.

A drawing of a steel square applied to a 2x4, giving the foot and plumb cuts of a one-half pitch roof, is shown by Fig. 356. The figures used on the square are 12 on the body of the square and 12 on the tongue. The body gives the foot cut along the edge pointed out at *A,* while the plumb cut is obtained by marking along the edge of the tongue, indicated at *B*. This being a half pitch, the foot and plumb cuts are both on a 45-degree angle. If the rafter has a tail, or lookout, the seat cut becomes what is known as a bird's mouth. The dotted line indicated at *a,* shows the plumb cut of the seat.

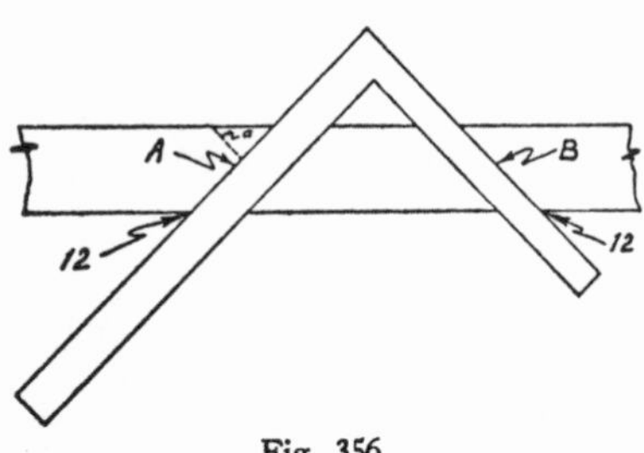

Fig. 356

Fence and Guides.—Fig. 357 shows two views of a job-made fence for a steel square. The holes pointed out at *a, a,* receive the thumbscrew bolts with which the fence is clamped to the square. At *b, b* are pointed out the slots into which the square is slipped. This fence gives satisfactory results if the pattern material is perfectly straight. But in cases where small humps are on the edge of the timber the fence will cause the square to rock, which renders the marking unreliable. There are available on the market, guides that are free from the objection just pointed out.

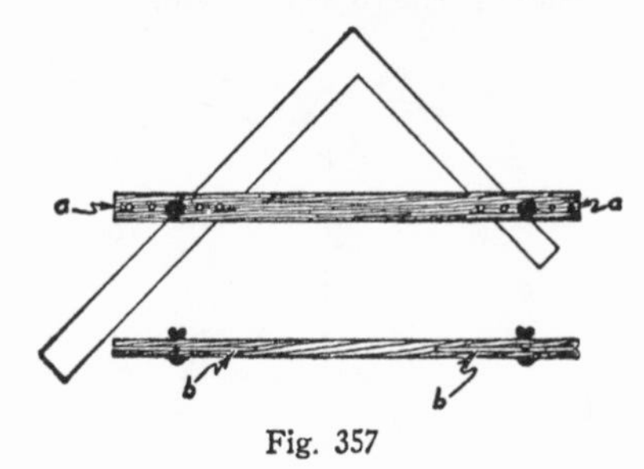

Fig. 357

These guides are made in pairs, much on the order of what is shown by Fig. 358. The upper drawing shows a side view, while an edge view is shown at the bottom. The dotted line shows the relationship of the guides to the timber, when they are set at 12 and 12 on the square.

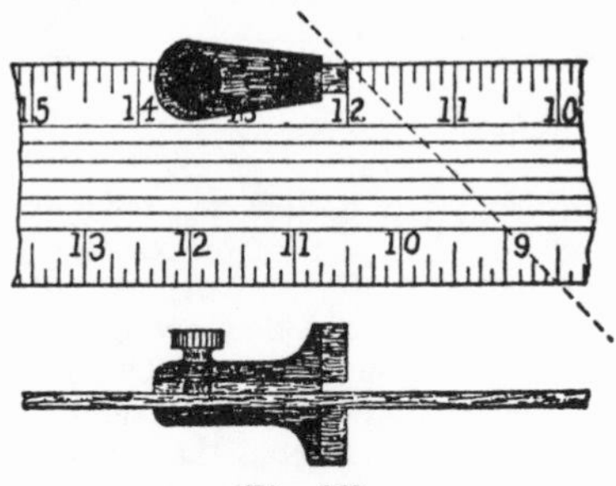

Fig. 358

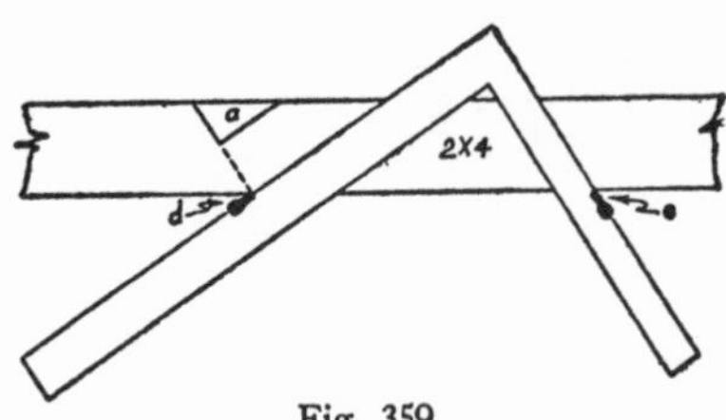

Fig. 359

Measuring Line.—Fig. 359 shows a square applied to a timber with the guides set for a one-third pitch roof. The guides are pointed out at *d* and *e*. At *a* is shown the seat cut marked for cutting. In this case the edge of the timber is taken for the measuring line. When this is done, the plumb cut of

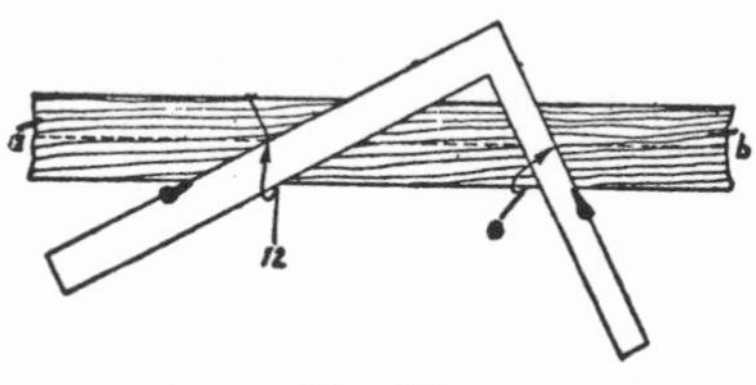

Fig. 360

the seat must be extended to the edge, as indicated by the dotted line. Where this dotted line intersects the edge of the timber, is the point where the stepping off must begin. Fig. 360 shows the guides set for a one-fourth pitch roof, and the measuring line is indicated between *a* and *b* by dotted line. In this method the stepping off begins at the corner of the seat cut and the steps are made on the measuring line.

Stepping Off.—At *A*, Fig. 361, is shown a common rafter with a full-width lookout. At *a* is shown the seat cut. At *B* is shown how to get the length of a common rafter for a third pitch roof by the stepping-off method. The run of this rafter is 12 feet and the rise is 8 feet. The number of feet in the run of the roof determines the number of steps to be taken on the timber to get the length of the rafter. The square, as shown, is applied for the first step. This application is repeated for each step. For example, the first step is numbered *1*, the second *2*, the third *3*, and so on up to *12*. Before removing the square from the 12th step the plumb cut should be marked on the timber as indicated by the dotted-line square. Here the edge of the timber is taken as the measuring line,

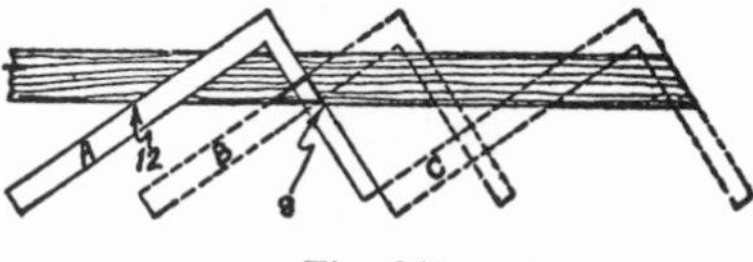

Fig. 362

while on the drawing shown at *C*, the measuring line is indicated by the dotted line. The tail of the lookout is stepped off in the opposite direction. In this case two steps are necessary, as the figures, *1* and *2*, indicate. Fig. 362 shows by dotted lines the application of the square (marked *C*) for the tail cut. Square *A* and *B* are in position, respectively, for marking the plumb and horizontal cuts of the seat. The heavily shaded triangle indicates the seat of the rafter.

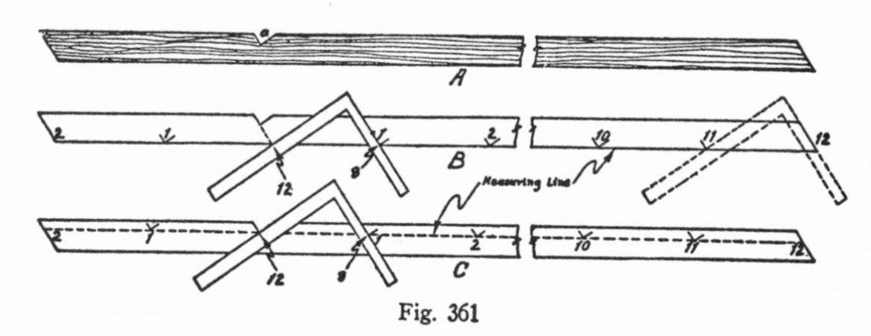

Fig. 361

Finding Length of Rafter.—The most accurate, but the least practical, method of finding the length of rafters is the square root method, shown by Fig. 363. The diagram gives a run of 4 feet, a rise of 3 feet, and a rafter of 5 feet. Here is the rule: *The base squared plus the altitude squared equals the hypotenuse squared.* In simple language, the square root of the sum of the squares of the run and the altitude equals the length of the rafter.

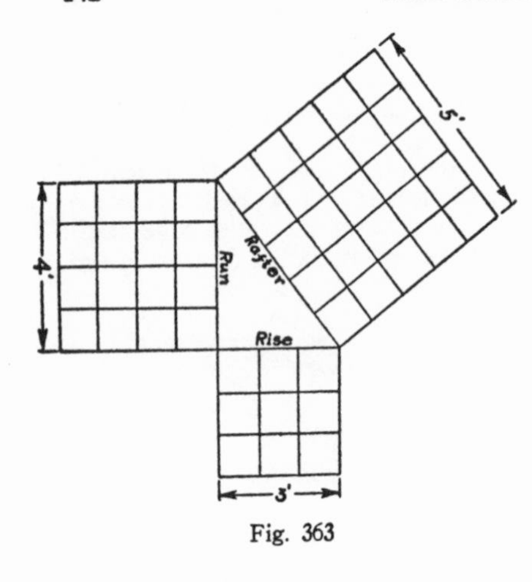

Fig. 363

Diagonal Method.—Fig. 364 shows another method of obtaining the length of rafter patterns. Here a pocket rule is shown in position for measuring the diagonal distance between 12 on the body of the square and 9 on the tongue. The diagonal distance between these two points is 15 inches, or the length of the rafter per foot run. To get the full length of the rafter multiply the length of the rafter per foot run by the number of feet in the run.

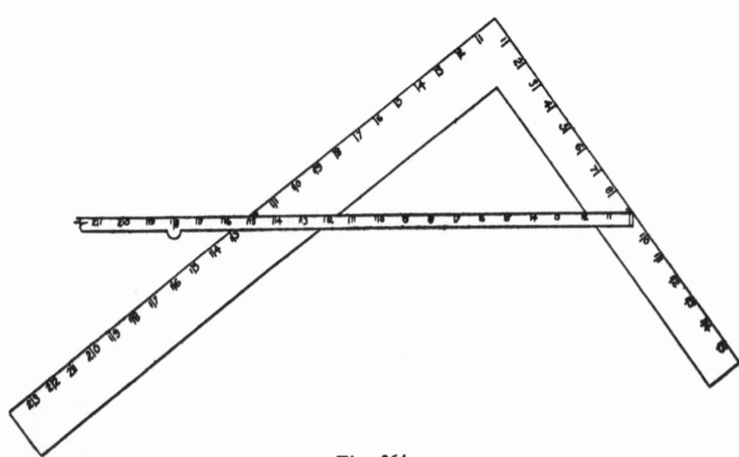
Fig. 364

Fraction of Step. — Fig. 365 shows to the left the square in position for the first step. The steps for this stepping off are numbered from *1* to *12* and marked as shown. But there is a 6½-inch fraction of a step to be added to the 12 steps. This is done after the last step has been taken, by marking along the blade of the square and then pushing the square forward 6½ inches, as shown by the dotted lines. The plumb cut is then marked along the edge of the tongue, as indicated. A detail of this operation is shown by Fig. 366. The square is first placed on the timber, using 12 and 8, as shown. Then the mark along the edge of the blade is made, as the "Mark Here" indicators point out. This done, push the square forward, as shown and mark the plumb cut. In the same way, any other fraction of a step can be marked. It should be mentioned here, that the principle

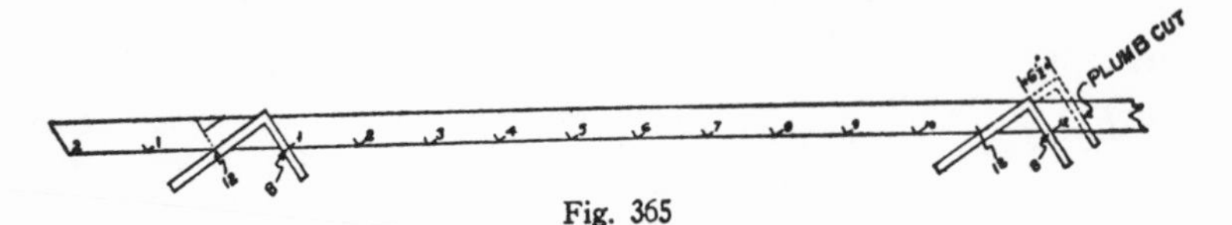

Fig. 365

involved in any of these illustrations for a certain pitch, will work equally well on any other pitch. The same results can be obtained by taking one extra full step, as shown in Fig. 367, marking along the blade of the square, and pulling the square back 5½ inches, as shown by the square marked *A*. It can also be done by taking the extra step, placing a mark at point *c*, 5½ inches from point *d*, or the heel of the square; or, on the other hand, 6½ inches from the base figure 12, as shown by figures. Then take 12 and 8 on the square, and apply the square in the position shown by the dotted-line square at *B*—be sure that the edge of the tongue will contact point *c* and mark along the edge of the tongue.

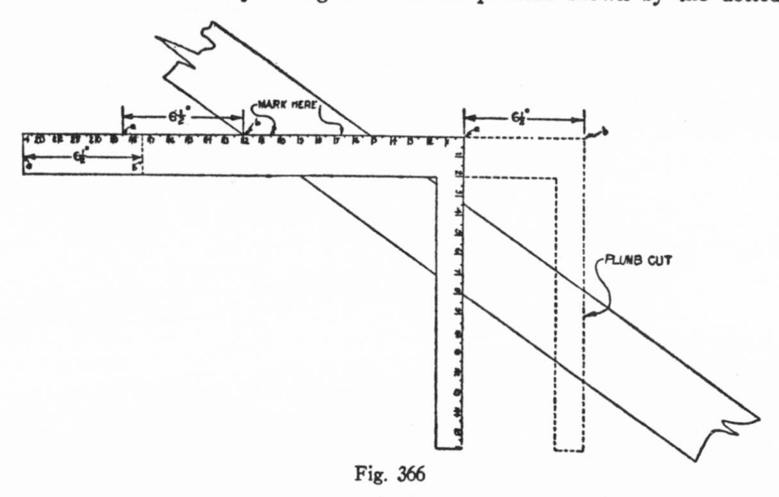

Fig. 366

Fig. 367

Figures for Unknown Cuts.—Fig. 368 illustrates how to get the foot and plumb cuts of a rafter when only the full run and full rise are known—for example, a run of 17½ feet, and a rise of 5½ feet. Apply the square to the timber, letting inches equal feet, which would mean that you would use 17½ on the blade of the square and 5½ on the tongue. Now these figures will give both the foot and the plumb cut. But if you want to make the matter still simpler, mark along the blade of the square and pull the square back until the figure 12 on the blade will intersect the edge of the timber, or from position *A* to position *B*. Then use point *b* on the tongue of the square and 12 on the blade, which will give you the foot and plumb cuts. You can also push the square forward until the figure 24 intersects the edge of the timber, as from position *A* to position *C*. Then take point *c* on the tongue of the square and 24 on the blade, this will again give the cuts.

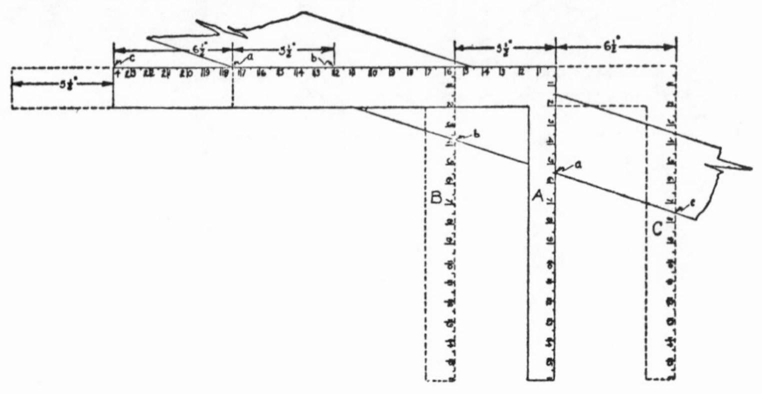

Fig. 368

Chapter 41

CUTS FOR HIPS, VALLEYS, AND JACKS

Achieve by Thought and Study.—Occasionally one meets a carpenter who understands the framing of common rafters, but when it comes to hips and valleys and jacks, he is puzzled. He goes ahead on these rafters more nearly on a trial and error basis, which is to say, that by cutting, trying, recutting and fitting he manages to get the roof framed, but never knowing just how it was accomplished. The principal reason for this lack of roof framing knowledge lies with the man himself.

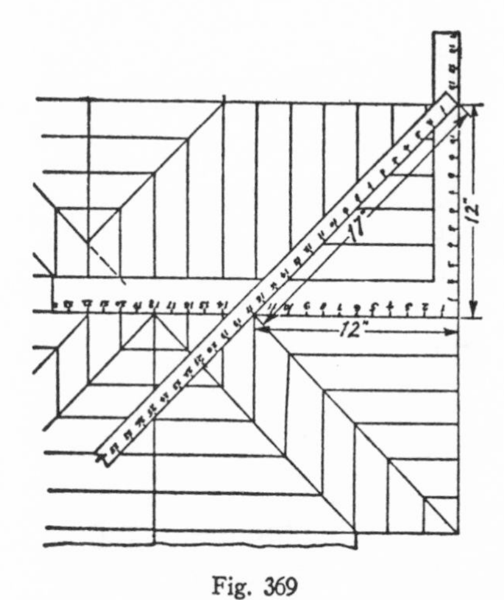

Fig. 369

Roof framing is an achievement—it must be acquired by practice and experience—it is not inherited. So let every apprentice get this thoroughly into his thinking, that while the framing of hips and valleys isn't any harder than the framing of common rafters, it is necessary that one give this part of roof framing more thought and study, because one gets less of it to do.

Run of Hips and Valleys.—The principal run of the roof, shown in part by Fig. 369, is 12 feet. The large-scale square shown by the drawing indicates that inches equal feet. To get the full run for the hip or valley rafters of a regular pitch roof, measure the diagonal distance between 12 and 12 on the square, which will give you 17 inches, less. (The exact figure is 16.97.) But for all practical purposes, 17 inches will answer.

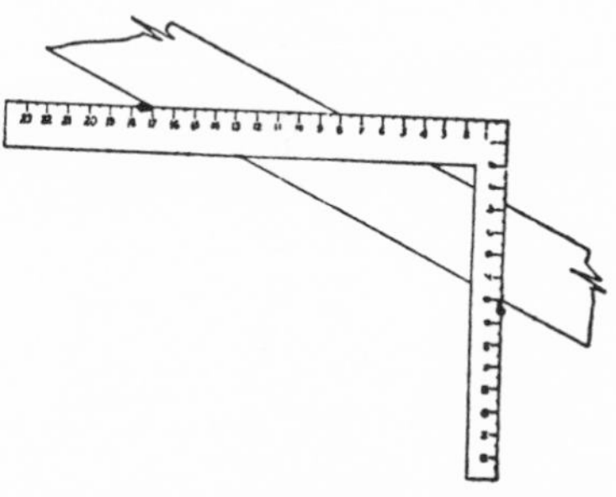

Fig. 370

Application of Square.—Fig. 370 shows the application of the square for stepping off the length of a hip or valley rafter. While the run for hips and valleys is 17 inches for every foot of common rafter run, as was shown by the previous illustration, the rise is the same as the common rafter rise.

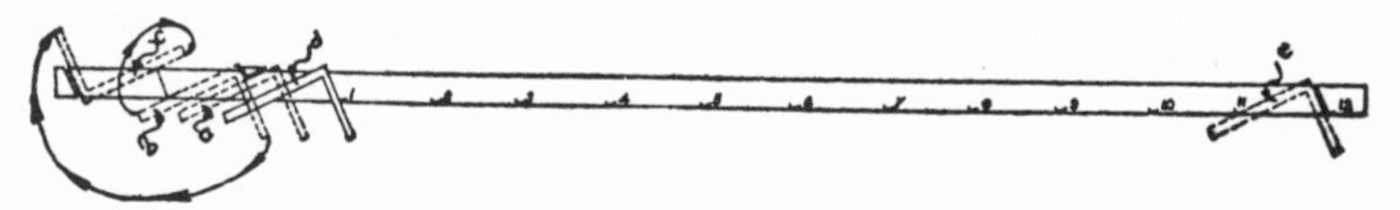

Fig. 371

Stepping Off Hips and Valleys.—Fig. 371 shows to the left four applications of the square. The figures used for all of them is 17 on the blade of the square and 8 on the tongue. The square pointed out at *a* is in position for marking the horizontal cut of the seat, the square at *b* is in position for marking

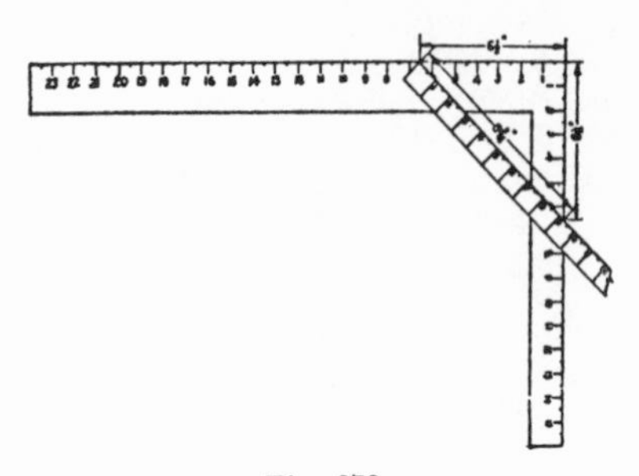

Fig. 372

the plumb cut of the seat, while the square at *d* is in position for the first step. At *e,* to the extreme right, the

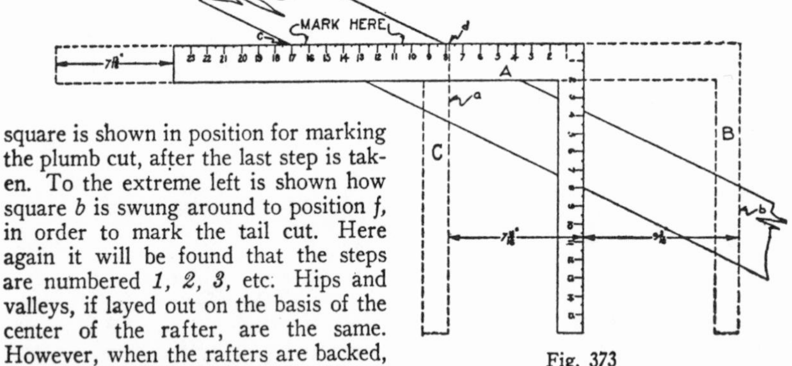

Fig. 373

square is shown in position for marking the plumb cut, after the last step is taken. To the extreme left is shown how square *b* is swung around to position *f,* in order to mark the tail cut. Here again it will be found that the steps are numbered *1, 2, 3,* etc. Hips and valleys, if layed out on the basis of the center of the rafter, are the same. However, when the rafters are backed, then the hip is backed so that it will have a ridge, while the valley is backed so that it will have a V-shaped groove.

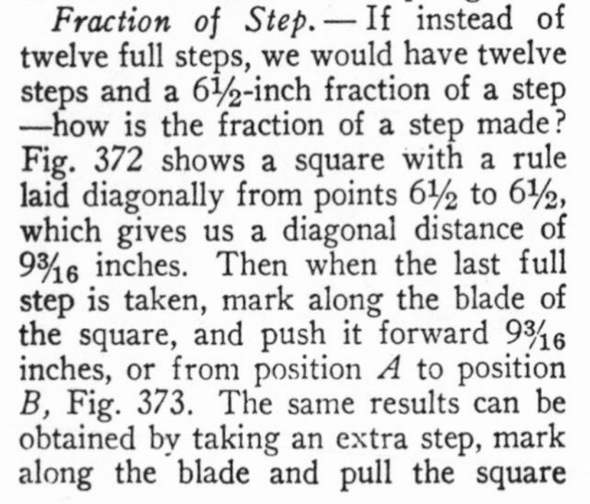

Fraction of Step.—If instead of twelve full steps, we would have twelve steps and a $6\frac{1}{2}$-inch fraction of a step—how is the fraction of a step made? Fig. *372* shows a square with a rule laid diagonally from points $6\frac{1}{2}$ to $6\frac{1}{2}$, which gives us a diagonal distance of $9\frac{3}{16}$ inches. Then when the last full step is taken, mark along the blade of the square, and push it forward $9\frac{3}{16}$ inches, or from position *A* to position *B,* Fig. 373. The same results can be obtained by taking an extra step, mark along the blade and pull the square

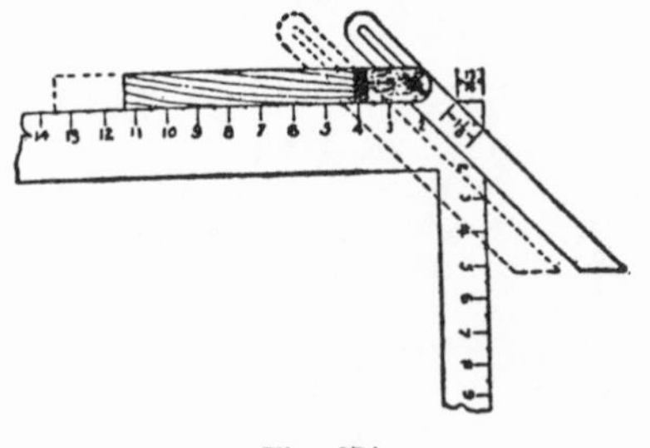

Fig. 374

back $7\frac{13}{16}$ inches, or from position *A* to position *C*. If you will subtract $9\frac{3}{16}$ from 17, it will give you $7\frac{13}{16}$, which is the distance the square is pulled back.

Deduction for Ridge Board.—If the rafters are framed to a $1\frac{5}{8}$-inch ridge board, then you will have to find the diagonal distance of $\frac{13}{16}$ and $\frac{13}{16}$ on the square. This is done by setting the bevel square to a 45-degree angle, or from 3 to 3 on the square, as shown by

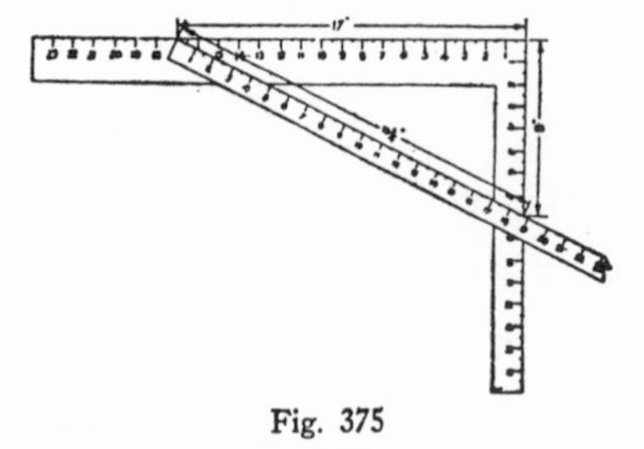

Fig. 375

dotted lines, Fig. 374. Then push it forward until it contacts the two 13/16-inch points, and measure the diagonal distance, which as shown by the drawing is 1 1/8 inches, or the amount to be

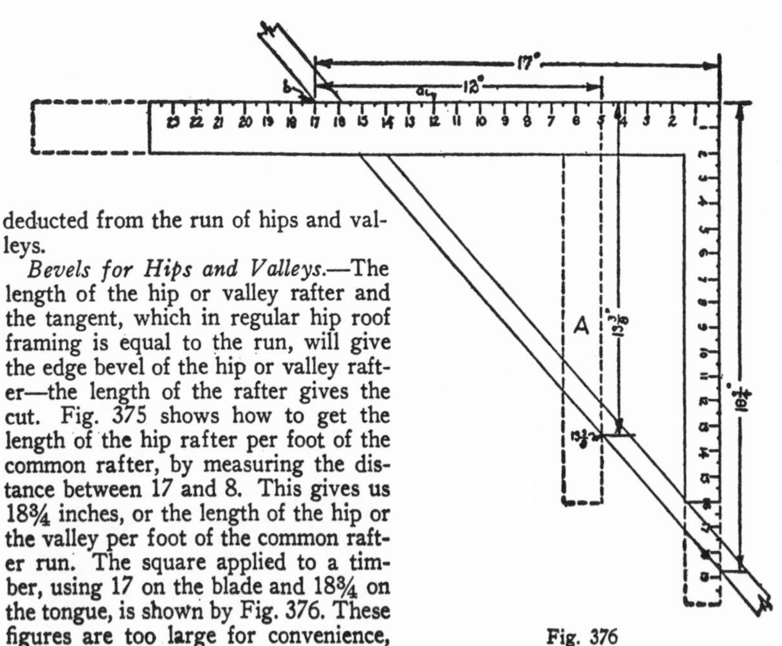

Fig. 376

deducted from the run of hips and valleys.

Bevels for Hips and Valleys.—The length of the hip or valley rafter and the tangent, which in regular hip roof framing is equal to the run, will give the edge bevel of the hip or valley rafter—the length of the rafter gives the cut. Fig. 375 shows how to get the length of the hip rafter per foot of the common rafter, by measuring the distance between 17 and 8. This gives us 18 3/4 inches, or the length of the hip or the valley per foot of the common rafter run. The square applied to a timber, using 17 on the blade and 18 3/4 on the tongue, is shown by Fig. 376. These figures are too large for convenience, so the square is pulled back until the

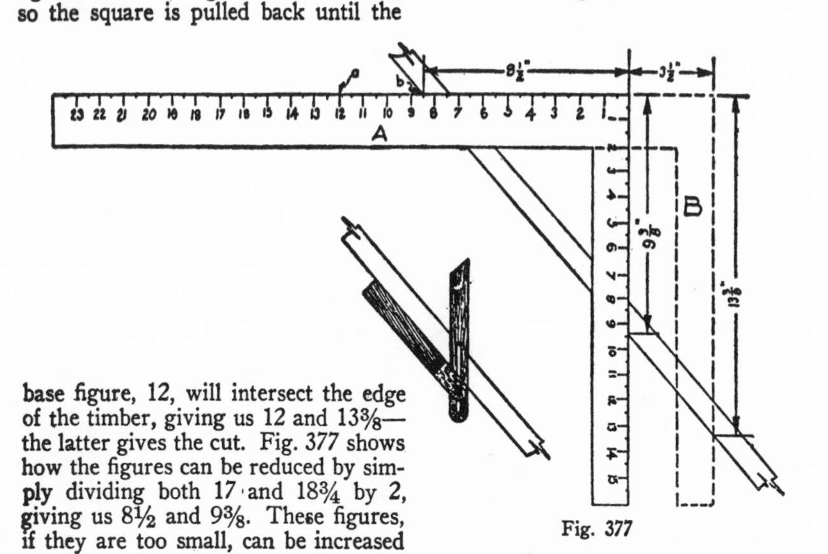

Fig. 377

base figure, 12, will intersect the edge of the timber, giving us 12 and 13 3/8—the latter gives the cut. Fig. 377 shows how the figures can be reduced by simply dividing both 17 and 18 3/4 by 2, giving us 8 1/2 and 9 3/8. These figures, if they are too small, can be increased by pushing the square forward from position *A* to position *B*, giving us the same figures that we had in Fig. 376. The steel square is used only for obtaining the bevel. Then the bevel

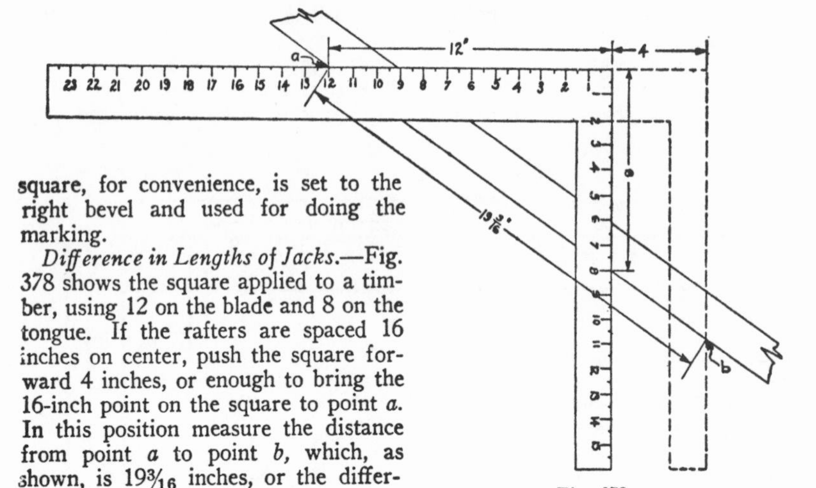

square, for convenience, is set to the right bevel and used for doing the marking.

Difference in Lengths of Jacks.—Fig. 378 shows the square applied to a timber, using 12 on the blade and 8 on the tongue. If the rafters are spaced 16 inches on center, push the square forward 4 inches, or enough to bring the 16-inch point on the square to point *a*. In this position measure the distance from point *a* to point *b*, which, as shown, is 19$\frac{3}{16}$ inches, or the difference in lengths of the jack rafters.

Fig. 378

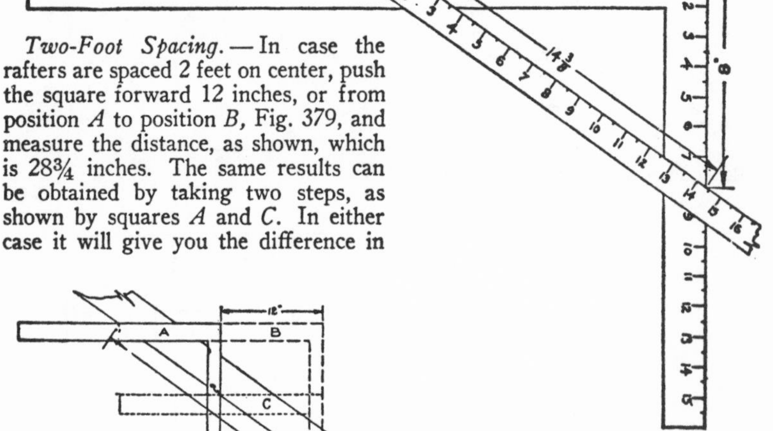

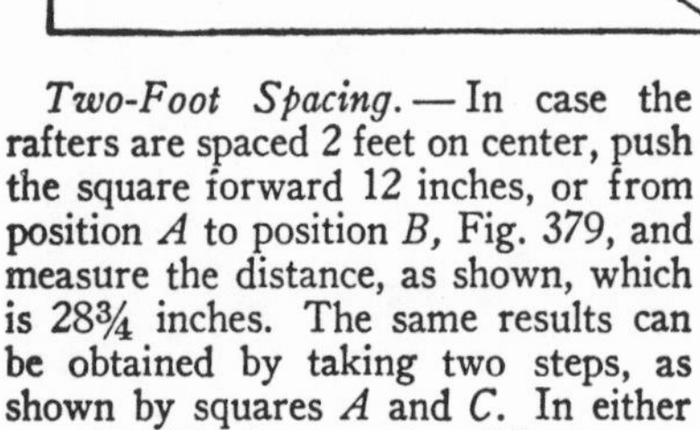

Two-Foot Spacing. — In case the rafters are spaced 2 feet on center, push the square forward 12 inches, or from position *A* to position *B*, Fig. 379, and measure the distance, as shown, which is 28¾ inches. The same results can be obtained by taking two steps, as shown by squares *A* and *C*. In either case it will give you the difference in

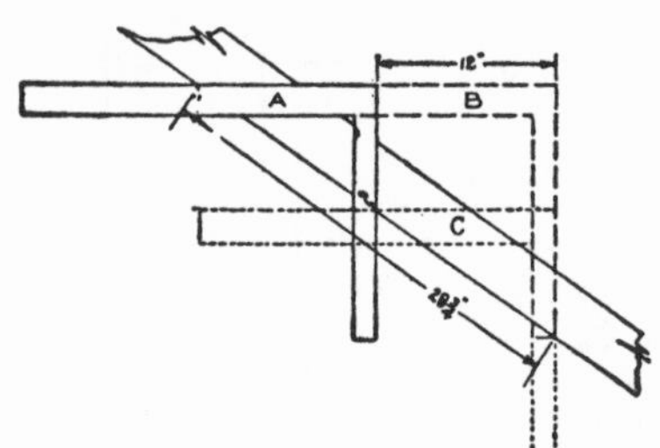

Fig. 379

the lengths of jack rafters spaced 2 feet on center, or 28¾ inches. These figures are for a third pitch roof. The principle is the same for any other pitch.

Fig. 380

Edge Bevel for Jacks.—The rule for obtaining the edge bevel for jack rafters is this: *Take 12 on the body of the square and the length of the common rafter per foot run on the tongue—the tongue gives the cut.* The length of the common rafter per foot run will have to be determined by taking the diagonal distance between 12 and 8 on the square, as shown by Fig. 380,

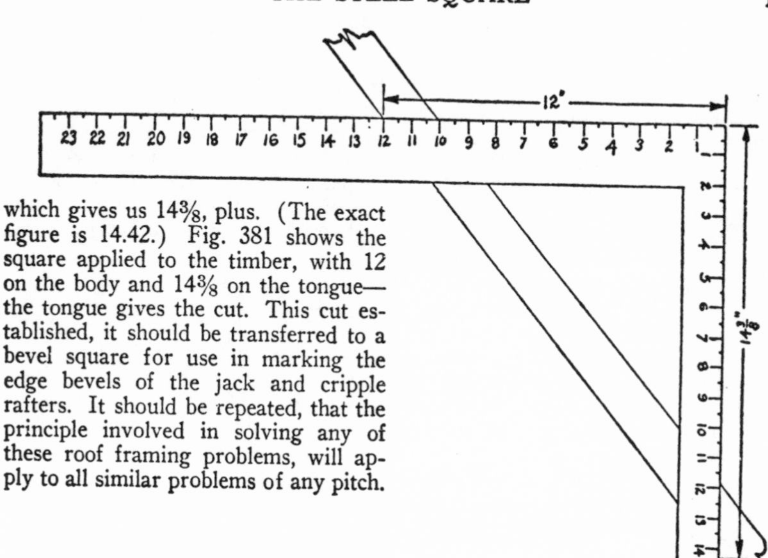

which gives us 14⅜, plus. (The exact figure is 14.42.) Fig. 381 shows the square applied to the timber, with 12 on the body and 14⅜ on the tongue—the tongue gives the cut. This cut established, it should be transferred to a bevel square for use in marking the edge bevels of the jack and cripple rafters. It should be repeated, that the principle involved in solving any of these roof framing problems, will apply to all similar problems of any pitch.

Fig. 381

Chapter 42

MISCELLANEOUS ROOF FRAMING PROBLEMS

Laying Out Bay Window.—The simplest bay window is the one called the octagon bay window — that is to say, the angles are the same as the angles of a true octagon. Fig. 382, at the top, shows a plan of an octagon bay window with 2x4 plates in place. The degrees of two angles are shown on the drawing. The butt joints are made on a 45-degree angle, or by using 12 and 12 on the square. The miter joints are made by using 12 and 4.97. Sometimes 12 and 5 are used, which is nearly enough correct for most practical purposes. The bottom drawing shows the

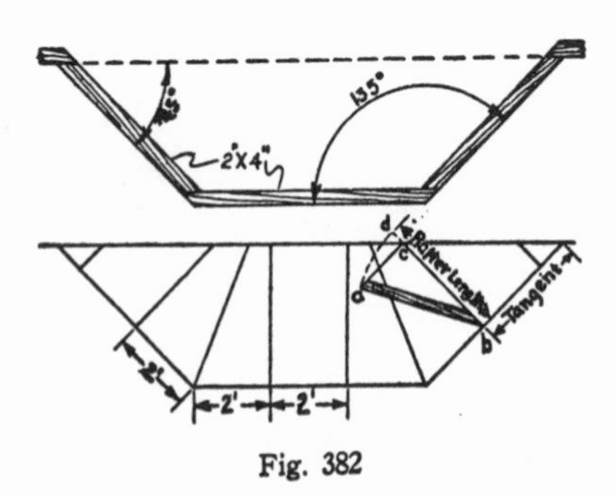

Fig. 382

roof plan. To the right is shown how to frame the jack rafters. Here a rafter is shown on the side, in which *a-b* is the rafter, *b-c* the run, and *c-a* the rise. The rafter length is transferred with a compass from *b-a* to *b-d,* as shown by the dotted part-circle. Now the rafter length and the tangent taken on the square, will give the edge bevel that fits against the side of the main building—the rafter length giving the bevel. The run and the rise, of course, give the plumb and level cuts.

Fig. 383 shows how to get the edge bevel for the hip rafter, which is shown on the side. Here *b-c* is the run, *c-a* the rise, and *a-b* the rafter. The rafter length is again transferred from *b-a* to *b-d,* as indicated by the dotted part-circle. Now the rafter length and the tangent gives the edge bevel—the rafter length giving the bevel.

Fig. 384, top, shows a plan of a hexagon bay window, with the 2x4 plates in place. The degrees of two angles are given on the drawing. To mark both the butt and the miter joints use

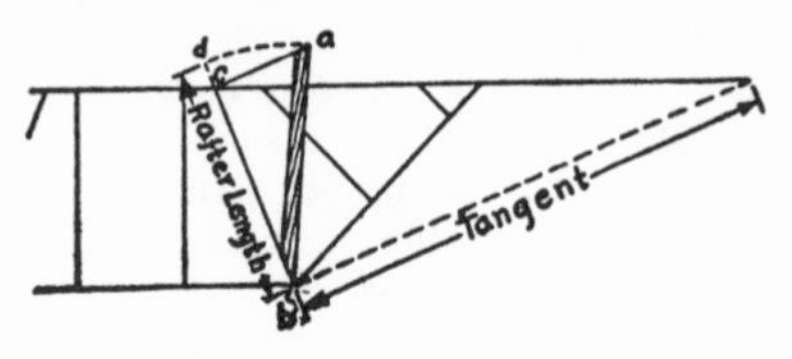

Fig. 383

12 and 6.93. Sometimes 12 and 7 are used for rough work. The bottom drawing shows a plan of the roof, which is framed on the same principle used for the octagon roof. Remember that in roof framing problems it is often necessary to read between the lines.

Length of Unequal Pitch Valley Rafter. — Since unequal pitches are used much less than regular pitches, the use of a diagram is suggested, somewhat on the order of the one

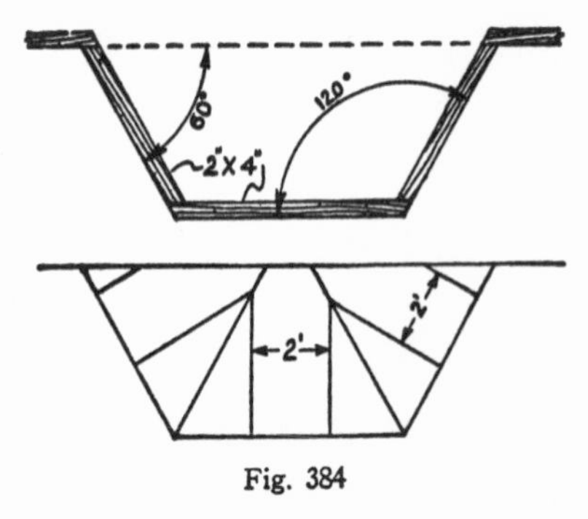

Fig. 384

shown by Fig. 385. In this diagram one roof has a third pitch, and the other a half pitch. There is shown to the upper left a hip rafter, and toward the right a valley rafter, hinged on the run and lying on the side. The reason

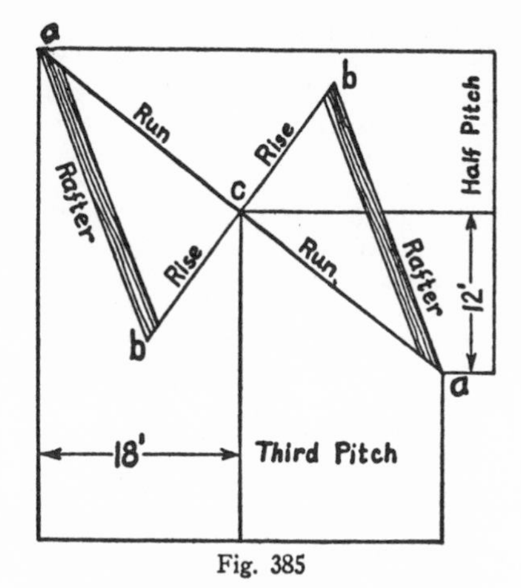

Fig. 385

both a hip and a valley are shown, is that the principle is the same in both rafters. The explanation of the valley rafter will apply to the hip rafter also, because the reference letters are the same in both cases. At a right angle to line *a-c*, the run, draw the line *c-b*, making it equal in length to the full rise of the roof. Now join point *a* with point *b*, which gives the rafter length. To make the drawing a little easier to understand, a side view of the rafter

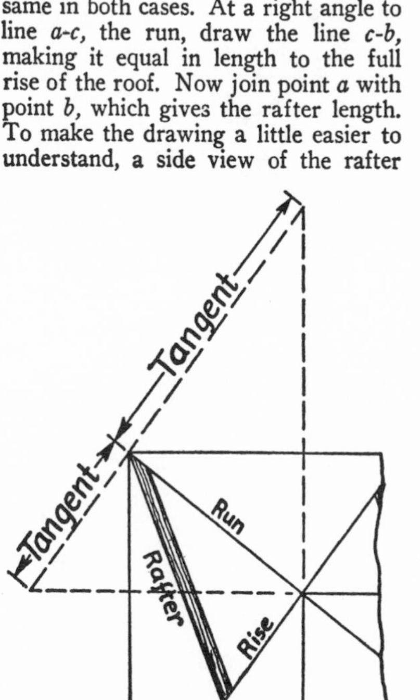

Fig. 386

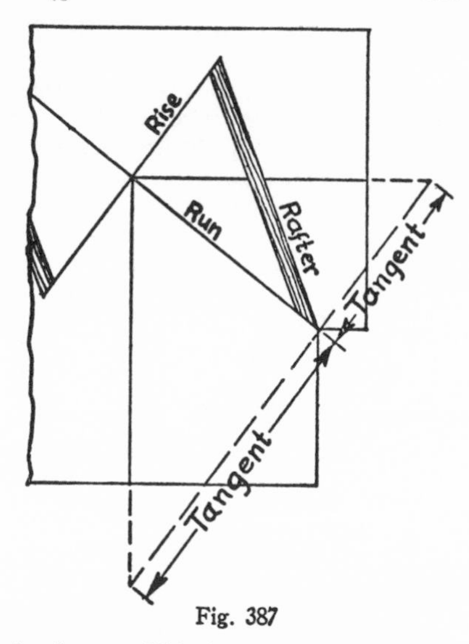

Fig. 387

is shown. This is not necessary after you understand the principle. In making a diagram, take a convenient scale; 1 inch equals 1 foot gives good results, but if the building is rather large, it might be more convenient to use a smaller scale.

To get the edge bevel of the hip or valley rafters, take the rafter length on one arm of the square, and the tangent on the other—the rafter length gives the bevel. Fig. 386 shows the hip rafter shown by Fig. 385 with the two tangents laid out. In the same way, Fig. 387 shows the tangents for the valley rafter.

Edge Bevels for Jacks. — Fig. 388 shows a roof plan, in which two pitches

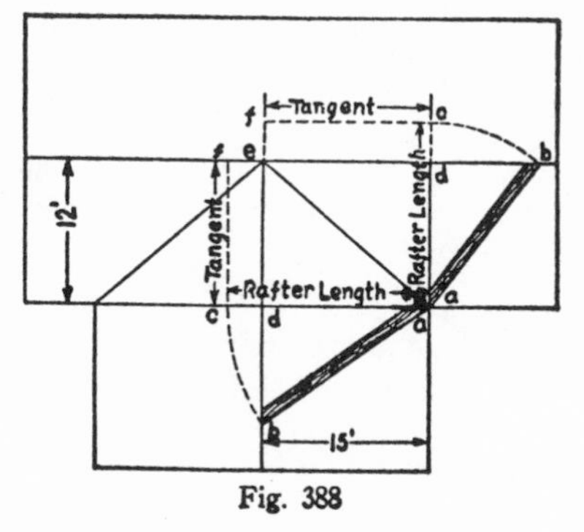

Fig. 388

are used. A common rafter of the 12 and 10 pitch is shown to the right of the main roof, and toward the bottom is shown a common rafter of the roof with the 12 and 8 pitch. If the reader will imagine that these two rafters are hinged on the run, or on line *a-d,* and lying on the side, he will have a good idea of the diagram. The reference letters for both rafters represent relatively the same points; namely, *a-d* represents the run; *d-b,* the rise, and *a-b,* the rafter. To develop the diagram further, set the compass at point *a,* and

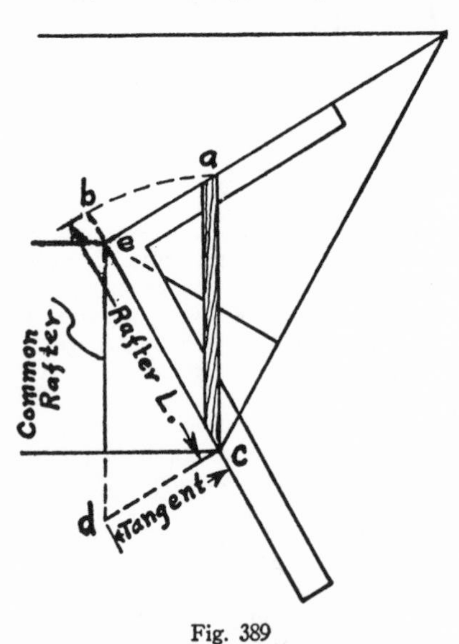

Fig. 389

transfer the rafter length, *a-b,* to *a-c,* as indicated by the dotted part-circle, *b-c.* Extend line a-d to *c,* and *d-e* to *f,* as shown by the dotted lines. This done, draw the dotted line *f-c* making it parallel with *e-d,* and the diagram will be completed. The diagram for the other rafter is made in the same way. Now to get the edge bevel for the jacks of either of the pitches, take the rafter length on one arm of the square, and the tangent on the other — the rafter length giving the bevel for the pitch you are working on.

Edge Bevel for Hip. — Fig. 389 shows a diagram of one end of an irregular plan hip roof. The blade of the square is on the seat of the hip rafter for the dull angle of the plan. To obtain the length of the hip rafter, make the rise, *e-a,* equal to the rise of the common rafter, and draw in the hip rafter as shown from *a* to *c.* With the compass set at *c,* transfer the rafter length, *c-a,* to *c-b,* as indicated by the dotted part-circle, *a-b.* Extend the seat of the hip, *c-e,* to *b,* as shown by the dotted line. Also extend the seat of the common rafter to *d,* as the dotted line indicates. To mark the edge bevel, take the rafter length, *c-b,* on the blade of the square, and the tangent on the tongue. If the hip rafter is to join the last two common rafters, shown to the

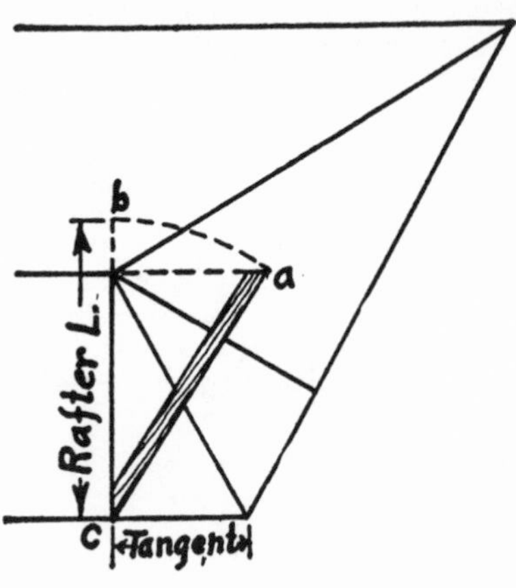

Fig. 390

left and right of the hip, then the blade will give the bevel. But if the hip is to saddle onto a ridge board or a deck, then the tongue will give the bevel.

Fig. 390 shows how to get the points for marking the edge bevel of the jacks that join the hip just explained. With the compass set at *c,* transfer the rafter length, *c-a,* to *c-b.* Now take the rafter length on the blade of the square, and the tangent on the tongue — the blade will give the bevel.

Cuts for Hood Braces. — Fig. 391 shows how to obtain the length of a brace for a hood by stepping it off with a steel square. The figures to be used on the square are 12 and 6, which also give the cut of the brace where it joins the building.

Fig. 392 shows how to get the cut of the brace where it joins the hood

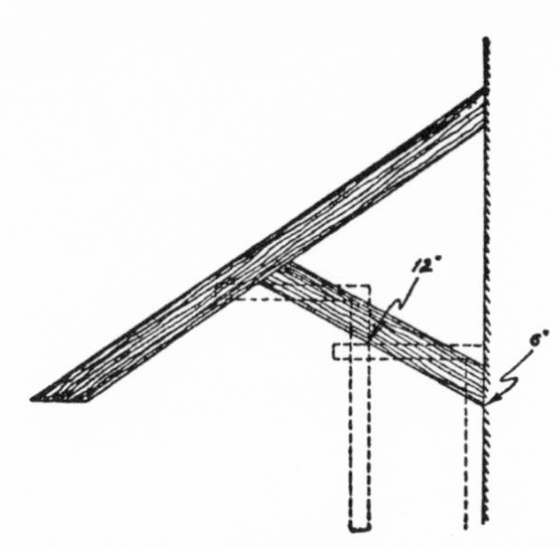

Fig. 391

rafter. Here square No. *1* is shown applied to the brace in a double-step position, using 24 on the body of the square and 12 on the tongue. The dotted outline of a square, numbered *2,*

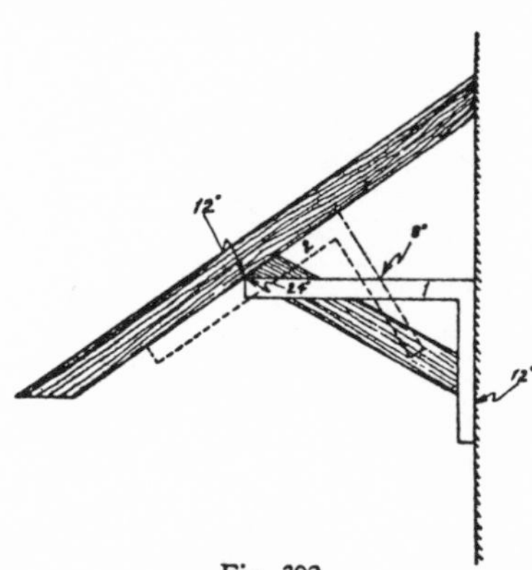

Fig. 392

shows how to get the cut by using 12 and 8, the figures used for framing the hood rafter, which has a third pitch.

The principles presented here will work for any pitch, when the figures used on the square conform with the pitch of the rafter and the angle of the brace, respectively.

Roof Framing Problem.—The problem is, how to join two equal-pitch roofs that have different spans. Fig. 393, to the left, shows perhaps the simplest solution. Here the wider building joins the narrower one with two regular valleys, and two small return hips that stop at the comb of the wider

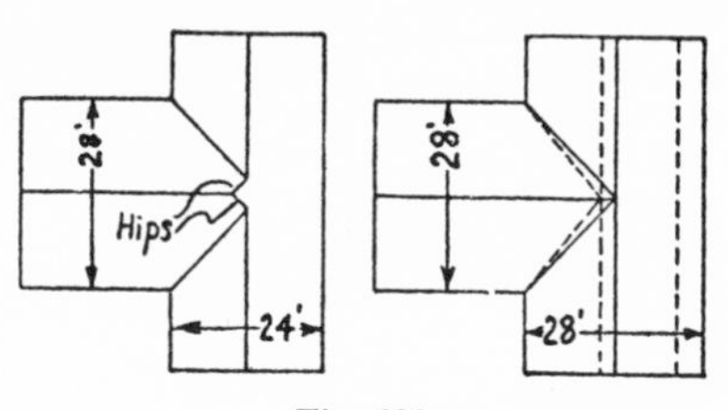

Fig. 393

roof. To the right are shown two solutions. The dotted lines show how the plan shown to the left could be framed with an irregular pitch roof, while the full lines show the narrower part of the building made the same in width as the wider part, which permits a regular pitch roof.

Fig. 394 shows to the left the original plan with a regular pitch roof, but the wider part is framed with a deck in order to keep it from extending beyond the comb of the narrower part. In the drawing to the right the full lines show a regular pitch roof with the wider part cut down to the width of the narrower one. The dotted lines

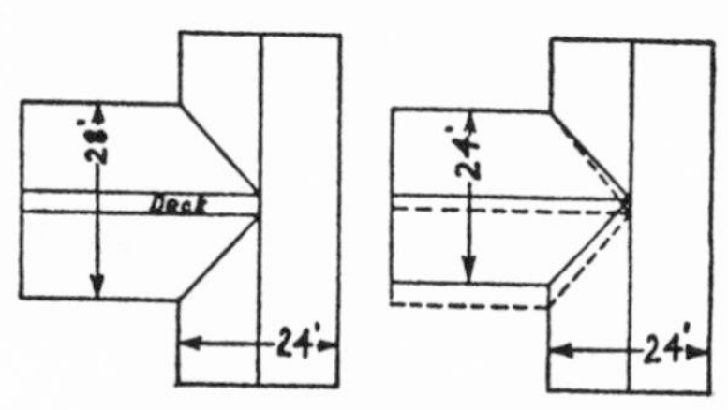

Fig. 394

show the original plan, but an irregular pitch roof.

Edge Bevel for Dormer.—This rule applies to all regular pitches: *Take 12 and the length of the rafter per foot run on the square—the larger of the two figures gives the bevel.* In this case we have a 12 and 9 pitch, as shown by Fig. 395. As given in figures the rafter per foot run is 15 inches long. Then to obtain the edge bevel for the dormer rafter, take 12 and 15 on the square—

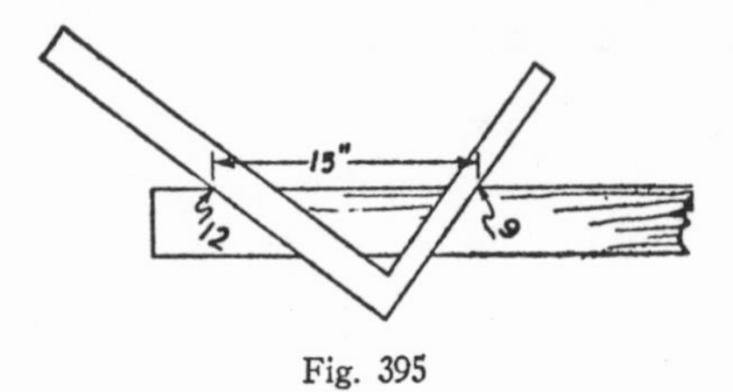

Fig. 395

the latter giving the bevel. A detail of the application of the square is shown by Fig. 396—the tongue giving the bevel. It is suggested that instead of nailing the rafters on the sheeting, that valley boards be used, such as are shown by Fig. 397, where a side of the main roof, looking straight at it, is

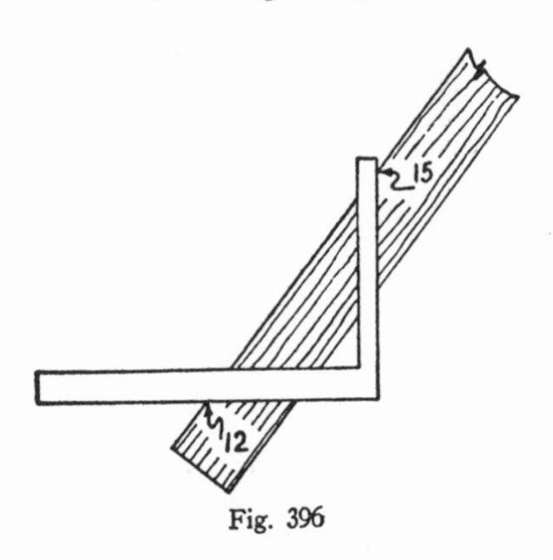

Fig. 396

shown with the valley boards for the dormer in place. The dormer in this case has a run at the gable of 6 feet. The distance from the base of the gable to where the comb of the dormer dies into the main roof is 7 feet, 6 inches, as shown to the right. To the left six applications of the square, giving the six steps for stepping off the length of the valley boards, are shown. The figures used for stepping off are 12 and 15, or 12 and the length of the rafter per foot run. Fig. 396 shows the application of the square for stepping off the valley boards. Here the blade gives the side bevel of the bottom cut of the

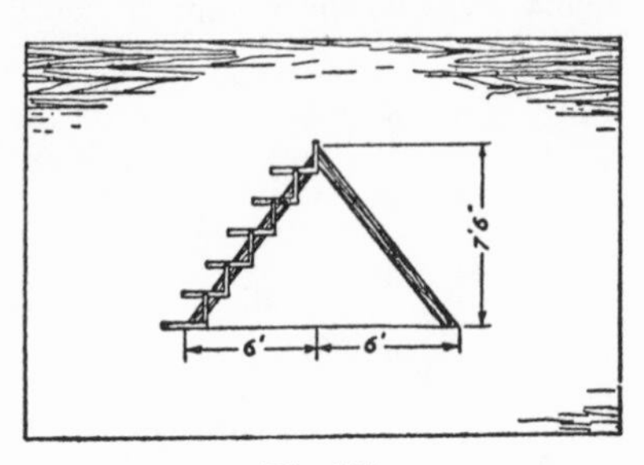

Fig. 397

valley boards, while the tongue gives the bevel of the top cut.

The edge bevel for the bottom cut of the valley boards is obtained by taking 17 and the diagonal distance between 12 and 15, or 19.21. Dividing both 17 and 19.21 by 2 will make the figures more practical, or 8½ and 9⅝, minus.

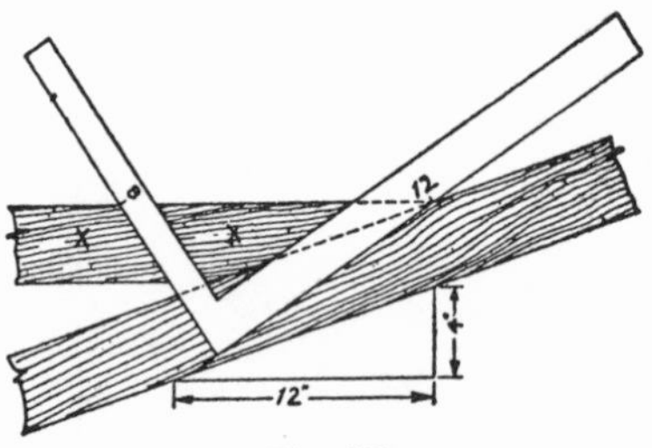

Fig. 398

The arm on which 9⅝ is used will give the bevel.

Fly-Rafter Cuts.—Let us assume that we have a main roof with a one-third pitch to which we are to join a dormer roof with a one-sixth pitch, or a 12 and 4 pitch roof. Fig. 398 shows how to obtain the cut. Take a short piece of material and give it a 12 and 4 cut and tack it to the rafter material in the manner shown by the piece marked *X X*. This, it will be seen, will put the upper edge of the piece in a horizontal position, if the rafter material were in the position of a rafter in place. To obtain the cut, take the figures for a one-third pitch, 12 on the body of the square and 8 on the tongue, and apply the square as shown in the drawing. Where the edge of the tongue intersects with the upper edge of the rafter material is the point to use with 12 to mark the cut. The blade will give the cut.

Fig. 399 shows how to get the cut when the same kind of dormer with

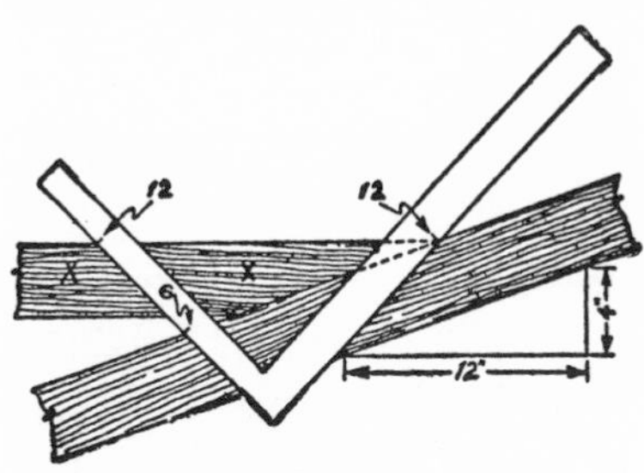

Fig. 399

the same pitch joins a one-half pitch main roof. Here the short piece, marked *X X*, is cut the same as in the other instance, but the application of the square is made to conform with the figures used in framing a one-half pitch roof, or 12 and 12. In this case as shown, 12 and 6 are the figures to use. The body of the square gives the cut. A little study of the two drawings will clarify the problem.

Stepping Off Odd Runs.—The problem that is taken up here is most frequently met on lean-to roofs. That is,

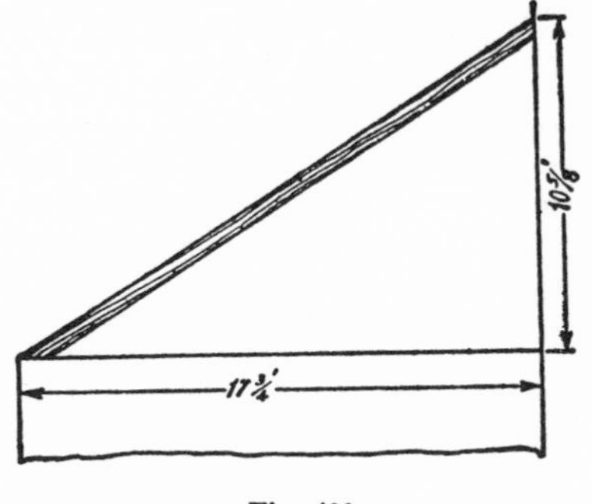

Fig. 400

single-pitch roofs that are framed against some building or some other object. Whether or not the runs and the rises of such roofs are of even or odd feet and inches, does not matter, they can be stepped off accurately with this simple method:

Fig. 400 shows a lean-to rafter in place. The run in this case is 17¾ feet and the rise is 10⅝ feet. With these figures to work with it would take some figuring to step off the rafter under the ordinary stepping-off rule. But the stepping off in this case is even simpler than the ordinary stepping-off method under favorable circumstances.

If you will let the number of feet and fractions of feet be represented on the square by inches, which would mean 17¾ inches on the body of the square and 10⅝ inches on the tongue,

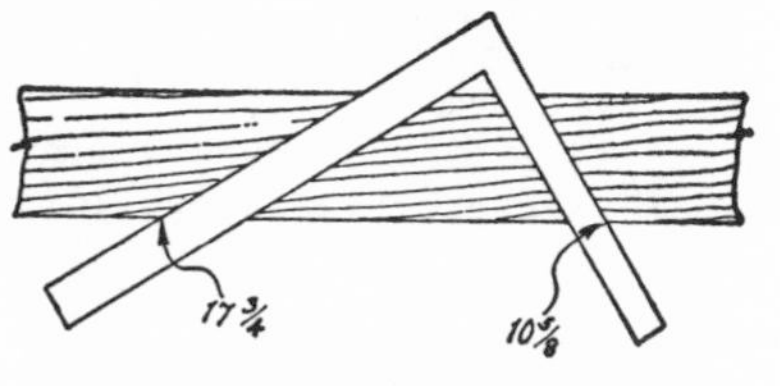

Fig. 401

as shown by Fig. 401, and take twelve steps on the rafter material, you will have the exact length of the rafter. Putting it in the form of a rule:

Let the run and the rise in feet be represented, respectively, on the body and the tongue of the square in inches, and take twelve steps, which will give you the length of the rafter.

Rafter Cuts.—Fig. 402 shows twenty-one different hip-roof pitches in the first column. The second column shows the figures to be used on the square for marking the edge bevels. The arm of the square on which the figures to the right in this column are used, gives the bevels. The third column gives the difference in the lengths of jack rafters spaced 16 inches, while the fourth column gives the difference in the lengths of jack rafters spaced 2 feet on center. The figures given in the table were ob-

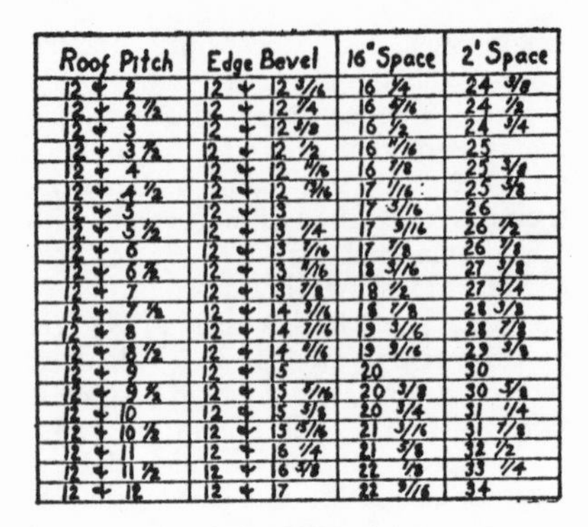

Roof Pitch	Edge Bevel	16" Space	2' Space
12 + 2	12 + 12 3/16	16 1/4	24 3/8
12 + 2 1/2	12 + 12 1/4	16 5/16	24 1/2
12 + 3	12 + 12 3/8	16 1/2	24 3/4
12 + 3 1/2	12 + 12 1/2	16 11/16	25
12 + 4	12 + 12 11/16	16 7/8	25 1/4
12 + 4 1/2	12 + 12 13/16	17 1/16	25 5/8
12 + 5	12 + 13	17 3/16	26
12 + 5 1/2	12 + 13 1/4	17 5/16	26 1/2
12 + 6	12 + 13 7/16	17 7/8	26 7/8
12 + 6 1/2	12 + 13 11/16	18 3/16	27 3/8
12 + 7	12 + 13 7/8	18 1/2	27 3/4
12 + 7 1/2	12 + 14 3/16	18 7/8	28 3/8
12 + 8	12 + 14 7/16	19 3/6	28 7/8
12 + 8 1/2	12 + 14 11/16	19 9/16	29 3/8
12 + 9	12 + 15	20	30
12 + 9 1/2	12 + 15 5/16	20 3/8	30 5/8
12 + 10	12 + 15 5/8	20 3/4	31 1/4
12 + 10 1/2	12 + 15 15/16	21 3/16	31 7/8
12 + 11	12 + 16 1/4	21 5/8	32 1/2
12 + 11 1/2	12 + 16 5/8	22 1/8	33 1/4
12 + 12	12 + 17	22 9/16	34

Fig. 402

tained by measuring the diagonal distances on the square, between the figures giving the various rises and 12. For the difference in the lengths of jacks for 2-foot spacing, the length of the rafter for a foot run was doubled, while for the 16-inch spacing, one-third was added.

Fig. 403, to the right, shows a square with the diagonal distances shown for

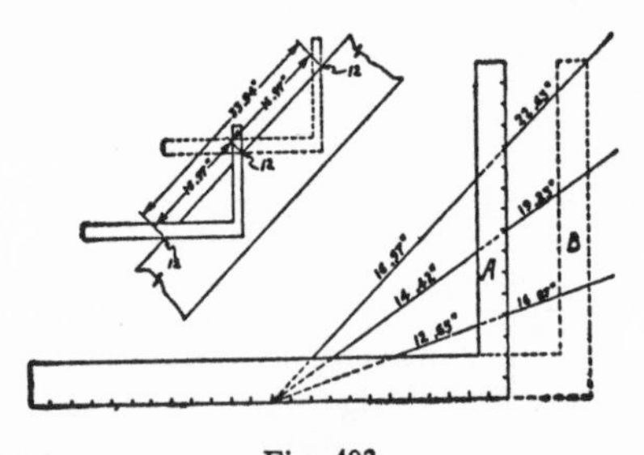

Fig. 403

these pitches: One-half, one-third, and one-sixth. The lengths of the rafters per foot run are given just above the diagonal lines for each of the pitches. The difference in the lengths of jack rafters spaced 16 inches for these pitches, is given between the tongues of the two squares marked *A* and *B*. To obtain these differences, apply the square for one foot run, mark along the blade and then slide the square forward from 12 to 16, or from position *A* to position *B*, keeping the blade on the line. Now, the diagonal distance between 16 and the point where the tongue intersects the line giving the pitch you are working on, is the difference in the lengths of the jack rafters spaced 16 inches on center.

To the upper left (Fig. 403) are shown two squares applied to a timber, using 12 and 12. Twice the diagonal distance between 12 and 12 is the difference in the lengths of jacks spaced 2 feet on center for a half pitch. The differences in the lengths of the jacks for the other pitches given in the table are obtained by applying the same principle.

Irregular Plan.—Fig. 404 shows two one-line drawings, or diagrams of a

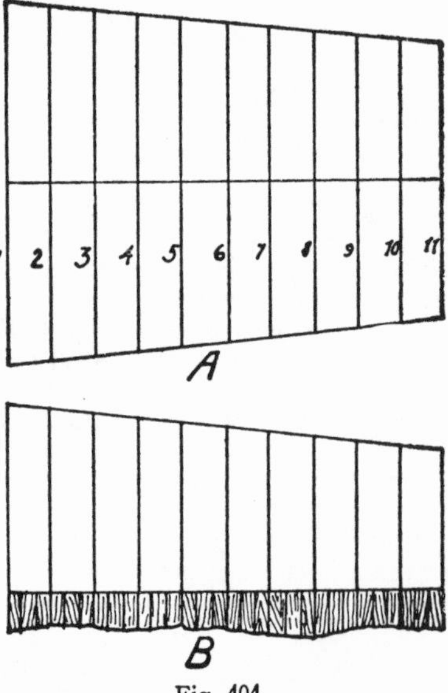

Fig. 404

double pitch roof that has a wider span on one end than on the other. At *A* is shown a plan of the roof with 11 pairs of rafters, and at *B* is a side view. The question is how to obtain the different lengths of the different rafters.

Fig. 405 shows a pair of rafters for the wide end in place, and by dotted lines the rafters for the narrow end are shown. To the right, shaded, a rafter cut for the narrow end is placed against the rafter of the wide end and the difference in the lengths has been divided into 10 equal spaces—the number of spaces there are for the rafters of the roof. The points that mark these

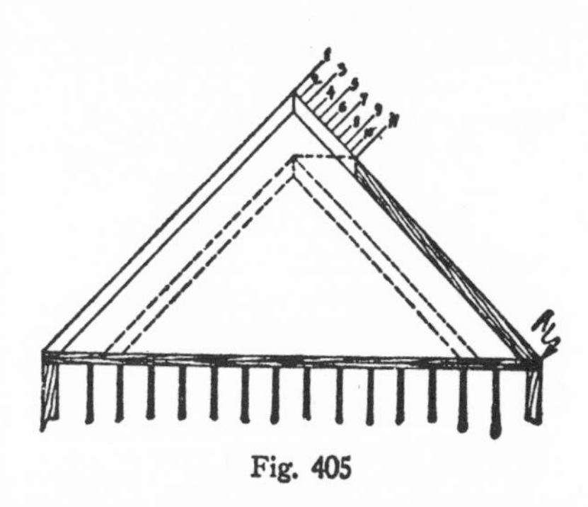

Fig. 405

spaces are numbered from *1* to *11*, or one more than the number of spaces. Now, the longest two rafters are cut as long as the distance between *1* and *A*, the next two are cut as long as the distance between *2* and *A*, the two following that, as long as the distance between *3* and *A*, and so on, *4-A*, *5-A*, *6-A*, until you come to the end rafters, which are cut as long as the distance between *11* and *A*. When these rafters are put in place in the order of their different lengths, the comb of the roof will be straight, but on an incline as shown at *B*, Fig. 404.

Practical Terms.—A reader once sent this writer a pencil sketch of a truss, giving the terms of the different members, including the roof joists and roof sheeting on one half of the truss, leaving the other half to be filled out, using practical terms.

Fig. 406 shows to the left in longhand, the terms that were on the sketch, and to the right, the terms that were supplied by the author before sending the drawing back, together with the following suggestions:

Truss rafter is a better term than principal, or principal rafter. Principal is used primarily as an adjective, as in, principal post, principal beam, principal wall and so forth. If a truss rafter is called principal, unless one knew from some other source what is meant, one could hardly be expected to know from the word itself. On the job, sim-

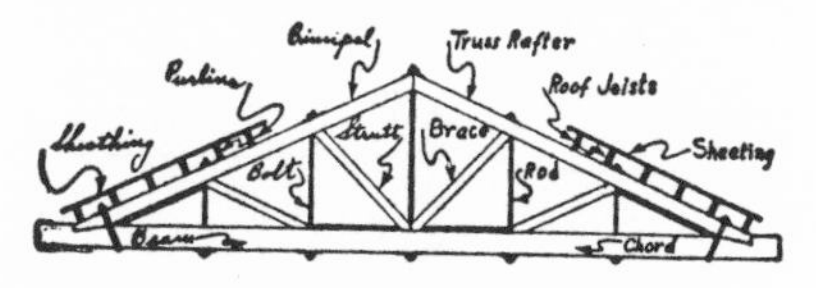

Fig. 406

ple terms that are readily understood should be used. . . . Sheathing is a poor choice of word, better say sheeting. . . . The timbers to which the roof sheeting is nailed, when they are placed as shown by the illustration, are roof joists. . . . A purlin is a timber that supports rafters about half-way between the seat and the comb of a roof. . . . Either brace or strutt is correct, but brace is commonly understood and therefore the better term. . . . A beam is a girder that supports joists or some other weight, while a chord is the bottom member of a truss, and in lattice trusses both the top and the bottom members are chords. . . . Bolt is a good word to use in the construction of beams, but rod is the better choice in this case.

Chapter 43

STAIR AND OTHER PROBLEMS

Laying Out Winders.—Fig. 407 is a plan of a quarter-turn flight of winders, with two steps of a straight flight of stairs to the left. Now to divide the space for three winders set the compass at point *a* and strike the quarter

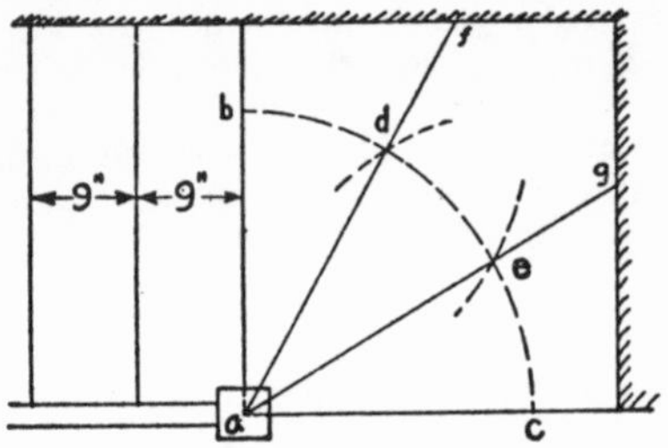

Fig. 407

circle from *b* to *c*, using any convenient radius. With the same radius set the compass at point *b* and cross the quarter circle, as at *e*. In the same way, set the compass at *c* and mark point *d*. This done, draw the line *a-f* in such a way that it will cross point *d*. Draw *a-g*, making it cross point *e*. This process is said to be trisecting a 90-degree angle with a compass. Fig. 408 shows how the trisecting can be accomplished by dividing the space into three 30-degree sections, making the

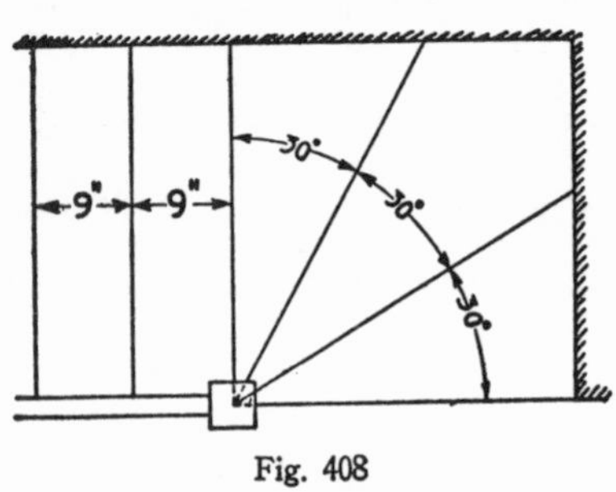

Fig. 408

lines radiate from the same corner. This method is as simple as the one just explained. Fig. 409 shows the same plan, giving the application of the square for obtaining the 30-degree divisions. The figures to be used on the square are 12 and 6.93. Sometimes 12 and 7 are used, but if you want to be exact, six and ninety-three one-hundredths should be used with twelve.

Narrow flights of winders are more

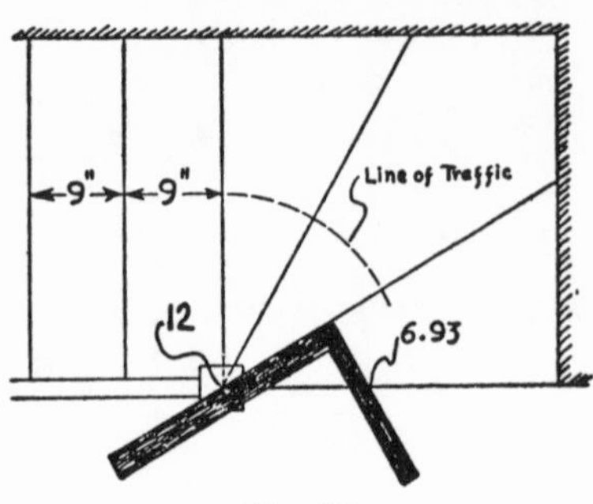

Fig. 409

satisfactory than the wider flights. Winders should be designed so the width at the line of traffic will be the same as the width of the straight steps.

Winding Stair Stringers.—Fig. 410 shows at the bottom, right, a plan of a three-step flight of winders. To the left are shown four straight steps. The winders are numbered *1, 2, 3,* and the

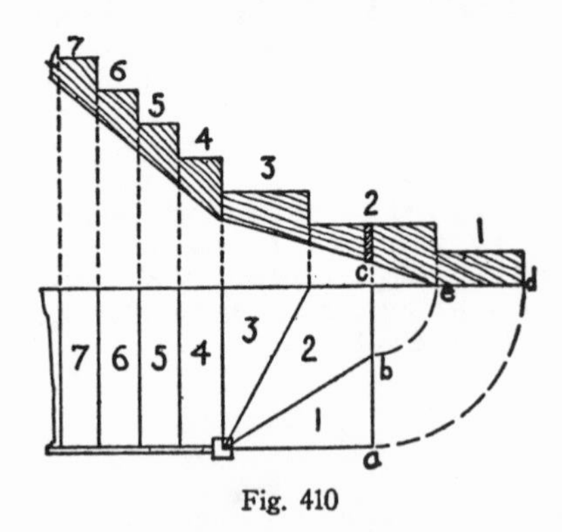

Fig. 410

straight steps are numbered *4, 5, 6,* and *7*. These numbers are the same on both the plan and the stretch-out of the rough stringers, or horses. First lay out the winders on the plan and then the straight steps. This done, set the compass at point *c* and carry point *a* to *d*, as shown by the dotted quarter circle. In the same way transfer point *b* to *e*. Now lay off the stringers. The first one supports the wide end of the first winder and half of the second winder. The second stringer supports one half of the second winder, and the third winder. The third stringer, shown in part, supports the four straight steps. The perpendicular dotted lines locate the risers, while the height of the steps is governed by the rise per step.

Quite frequently the skirt board, or finished stringer, for winders is not housed, but in the better stairs they usually are. Fig. 411 shows the same

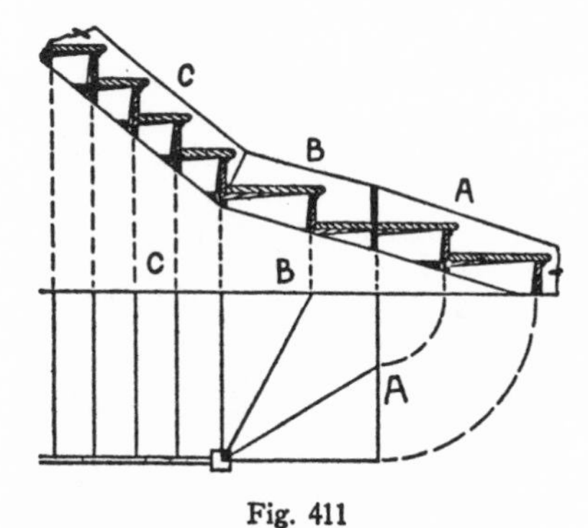

Fig. 411

plan of the flight of three winders and the four straight steps at the bottom. A stretch-out of the skirt boards is shown at the top. The board marked *A* on the stretch-out is for the side marked *A* on the plan. In relatively the same way, the board marked *B* goes on the side marked *B*, and *C* is for the side *C*. The perpendicular dotted lines again locate the rough risers, as a little study will show. Both the winders and the steps are shown wedged into the housing, which is as practical with winders as with straight steps, but when this is done the rough stringers must be held away from the wall, to let the bottom edge of the skirt board in.

Stair Problem.—A continuous handrail would have an awkward drop and twist at the turning point, marked *A* in Fig. 412. The solution is simple. All that is necessary is to change the design slightly. This is shown by Fig.

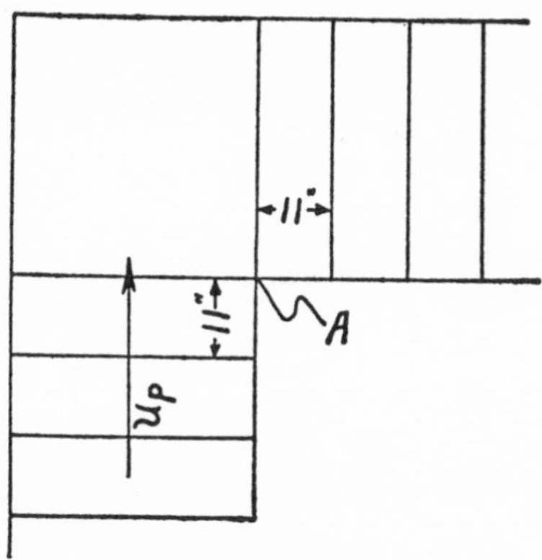

Fig. 412

413, where the two flights of stairs have been offset 7 inches from the original landing. By doing this the handrail will have a uniform fall, and the same number of balusters can be used

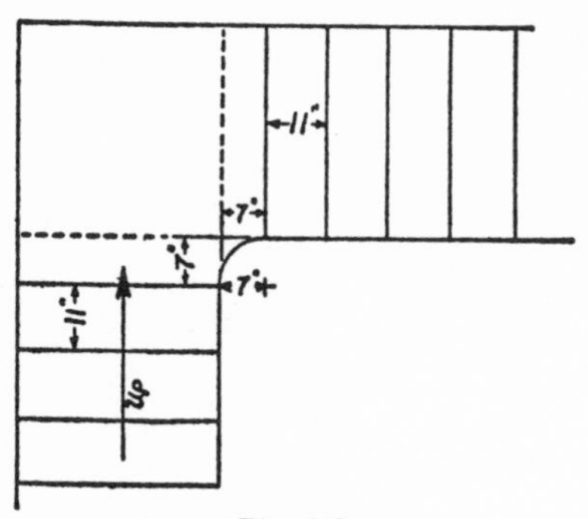

Fig. 413

on the curve at the angle as are used on the steps.

How the 7-inch offset for the two flights is arrived at is also simple. In order to make the balusters work out right on the curve at the angle, the curve will have to be as long as the width of a step, or 11 inches. If the quarter circle will have to be 11 inches long, then the circumference of the whole circle will have to be four times 11, or 44 inches long. To get the radius of a circle whose circumference is 44 inches, divide 44 by 3.1416 and divide the quotient by 2. This gives approximately 7 inches for the radius, which at the same time is the distance of the offsets indicated by dotted lines in Fig. 413.

Spacing for Porch Balusters.—Fig. 414 shows the layout. The distance between the porch posts is 122½ inches. To this must be added the thickness of the balusters, or 1½ inches. This is necessary because the spacing, technically, begins and ends one half the

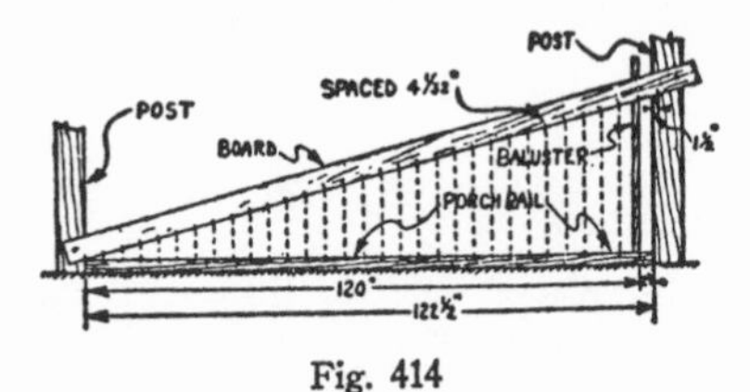

Fig. 414

thickness of the balusters back of the surface of the posts. The balusters are 1½ by 1½ inches, which are spaced 4 inches on center. By dividing 124 by 4, we get 31, or one more than the number of balusters needed. In this case figures were taken that will come out just right, in order to simplify the problem. In practice the figures seldom come out just even. Now take a light board with a straight edge, and step off 31 spaces, with the compass set at a little more than 4 inches, say, 4⅓₂ inches, as shown on the drawing. Fasten the board to the posts, bringing the starting point of the spacing to the angle between the porch rail and the post to the left. (The porch rail is shown shaded, placed between the bases of the posts for marking.) The right end of the board is to be fastened in such a way that the last spacing mark will be just the thickness of a baluster, or 1½ inches, past the corner of the post, as indicated on the drawing. Then take the square and drop the spaces on the board to the porch rail, and mark it. This will give you the practical spacing shown in Fig. 415. This spacing is as accurate as the center to center (*C* to *C*) spacing, and more convenient. The center to center spacing must be started one half the thickness of the balusters back of the surface of the

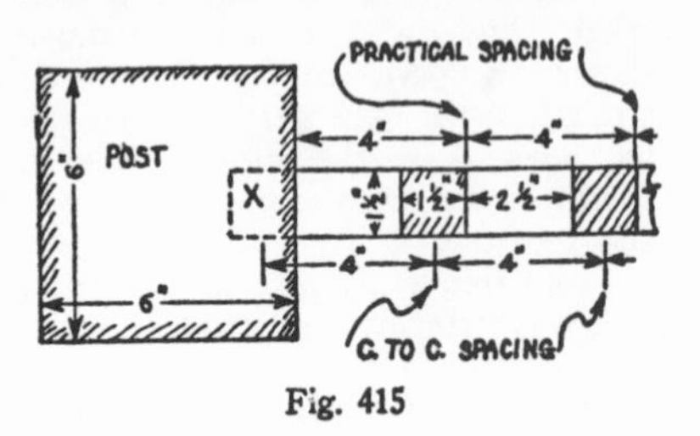

Fig. 415

post. This is necessary because the post takes the place of a baluster, as shown by the dotted lines at *X*.

Fig. 416 shows a plan of the post to the right. Here is shown how the practical spacing ends at the right of the blind baluster, marked *X* and shown by dotted lines. The center to center spacing is also shown which ends at the center of the false baluster. Study the three drawings, and remember that there is always something to be read between the lines.

Boring Angling Holes.—Fig. 417 shows a square applied to a timber on the 8 and 12 points, for the purpose of

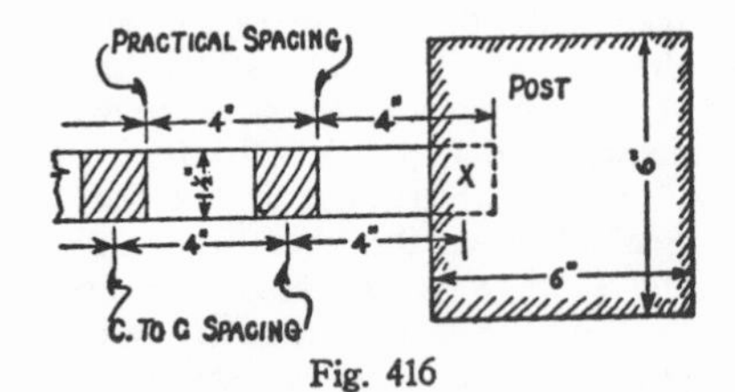

Fig. 416

showing how to bore angling holes so they will be on the angle desired. It does not matter whether the holes are to be bored in a piece of timber, railing, or on some flat surface; this method will apply. Start with the square in the position shown and mark along the outside edge of the body and tongue.

This gives the two important angles. The dotted lines at *A* show where a hole is bored at a right angle to the edge of the timber. This hole should be bored with the bit that will be used in boring the angling hole shown by dotted lines at *B*. Having done all of

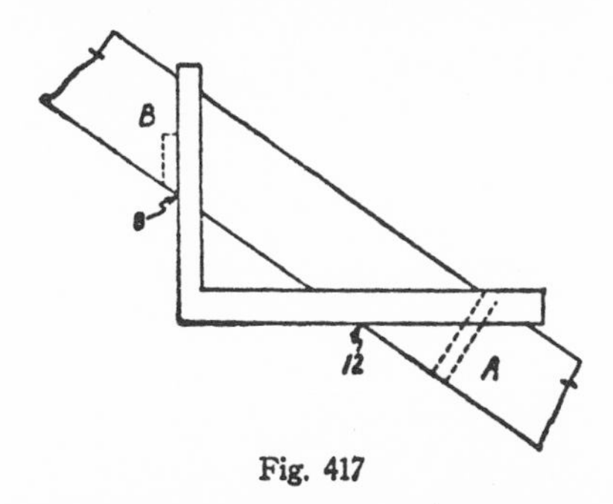

Fig. 417

this, cut the timber as marked along the edge of the body of the square to obtain the block marked *A*. This block is then transferred to the point where the angling hole is to be bored, in this case, to the dotted lines shown at *B*, and put together as shown by the detail in Fig. 418. The two pieces should

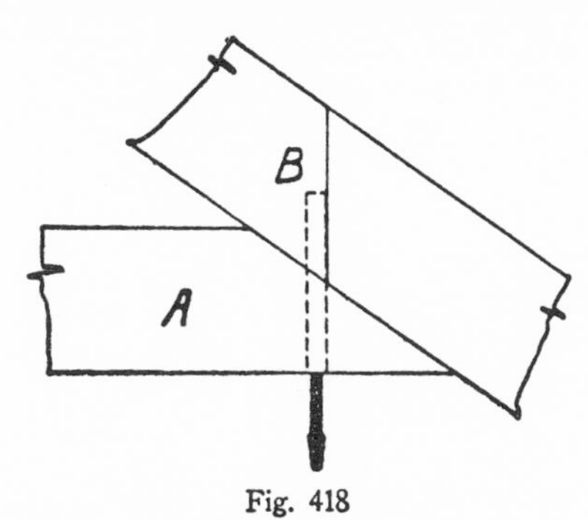

Fig. 418

be securely fastened together before boring of the angling hole is started. The shaded shank and tang of an auger bit, shows the position of the auger when the boring is done. Whenever a hole has been bored the block is removed and fastened again for boring the next hole. This process is repeated until the holes have all been bored.

Cuts for Braces.—A very practical and simple way is illustrated by Fig. 419. The panel is 32 inches by 48 inches. The two cuts can be obtained with the square by taking 48 inches, less the 4-inch bearing, or 44 inches; and 32 inches, less the 4-inch bearing,

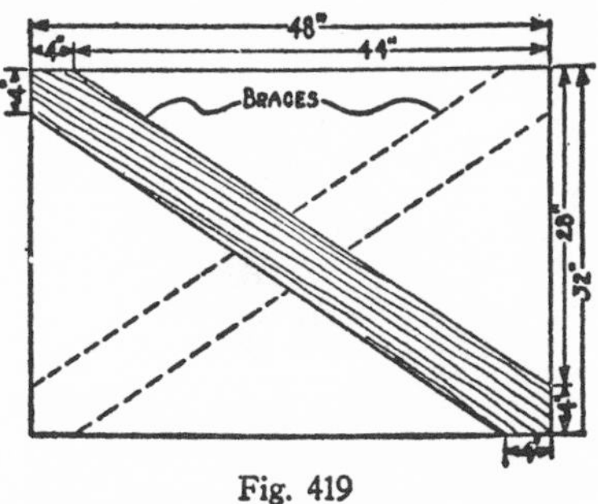

Fig. 419

or 28 inches, as shown by the figures on the drawing. But these figures are too large to be taken on the square, so 44 and 28 will have to be reduced by dividing both figures by 2, which will give 22 and 14. Now take 22 on the blade and 14 on the tongue of the square—the blade will give the cut that joins the side of the panel, while the

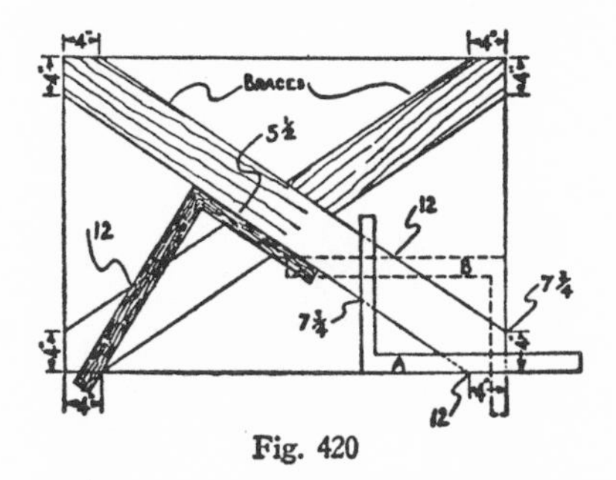

Fig. 420

tongue will give the cut that joins the end of the panel.

Fig. 420 shows the same layout with the braces in place. Here 12 is taken as the base figure, while the key figure that is used on the tongue, must be found by applying the square as shown. Square *A* is in position for marking the cut that joins the side of the panel, while square *B* gives the cut that joins the end of the panel. The same figures are used in both applications, but in *A* the blade gives the cut, while in *B* the tongue gives the cut. Square *C* gives the application of the square to find the cut for the center joints.

Fig. 421 shows a rather long panel with three braces on top of each other, joining the same two corners. The

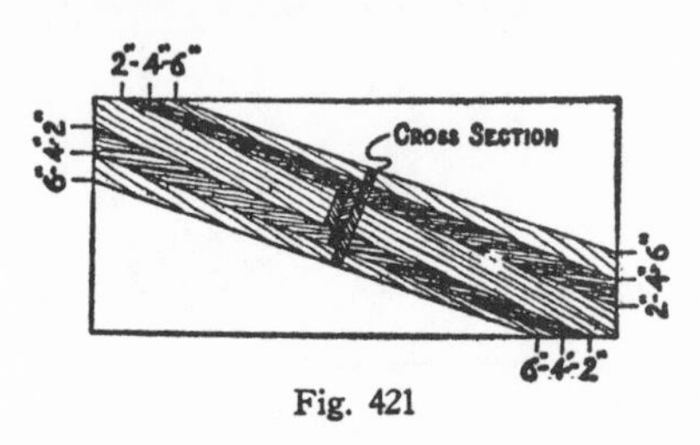

Fig. 421

cross section at the center shows the position they hold. The first brace has 6-inch bearings, the second brace has 4-inch bearings, and the third brace has 2-inch bearings. Although the three braces join the same two corners, they do not hold the same position, as to angles, slopes, and cuts. Compare this with the braces shown in Fig. 420.

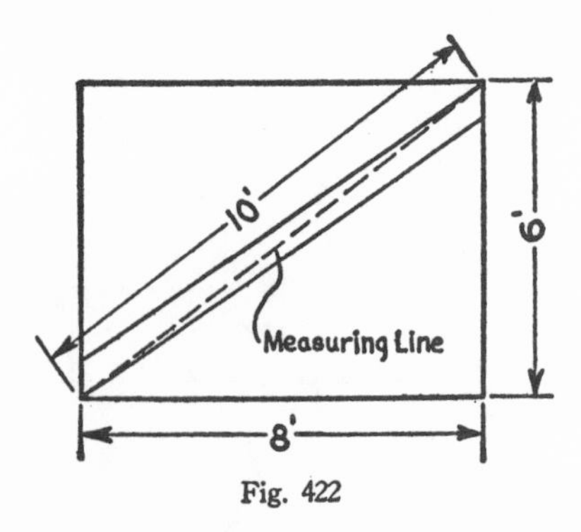

Fig. 422

Remember that a difference in the widths of braces, changes their position when in place.

Bevel for Braces.— Fig. 422 shows an oblong, 6 feet by 8 feet, that is to be held in a square position with a brace cut as shown. In order to get the right bevel, a measuring line (chalk line will do) should be made on the material, as shown by dotted line on the drawing. To do this it is necessary to know the diagonal distance, which in this case is 10 feet. This distance can be obtained by taking 6 squared and 8 squared and extracting the square root. A more practical method

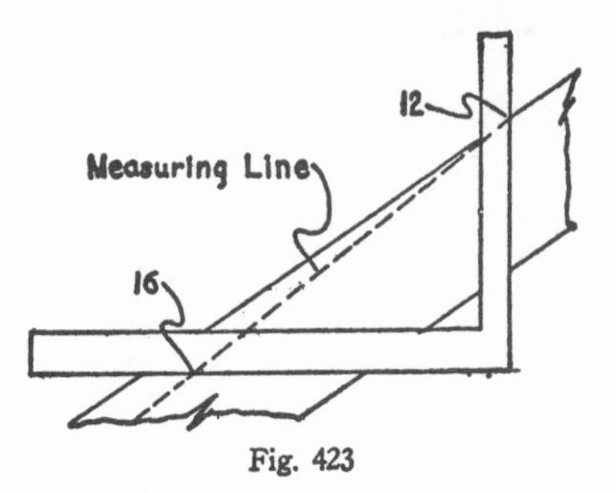

Fig. 423

would be to transfer the distance with a steel tape to the brace material, marking one end on one edge of the material and the other end on the other edge, as indicated by the dotted measuring line on the drawing.

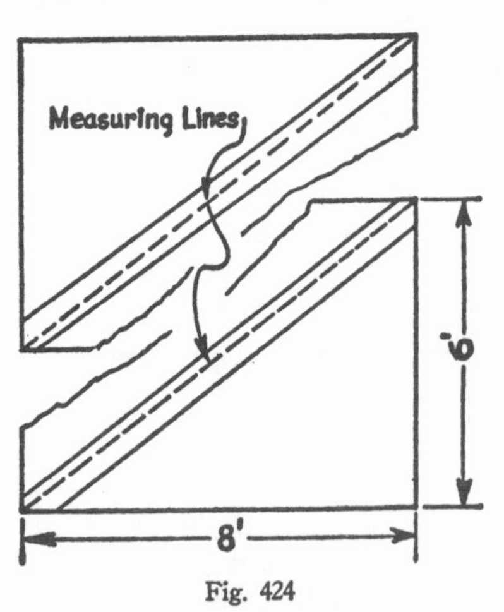

Fig. 424

Fig. 423 shows the square applied to the measuring line of the brace, shown in part, for obtaining the bevel of one end of the brace. The other bevel is obtained in the same way. In this case 16 is taken on the blade and 12 on the tongue. These figures were obtained by multiplying both 8 feet and 6 feet by 2, which resulted in 16 and 12.

Fig. 424 shows two braces with double bevels on the ends. To get these bevels the principle is the same as in the single bevel. The square must be applied to the measuring line, using either the original figures 8 on the blade and 6 on the tongue, or doubling them and use 16 and 12, as explained under Fig. 423. The figures 6 feet and 8 feet were taken for convenience, in practice the figures would probably come out in fractions.

Sways for Hips and Valleys.—Fig. 425 is a diagram of a common circular

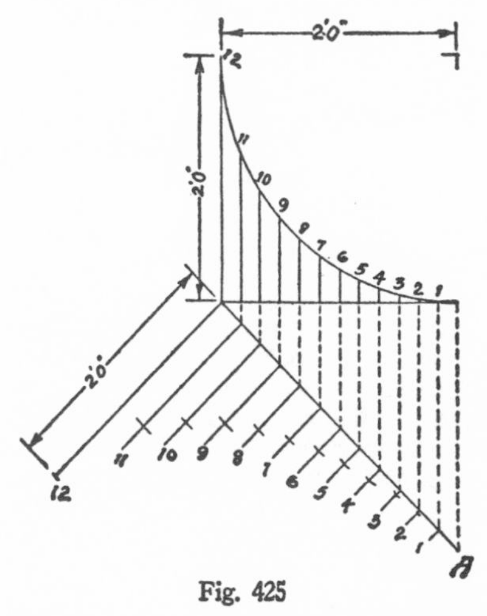

Fig. 425

rafter, the run of which is divided into 12 equal parts. (Any number of spaces can be used, and they do not have to be equal.) These points have been raised perpendicular to the base line, until they intersected the common-rafter curve. At the bottom of the diagram is shown the development of the hip curvature. First, draw the base line of the hip on a 45-degree angle and drop the division points of the common rafter until they strike this line, as indicated by the dotted lines. At each of these intersections draw a line perpendicular to the base of the hip, as shown on the drawing. Then with a compass transfer the distance between the base line and curved line of each perpendicular line of the common rafter, to the respective perpendicular line of the hip. In other words, make the perpen-

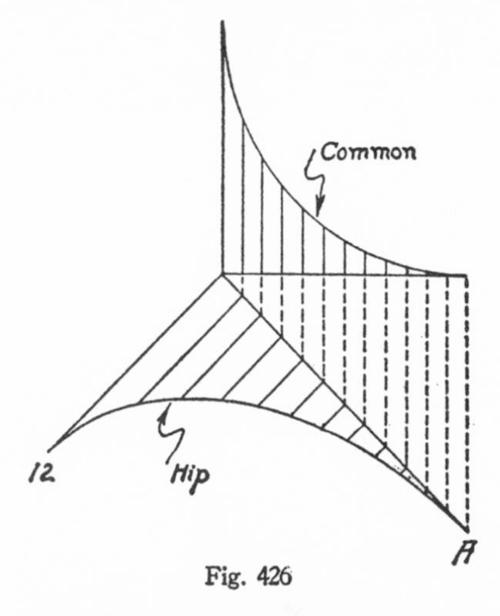

Fig. 426

dicular lines of the hip as long as the respective perpendicular lines of the common rafter. Now draw a line from point *A*, crossing these marks, and stop at *12*. The curvature between *A* and *12* is the curvature of the hip rafter.

Fig. 426 shows the same diagram with the hip rafter fully developed.

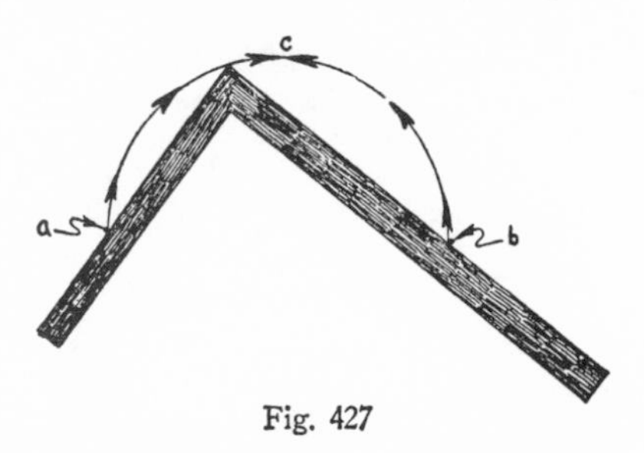

Fig. 427

You will notice that the curvature of the hip is not a true circle curve, and therefore cannot be obtained with a radius.

Circles with Square.— Stick two nails, as at *a* and *b*, Fig. 427, as far apart as the diameter of the circle you want. Place the square in somewhat the position shown, and with a pencil at the heel mark the circle as the square

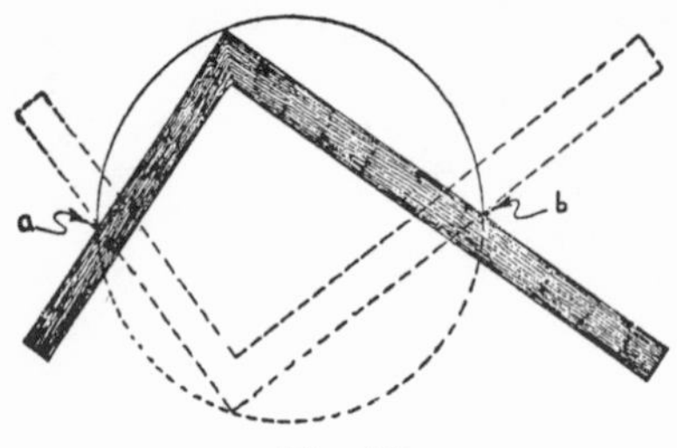

Fig. 428

is moved in the direction of the arrows. When you have one half of the circle marked, place the square in the position shown by dotted lines in Fig. 428,

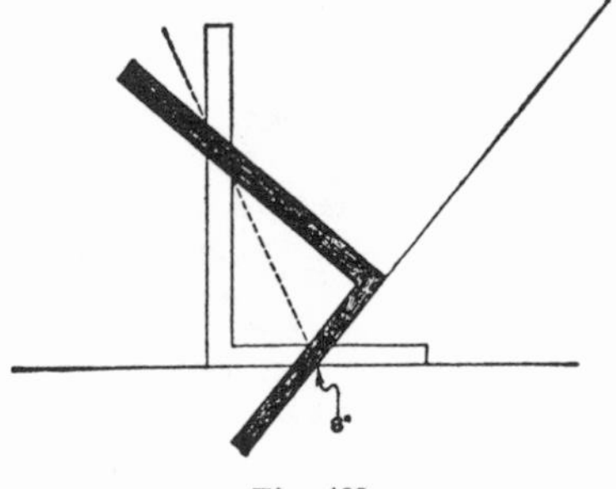

Fig. 429

and proceed to mark the other half of the circle. In case you want a circle that is larger than what the steel square will make, make a wooden square as large as needed to make the circle.

Member Cuts.—Fig. 429 shows how a member cut can be found with a square. The square is shown applied to an obtuse angle, but it does not matter what kind of angle you might have, the principle is the same. Shown are two squares, one shaded and the other not. The unshaded square is placed in such a position that 8 on the tongue intersects the angle, bringing the blade in a perpendicular position with the base line. The shaded square is so placed that the tongue rests on the inclined line with the figure 8 also inter-

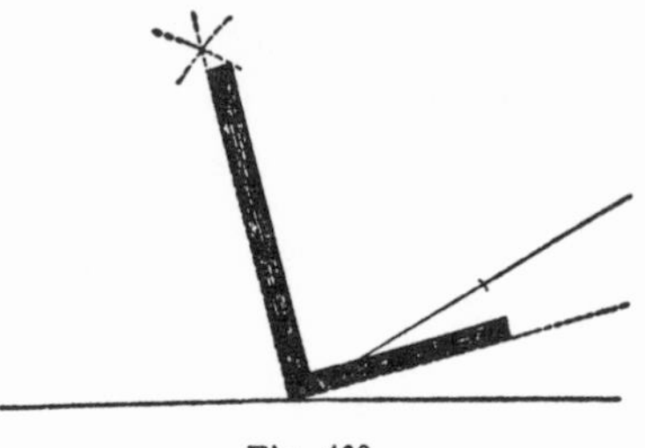

Fig. 430

secting the angle. Having this, the member cut is obtained by striking a line from the point where the blades of the square cross, to the point of the angle, or the point where the tongues cross. The dotted line represents the line that will make the member cut. Such cuts must often be made in finishing around open stairs. After the dull angle has been bisected, either with a square or with a compass, the sharp angle can be bisected with a square by placing it as shown in Fig. 430.

Chapter 44

MITER AND SQUARING PROBLEMS

Mitering by Reflection in Saw.—Cutting miters on small moldings by the reflection of the molding in the saw blade is a trick that every carpenter should practice. It is especially suit-

Fig. 431

able for use in cutting base shoe and quarter rounds. Any carpenter with good judgment and an accurate eye can miter small moldings with it so

Fig. 432

that the joints will fit perfectly—rarely will he have to do recutting.

Fig. 431 shows a saw applied to a half round for making a square cut by means of the reflection in the saw blade. Fig. 432 shows the same saw applied for cutting a true miter, also by means of the reflection, while Fig. 433 shows four samples of miters that are not true miters. The half round molding is used in all of these illustrations, but the principle is the same in cases of quarter rounds and other small moldings. In using the reflection in the saw blade, the workman's judgment and his eye must be trained so that when he looks at the angle that the molding must fit, he can apply the saw to the molding and adjust it in such a manner that when the reflection shows the same angle, he can cut the molding

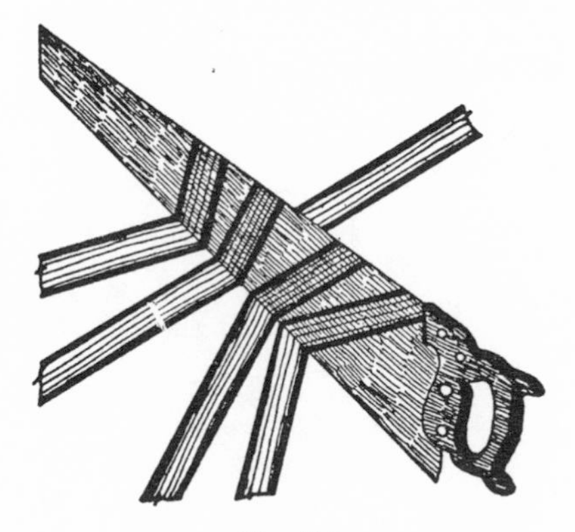

Fig. 433

and it will fit. This trick is a time saver for the carpenter who acquires the ability to do it skillfully.

Hunting Miter.—To get the cut for fitting a straight molding to a curved molding of the same design, make a drawing on the order of the one shown by Fig. 434, with a cross section of the molding. Divide the width of the molding into convenient spaces (they need not be equal) as indicated by the lines *1, 2, 3, 4,* etc. Now set the compass at point *X,* and with a radius that will give the outside line of the curved molding, strike the curved line that is numbered *1.* This done, transfer the numbered spaces shown to the left, to the upper right, as indicated by numbers. With the compass again set at point *X*, draw in the other curved lines from the different points, or *2, 3, 4, 5,* etc. Where the curved lines meet the straight lines, draw a line, as shown in Fig. 435, crossing points *a, b, c, d,*

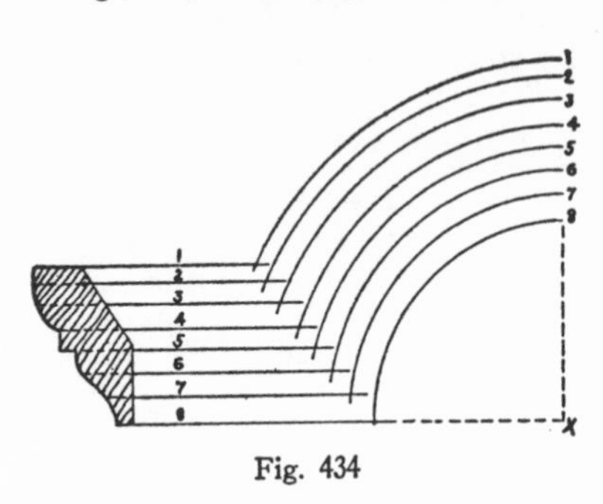

Fig. 434

etc. This, you will find, will produce a slightly curved line. To make the joint, cut the moldings straight, as indicated by the dotted line, Fig. 435, between points *a* and *h*. Then file the slight curve onto the cuts with a rasp, or perhaps a block plane could also be used in making the joint fit.

Rake Molding Miter. — Fig. 436 shows in part a bed molding in place on the rake and eaves of a cornice.

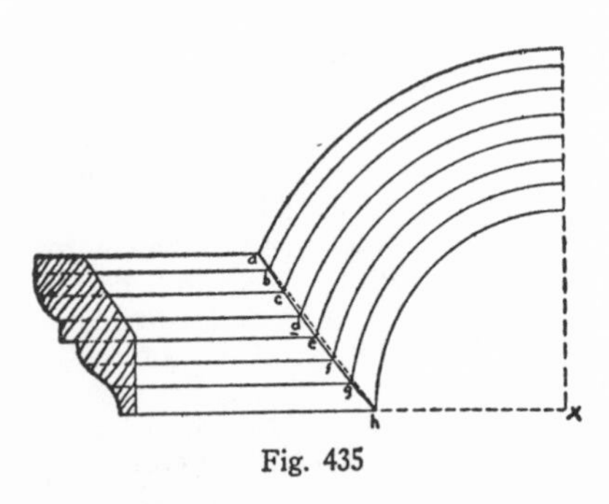

Fig. 435

The outline of the back of the molding under the eaves is shown by dotted lines. This molding fits the plancher with a full bearing, but joins the frieze by a mere corner contact. By placing the bed molding under the eaves in this way, a true miter cut will make the joint at the corner. The rake molding fits into the angle in the regular way. Fig. 437 shows the same treatment for a crown molding. This is especially

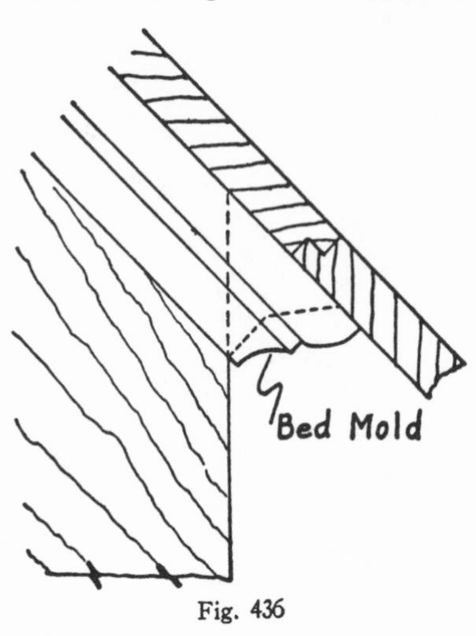

Fig. 436

suitable for a dehorned cornice, because it gives the eaves of the roof just a little more drip.

In case the eaves molding must be

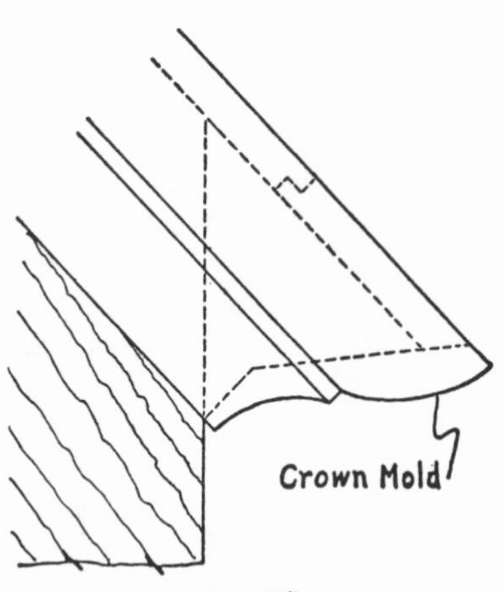

Fig. 437

set in a plumb position, that is, set so it will have a full bearing against the frieze, then, to make the moldings member at the corner, the rake molding will have to be especially designed for the purpose. This is true of both bed and crown molding.

Marking Cuts for Sprung Moldings.—Fig. 438, to the left, shows a cross section of a large molding in place. The part of the back dealt with here is shown to the right, shaded, looking straight at it. This is indicated by the

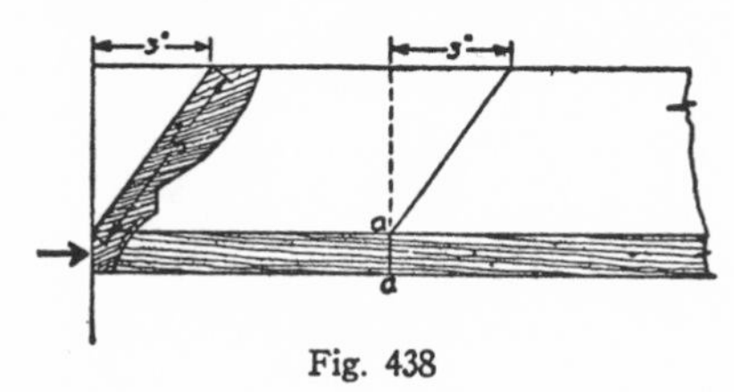

Fig. 438

arrow to the left, which points to the bottom of the cross section. From the starting point make the first mark, as between *a* and *a*, which is a square mark. Now turn to Fig. 439, which shows a cross section, upper left, and a drawing of the back, sloping down toward the right. Again looking

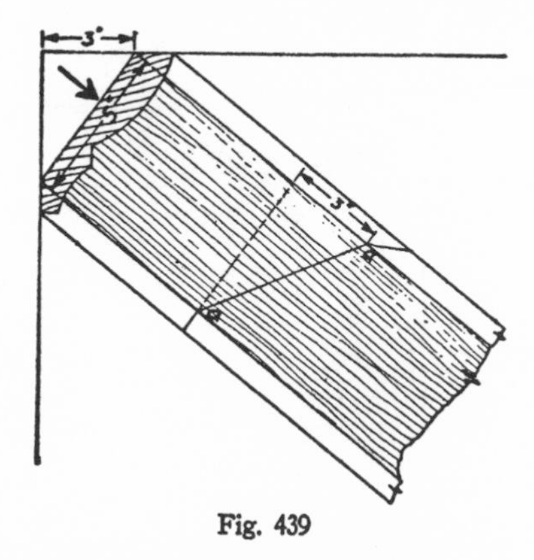

Fig. 439

straight at the back, as indicated by the arrow to the left of the cross section. Here make the mark between *a* and *a*, by squaring across the back, the shaded part, as shown by the dotted line. At a right angle to this line measure 3 inches, the spring of the molding, to get the second point, and draw a line from *a* to *a*. This marking can also be done with the steel square, by taking 3

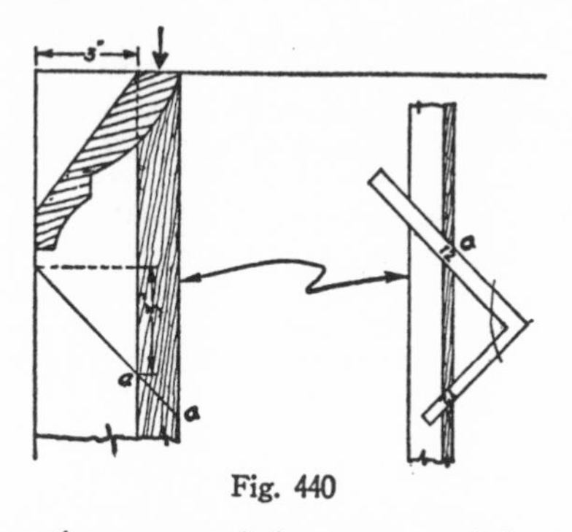

Fig. 440

on the tongue of the square and 5 on the body, the tongue giving the cut. Those figures are given with the cross section at the upper left. The cut is the same as a sheeting or plancher cut for a hip roof.

Fig. 440, to the left, again shows the cross section of the molding, and the arrow shows the direction of the view, which is a top view. The shaded part below the cross section is the upper bearing of the molding, which takes a true miter cut. This is shown again between *a* and *a*. The square (at a much smaller scale) applied to the molding to get this cut is shown to the right. The figures used on the square are 12 and 12. With the back of the molding marked in this way, take a well-sharpened fine saw and cut the molding from the back so as to cut away the three marks. If this is carefully done you will have a true miter cut on the molding.

Squaring with Tape.—Fig. 441 shows how the steel tape is used in

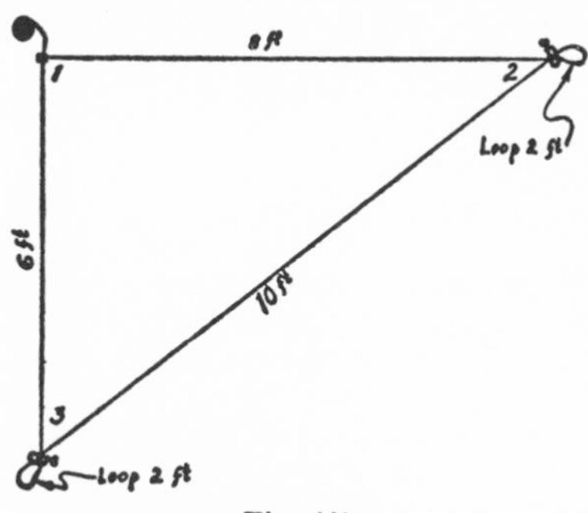

Fig. 441

squaring by using the 6-8-10 method of squaring. At number *1* the end of the tape is hooked on a nail that has been stuck at the established corner of the building. At number *2*, the two feet of the tape between 8 feet and 10 feet, is formed into a loop, and the tape is clamped together in such a manner that the 8-foot and the 10-foot points will meet. Then 10 feet more of the tape is unrolled, which reaches to number *3*. Here again a loop of 2 feet is formed, as shown. From number *3* the tape is run to the established corner, number *1*. If the 8-foot and the 6-foot sides of the triangle are kept on the building line, while all the sides are stretched tight, you will have a square corner at number *1*.

Six-Eight-Ten Method.—Fig. 442 shows how the principle of the 6-8-10 squaring method can be used with a

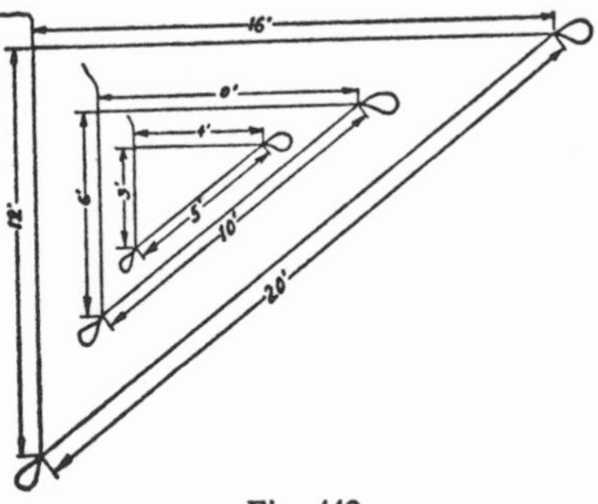

Fig. 442

small triangle as well as with a large triangle. At the center of this figure 3, 4, and 5 are used to make the right angle. These numbers were obtained by dividing 6, 8, and 10 by 2. The large triangle shown is made by multiplying 6, 8, and 10 by 2, resulting in 12, 16, and 20. The loops at the two angles show how the tape is held, either with a clamp or with the fingers. Clamping the tape together where the loops are at the corners insures accuracy. On the other hand, when there are three persons, each one holding a corner, if they are careful, good results can be obtained by holding the tape with the hands.

Squaring a Building.—A simple way to square a building, say for example, 32 feet by 40 feet, when it is staked out, is illustrated by Figs. 443 and 444. Start with Fig. 443 by stretching line

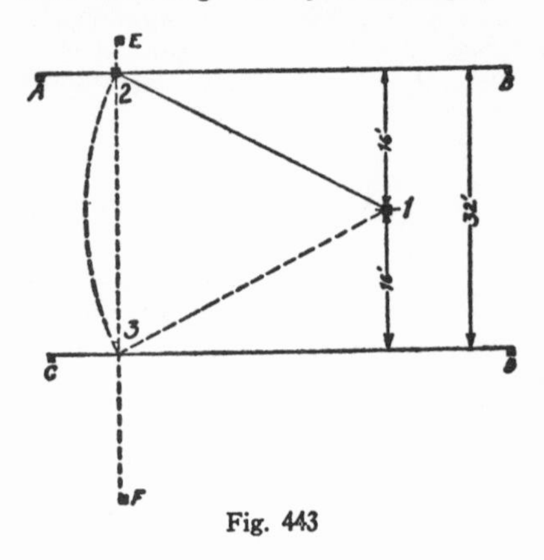

Fig. 443

A-B on the building line, making it cross the established corner of the building, as shown at *2*. Then set line *C-D* parallel to, and 32 feet from, line *A-B*. At a convenient point, exactly

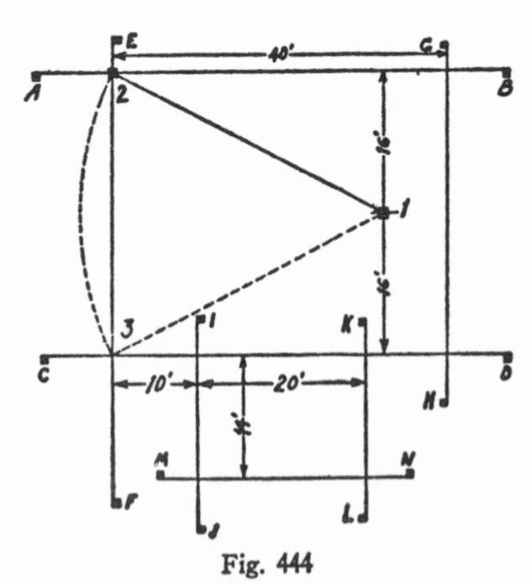

Fig. 444

halfway between the two lines, drive a stake, as at *1*, and stick a nail in the top of the stake in such a way that it will be exactly halfway between the two lines. Hook the tape on this nail and get the distance from *1* to the established corner, number *2*. Then carry this distance to number *3*, and stretch a line from *E* to *F*, crossing points *2* and *3*, and you will have two of the corners squared. To finish the squaring of the main part of the building, set line *G-H*, Fig. 444, parallel to and 40 feet from line *E-F*. Now set line *I-J* parallel to and 10 feet from line *E-F*, and *K-L* parallel to and 20 feet from *I-J*. Finish the staking out by setting line *M-N* parallel to and 14 feet from line *C-D*, and the staking out is completed.

Origin of Six-Eight-Ten.—How the 6-8-10 squaring method originated is illustrated by Fig. 445. Here are shown a square and a circle both of which are involved in every squaring method.

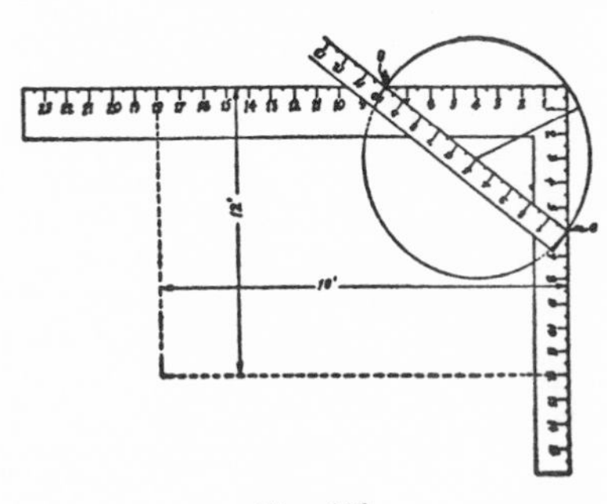

Fig. 445

The drawing shows a full-sized square giving the inches and a part of a rule. The rule contacts the outside edge of the tongue of the square at 6 inches and the outside edge of the blade at 8 inches, thus the rule shows the diagonal distance between these two points to be 10 inches. These three parts, it can readily be seen, form a right-angle triangle with a ratio of 6-8-10. To find where the circle comes in, bisect 10 inches and set one leg of the compass at 5. With the compass set in this manner, strike a circle with a 5-inch radius, intersecting the three important points of the triangle. By changing inches to feet, and adding the two dotted lines, a rectangular figure is formed, measuring 12 feet by 18 feet.

Circle with Square.—Fig. 446 shows the oblong figure given in Fig. 445. Now if a nail were set at point *A* and another nail at point *B*, the framing square would describe a circle by keeping both the tongue and the blade of the square in contact with the nails while the heel of the square is moved from *A* to *1*, *2*, *3*, and on to *B*, as indi-

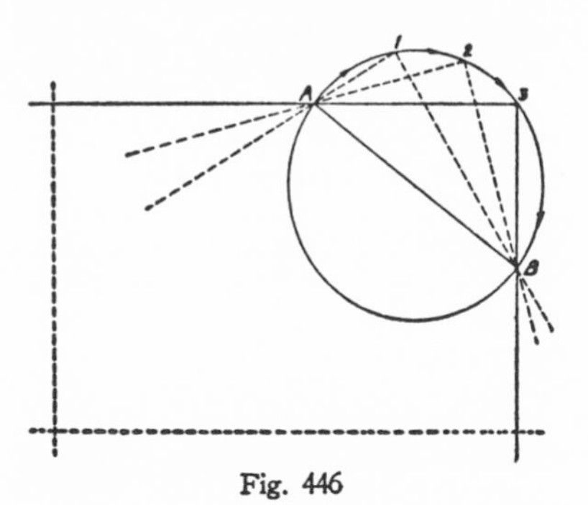

Fig. 446

cated by arrows. The marking is done with a pencil held at the heel of the square as it is moved from *A* to *B*. The other half of the circle is marked in the same way.

Squaring Building.—Knowing that with a square a circle can be described,

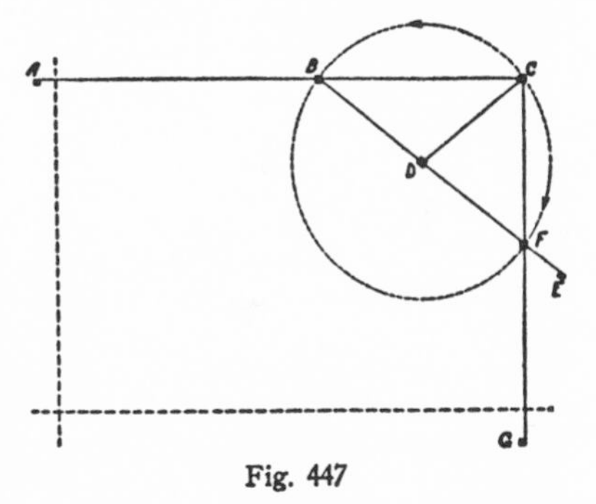

Fig. 447

and with a circle a perfect right angle can be produced, it can be shown by Fig. 447 how simple a matter it is to square a building when it is being staked out. Proceed by stretching a line from *A* to *C* on the established building line, the point at *C* being the established corner. At any convenient point, approximately between a 40 and

a. 50-degree angle, drive a stake, as at *D*. Stick a nail on top of this stake and hook the tape on it. With the radius, *D-C*, move from *C* to *B*, as indicated by the dotted part-circle and arrow, and drive a stake there. Establish this point by driving a nail in the stake where the circle crosses the building line. This done, stretch a line from *B* to *E*, crossing point *D*. With the radius, *D-C*, move from *C* to *F*, and drive a stake there, establishing the point by driving a nail where the circle crosses line *B-E*. A line run from *C* to *G*, crossing point *F*, will form a perfect right angle at *C*, which is the corner of the building. The two dotted lines that complete the oblong figure, can be located by measurements.

Making Louver.—To make a louver with a circular casing so that the wind will not blow in rain, and that otherwise it will not leak, proceed as follows:

Space the sloping boards so that the back edge of one board will be as high or higher than the front edge of the board directly above it. The width of these boards will also have something to do with keeping the wind from blowing the rain to the inside. In any case, it is suggested that tin flashing be used somewhat on the order pointed out by Fig. 448. The drawing shown to the right gives a cross section of the louver, while the drawing to the left is a face view. Here is shown by dotted lines an octagon frame into which the louver boards are fastened. Then the circular casing is put on, as shown. A circular frame can be used, instead of the octagonal frame, but the expense will be much higher. In either case, the tin flashing should be used to carry off rain water that might blow or leak in. Study the drawing.

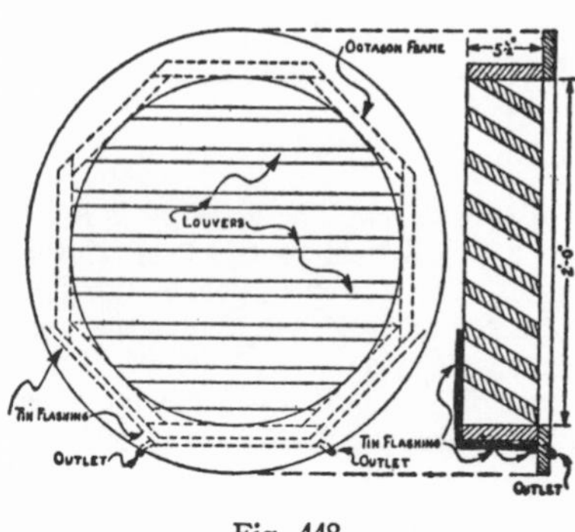

Fig. 448

Framing Gable-Shaped Louvers.—The problem is the same as a roof framing problem. Let us say that we are framing a third pitch dormer roof against a main roof. This would have two valleys. Fig. 449, upper drawing, is a diagram of such a dormer gable. The square is shown applied so that the figures will intersect the lower edge of the rafter, as it would have to in a louver frame. The figures to use on the square are 12 and 8, or the run and the rise. If the square is applied to the rafter as shown, the part of the groove marked *A* and *B*, bottom drawing, can be marked with the blade of the square. The bevel marked *C* of the louver, can be marked with the same figures, the blade giving the cut. These are all edge bevels. To make the edge bevel come out right, the louver board will have to be beveled on the front edge first. How to mark the board for the front bevel, is shown by Fig. 450, where an end view of the louver board is given, with the square shown at a smaller scale. Here again, 12 and 8 are the figures to be used, the tongue of the square giving the cut. Fig. 451

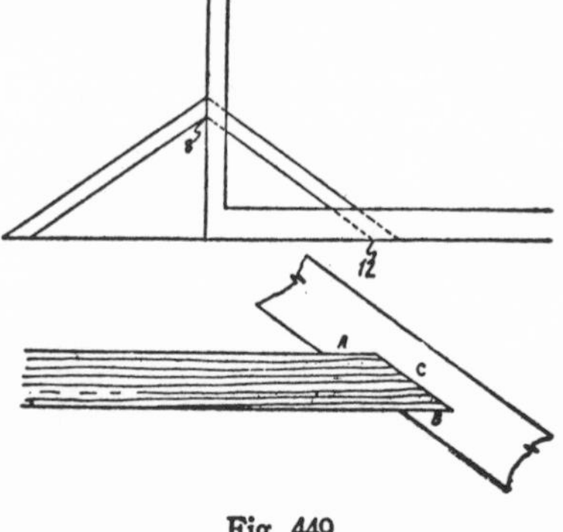

Fig. 449

shows the square applied to a board, using the run, 12, and the length of the rafter per foot run, 14⅜, plus. In the position the square is shown, the tongue will give the side bevel of the groove that receives the louver boards, while the blade will give the side bevel of

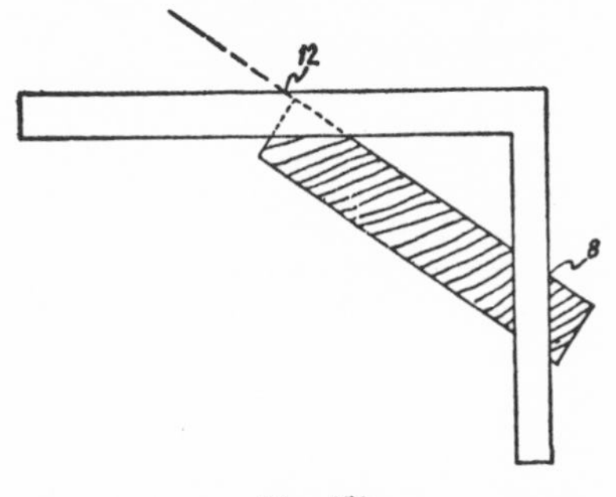

Fig. 450

the ends of the louver boards. In roof framing language, the blade of the square gives the sheeting cut for a third pitch roof, while the tongue will give the edge bevel of the jack rafters. The louver board will have to be sloped to the pitch of the rafter boards of the

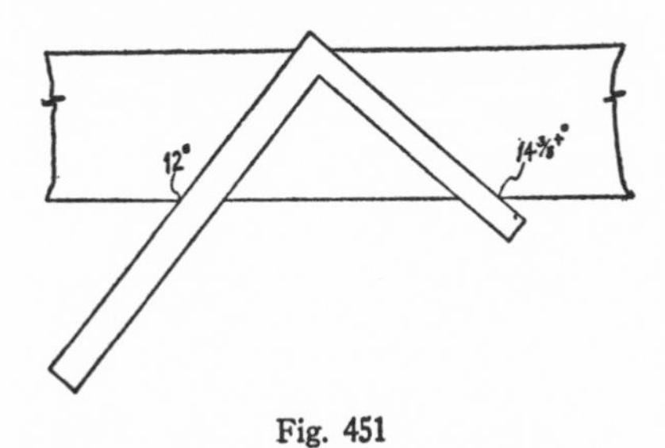

Fig. 451

louver frame, otherwise it will be a problem in uneven-pitch roof framing. What is said here about a third pitch, in principle, will apply to any other pitch.

It should be remembered that in dealing with roof framing problems, much reading between the lines must be done, to understand the solution; which is to say, the reader should be able to visualize the various cuts in such a manner that he will know in his own mind when they are right and also when they are wrong.

Describing Elliptical Arch.—Fig. 452 shows one half of an ellipse. The horizontal center line gives the half-way line of this ellipse. What is above this line is an elliptical arch, which is also divided into two parts with a perpendicular center line. To the left is shown the method that was used in describing the arch.

To describe one half of an elliptical arch, first draw two lines at a right angle, making the short arm of this right angle, as long as half the width of the arch, and the other as long as

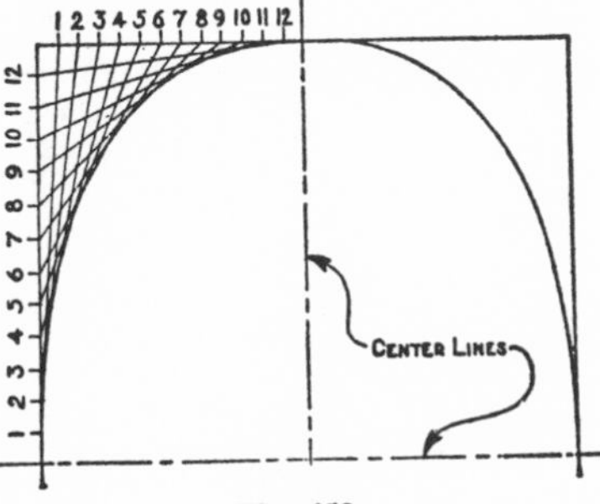

Fig. 452

the height of the arch. Then divide each of these arms into the same number of spaces, as shown. In this case there are 13 spaces and 12 points marking the divisions. The division points shown to the left are numbered *1, 2, 3,* etc., from the bottom up to *12.* In the same way the division points at the top are numbered *1, 2, 3,* etc., from left to right on to point *12.* With the division points so marked, draw a line from *1* to *1,* from *2* to *2,* from *3* to *3,* etc., up to *12* and *12.* The curve described in this way gives one half of the elliptical arch shown by Fig. 452. The other half of this arch is a reproduction of the curve shown to the left.

The number of spaces, as mentioned before, must be the same on the two arms. Any number of spaces can be used, but there must be enough to insure accuracy.

Chapter 45
SQUARES

Testing Steel Square.— Every carpenter should know how to test his square. A good way is to take a straightedge, as shown in Fig. 453,

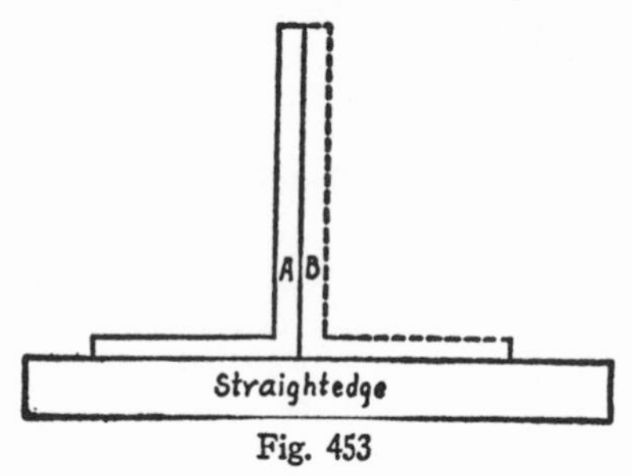

Fig. 453

place the square in position *A,* and mark along the outside edge of the blade. Then place it in position *B*—if the edge of the blade coincides with the mark, the square is true. But if the

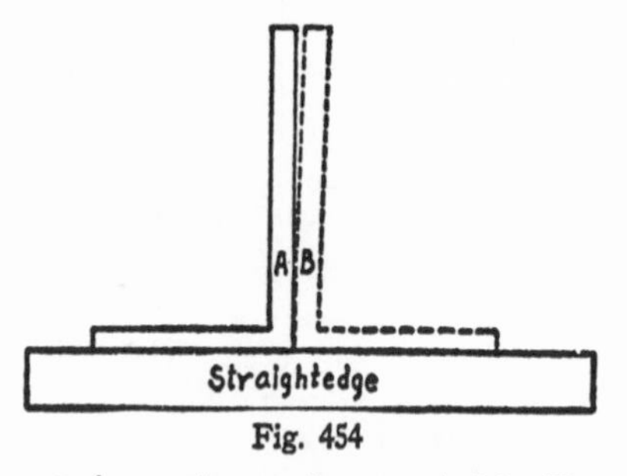

Fig. 454

test shows the results revealed in Fig. 454, it indicates that the tongue has been sprung inward. Should the results be as shown by Fig. 455, then the tongue has been sprung outward.

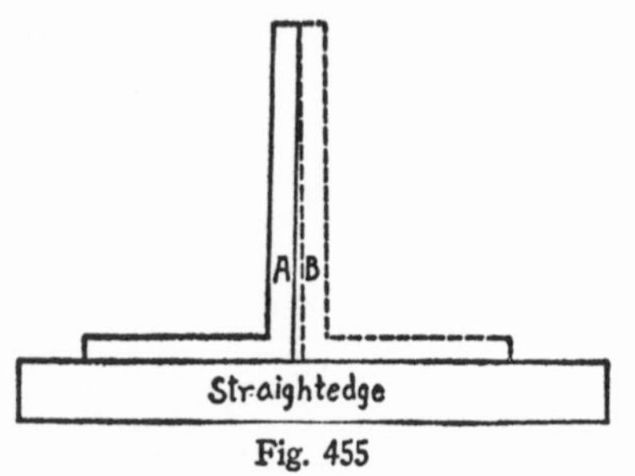

Fig. 455

Steel Square and Miters.—Fig. 456 shows four applications of the steel square for marking miters. At the top, to the right, the square is applied for a true miter, by using 12 on each arm of the square, while to the left it is also applied for a true miter, using 16 on each of the arms. The latter application gives the workman a chance to locate the point on the tongue by the feel of the hand, while he locates the point on the body of the square with the eye. This is especially suitable for marking boxing boards and rough flooring boards, when these are put on diagonally. At the bottom are shown

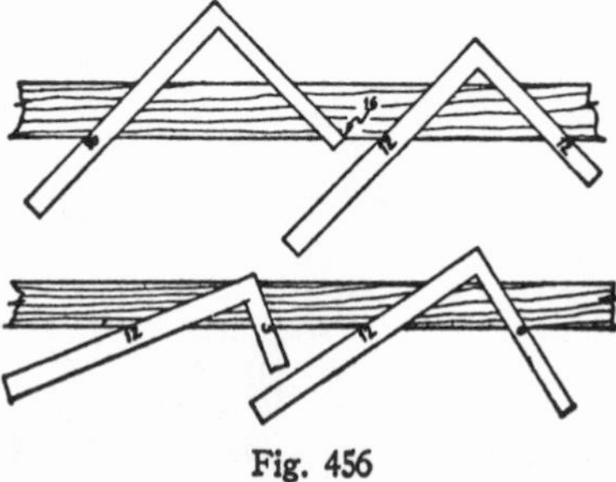
Fig. 456

two applications of the steel square for miters that are not true. Each of these applications gives a sharp bevel and a dull bevel. These are only six miters out of an unlimited number that can be marked with a steel square.

Difference in Length of Gable Studding.—As a rule the apprentice carpenter makes this problem harder than it really is—in fact, it is simple. Let's put it in simple terms: The roof, say, has a 12 and 8 pitch; that is, 12 inches run and 8 inches rise. Now if the studding were spaced 12 inches on center, the difference in the lengths of the gable studding would be 8 inches. And if the studding were spaced 24 inches on center, then the difference in the lengths of the studding would be twice 8, or 16 inches. But in case the studding are spaced 16 inches on center, because 16 is one third less than 24, the difference in the lengths of the studding would be one third less than for the 24-inch spacing, or 10⅔ inches.

Fig. 457 shows how what has just been explained, can be gotten by means of the steel square. The square in position *A,* shows a 12-inch run and an 8-inch rise—the rise is the difference in the lengths of gable studding spaced 12 inches on center. Position *B* shows

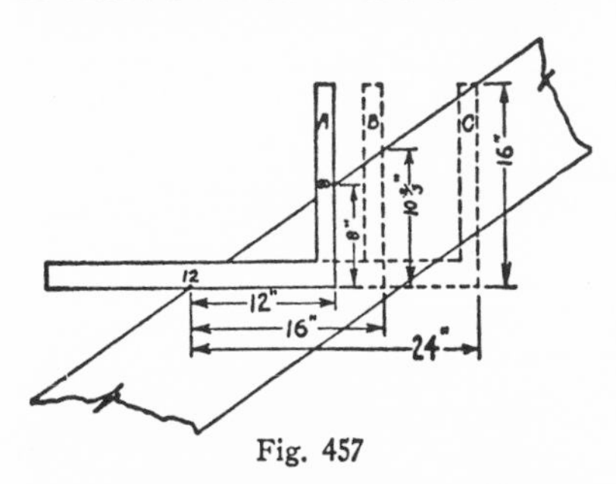

Fig. 457

the tongue of the square by dotted lines, giving the difference in the lengths of the studding for a 16-inch spacing, or 10⅔ inches, while position *C* gives the difference in the lengths of the studding for a 24-inch spacing, or 16 inches. The student should apply the square to a board as shown, and think through and compare the two explanations given here.

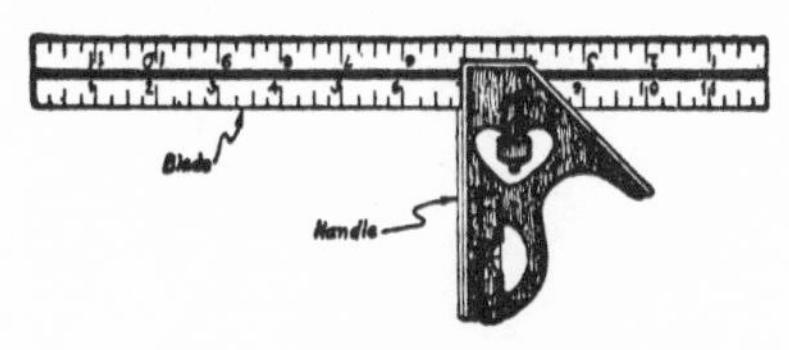

Fig. 458

Combination Square. — The combination square, a sample of which is shown in Fig. 458, is pushing the try square off the market. While the try square is still used in carpenter's shops, it is seldom seen on the job. The combination square, as has been pointed out, is only one of the reasons why the old fashioned try square is going out of use. The other reason is the advent of the power-driven saw, together with other power-driven machinery. These machines are taking over much of the shop work as well as much of the cross-cut sawing and ripping on the job. When material is cut and shaped with a machine, testing with a try square is unnecessary in many instances. When it does become necessary, the much better combination square is used instead.

Fig. 459 shows the combination square with the blade shifted to one side for marking miters — the dotted

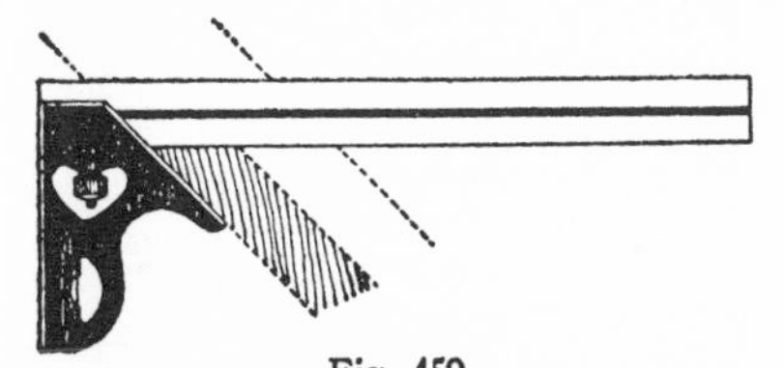

Fig. 459

lines represent possible timbers to be marked or tested. The shaded part shows how a square in this adjusted position can be used for trying an edge bevel of a board. If the blade were shifted to the left, then the square could be used for marking square across and also for trying square edges. Among the other things that can be done with a combination square are

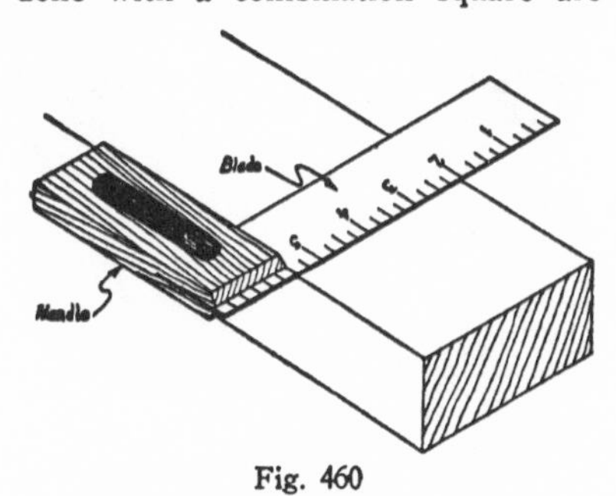

Fig. 460

leveling, plumbing, gauge marking, and depth gauging. The blade, if it is taken from the frame, will make an excellent bench rule.

Try Square.—Since the try square is still being used it is shown here, but with the understanding that any test or mark made with it can be made as well if not better with the combination square. Fig. 460 shows a try square with the blade and the handle pointed out. The square is in position for marking the face of a piece of material. The important thing is to work from a straight edge. If the edge is not straight, especially in accurate work, it should be jointed before the square is

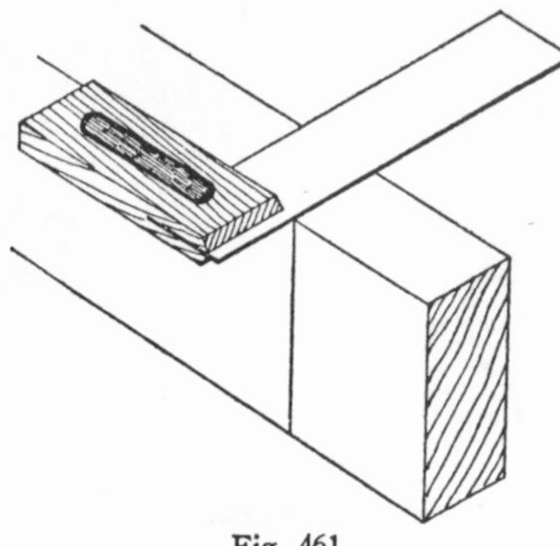

Fig. 461

applied. Fig. 461 shows the try square in position for marking the edge of a piece of timber, the face side having already been marked. The experienced carpenter usually omits the edge marking, because he can saw the timber square enough without it. The apprentice should mark the edge as well as the face before he starts to do the sawing. This will not only give him practice in handling the square, but it will help him eventually to be able to do the sawing without marking the edge.

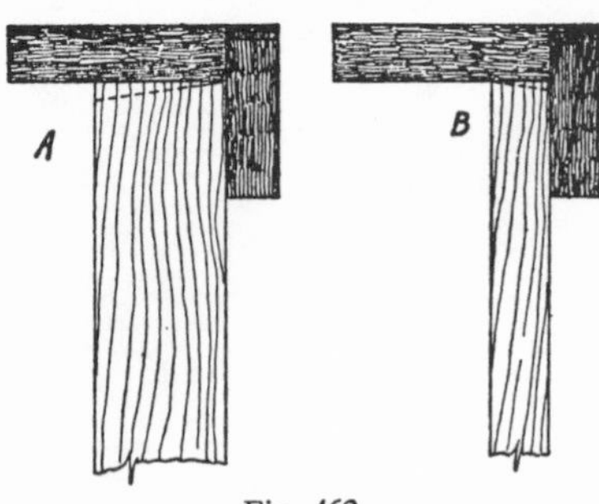

Fig. 462

Try Square Applications.—Fig. 462 shows two applications of the try square. At *A*, the square is applied for testing the squareness of the timber across the end. If the end were cut as indicated by the dotted line, or somewhat like it, this test would reveal a wedge-shaped opening between the end of the timber and the blade of the square. At *B*, the square is applied to test the squareness of the end, viewed from the edge of the timber.

Witness Marks.—Fig. 463 is a perspective view of a try square applied to the edge of a board for testing the squareness of it. There are also shown four different witness marks that good mechanics use to distinguish the face

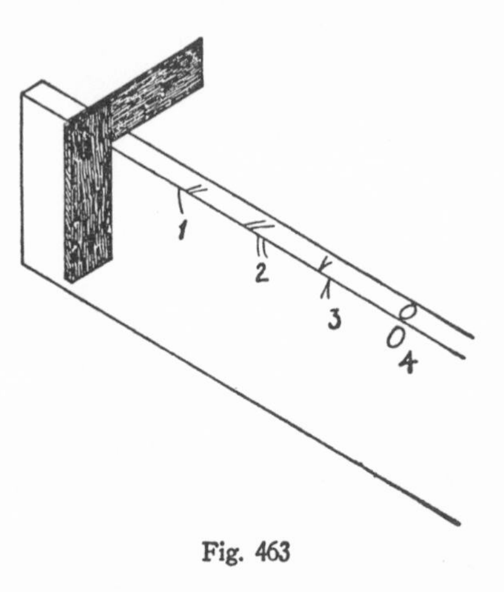

Fig. 463

edge and the face side. At number *1* the face edge is indicated by two marks, while the face side is marked with only one mark. This distinction in marking between the face edge and the face side is quite unnecessary, since the edge can always be distinguished from the side whether it is marked or not. At number *2* both the face edge and the face side of the board are marked with two lines. At *3* the face edge and the face side are indicated with check marks, while at *4* they are marked with circles. Any of these witness marks, or even other marks will answer the purpose. The important thing is that the face edge can readily be distinguished from the unworked edge, and the face side from the back side.

Bevel Square.—Fig. 464 is a drawing of a bevel square, giving the names of such important parts as the blade, the handle, and the thumbscrew. The bevel square is still an important tool for the carpenter, although it is not used as much as it used to be. It is indispensible in roof framing for marking the bevels of the different cuts.

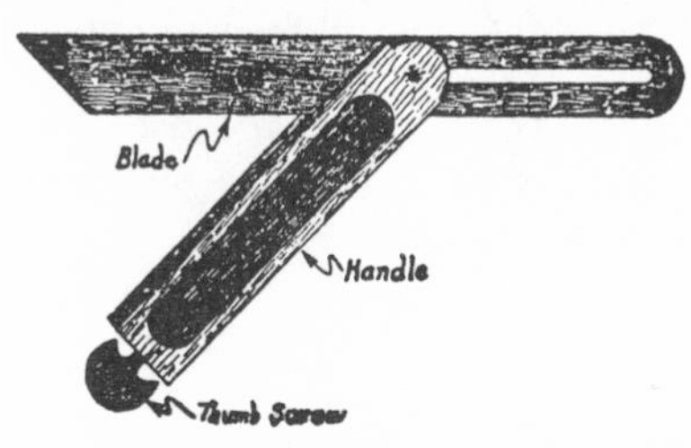

Fig. 464

The roof framer obtains the bevel with the steel square and then transfers it to the bevel square for marking the different cuts. Some roof framers use two and sometimes three bevel squares, each square set for a particular cut. This is especially true in framing irregular-pitch roofs. For most cases one or two bevel squares will serve all practical purposes in roof framing.

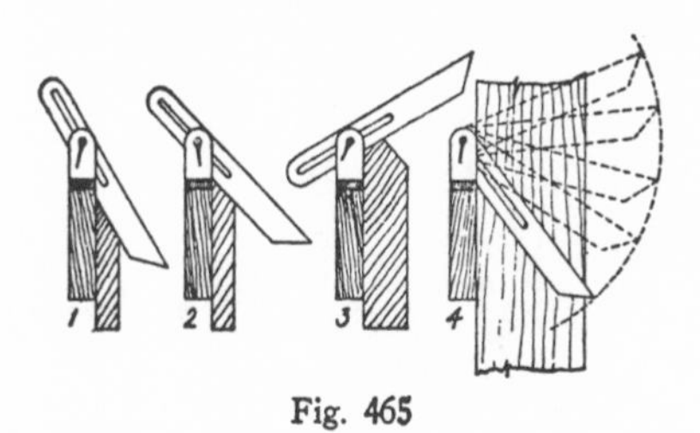

Fig. 465

Ways to Use Bevel Square. — Fig. 465 shows a number of ways a bevel square is used. At *1* it is used as a try square for a very sharp bevel on the edge of a board. At *2* a similar test is given on the edge of a board with a less-sharp bevel. At *3* the square is used for testing still another bevel. The tests shown here are only samples of hundreds of different bevels that might require testing for which the bevel square has no rival. At *4* is shown how the bevel square can be used for marking miter and other cuts on the face side of a board. The dotted lines show four different positions of the blade, which indicate that there is no limit to the kind of bevels that can be marked with a bevel square.

Divide and Conquer. — Fig. 466 shows a steel square applied to a straightedge with the figures 12 and 4½ contacting the edge. To the right is shown a guide nailed to the straightedge, which should be thin material. The figures used on the blade and tongue of the square, as shown on the diagram, are always read as inches, but the figures shown to the right between

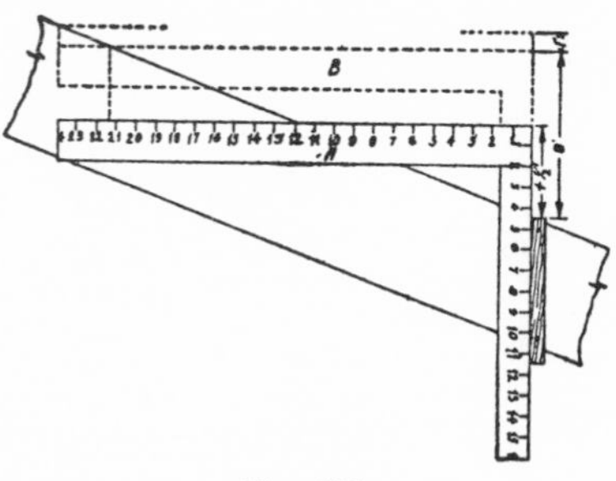

Fig. 466

the arrows, are always read so as to make inches represent feet.

The square as shown on the drawing is set for a problem in division. Reading the figures on the tongue of the square there are 4½ inches, but the figures between the arrows must be read 4½ feet. The figure on the blade would read 12 inches. This means that 4½ feet will have twelve 4½-inch spaces in it. This is the basic, or starting problem and is used in this way: If 4½ feet have 12, 4½-inch spaces, how many 4½-inch spaces are there in 8 feet? The problem is solved by moving the square from position *A* to position *B,* shown by dotted lines, and reading the figures at the intersecting points, which in this case would be 21⅓, or 21 and ⅓ spaces—the answer.

Another example: How many spaces of 4½ inches will there be in a distance of 9 feet? Move the square from position *A* to one inch past position *B*, indicated in part by dotted lines, and read the figure at the intersecting point on the blade, which is 24, or 24 spaces —the answer.

Figures that are easily divided were purposely chosen, so that the student could prove the examples quickly. But in practice the figures that must be used in spacing will, in most cases be fractions. For example: How many spaces of $3^{13}/_{16}$ inches are there in a distance of 5 feet $3^{7}/_{16}$ inches? The square is placed on the straightedge, using $3^{13}/_{16}$ on the tongue, and 12 on the blade. To get the answer move the square to the point on the tongue that represents 5 feet $3^{7}/_{16}$ inches, and read the answer on the intersecting point of the blade. (So that feet, inches, and fractions of inches can be easily determined on the square, use the side of the square that has the inches divided into 12ths, or use decimals, and measure the fractions with a compass.)

One more example: If 4½ feet will have 12 4½-inch spaces, how many 4½-inch spaces will there be in 2¾ feet? To get the answer pull back the square until the 2¾-inch point on the tongue intersects the edge of the straightedge. The answer will be found at the intersecting point on the blade.

The application of the square is important. It is suggested that the student take a steel square and a straightedge and practice with them until he thoroughly understands the process of dividing distances with the square.

Rapid Calculator.—Besides all of the other uses the steel square can be put to, it can be used as a board-foot calculator and the results will be quick and accurate.

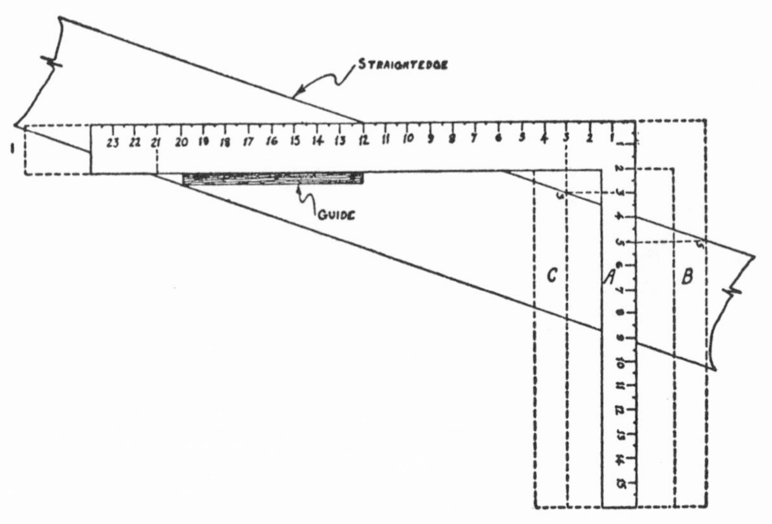

Fig. 467

If the reader will turn to Fig. 467, he will find the steel square placed on a straightedge with a guide under the blade. The figures that are used are 12 and 4. In this position the square is set for calculating the number of board feet in 1x4's of different lengths. The 1x4 board is taken here, because the board feet in it are easily figured, thus the reader can prove the accuracy of this calculating method. But the calculator will be equally accurate in giving the exact board feet for boards 5½ inches wide, 6⅞ inches wide or any other width that might need to be figured. The same is true with regard to the lengths of boards. They can be in even feet lengths, or they can be in lengths of feet and inches, including fractions, still the results will be accurate.

In the position the square is shown, Fig. 467, it gives the board feet for a board 12 feet long and 4 inches wide. The length of the board is shown on the blade of the square, where it intersects the edge of the straightedge. To find the number of board feet in the board, read the figure where the tongue intersects the edge of the straightedge. In this case it is 4, or 4 board feet. The original position of the square is marked *A*. Now, if the square is pushed forward to position *B*, shown by dotted lines, the intersection of the blade with the straightedge will read 15, and the intersection of the tongue will read 5, meaning that a board 15 feet long, 4 inches wide, will have 5 board feet of lumber in it. Or if the square were pulled back 3 inches, the intersections will read 9 and 3, meaning that a 1x4, 9 feet long, has 3 board feet of lumber in it.

It should be remembered that the 12 on the blade and the width in inches, including fractions, on the tongue give the starting position. Then fasten the guide and shift the square according to the lengths of the boards, and the intersecting points on the tongue will give the number of board feet of lumber in the various boards.

The Steel Square.—Speaking from the standpoint of the composition of the square, the name, The Steel Square, is correct. This term is probably used by a large majority of square users, and it is a good term. But in the building industry it is often called The Framing Square, and some insist that this is the only legitimate term. However, this term has support in cases where the square is used for framing purposes, but the square has a great many other uses. It is the conclusion of this writer, that common usage, both of the tool and of the term, lends its support to the name, "The Steel Square," which is the title of this book. A perspective view of the steel square is shown by Fig. 468, where the blade, tongue, and heel are pointed out.

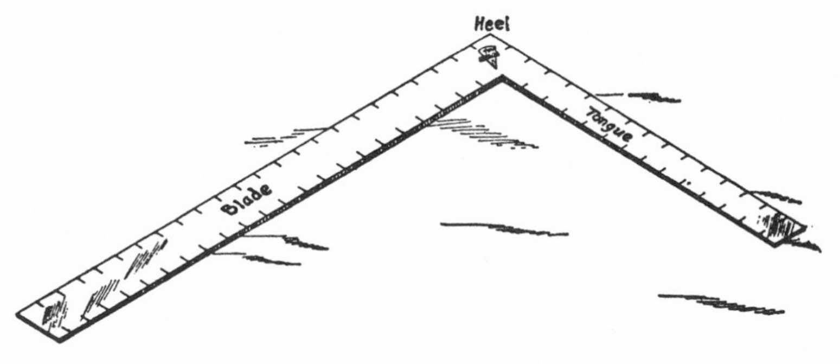

Fig. 468

Chapter 46

MISCELLANEOUS PROBLEMS

Gable Studding.—Gable studding can be framed in advance so that the studs will space themselves, as it were, without marking the plate. All that is necessary is that they be cut right and set plumb. The difference in the lengths of the different studding is the rise of the roof for the distance of one space from center to center. This difference in the lengths may also be obtained in a practical way with the steel square.

Fig. 469 shows a pair of gable rafters in place. The pitch is one-half. The

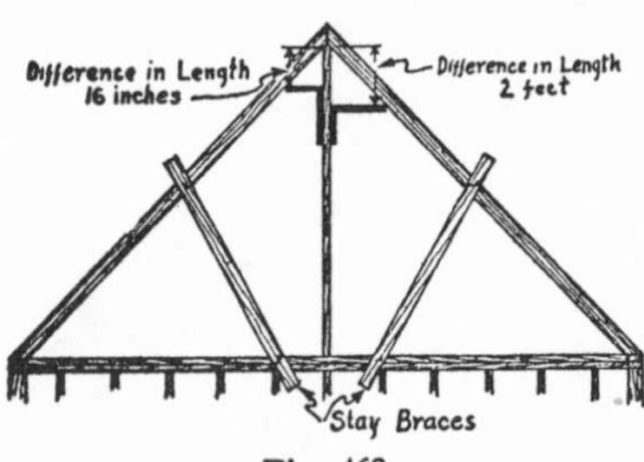

Fig. 469

first thing to do is to make sure that the rafters are perfectly straight and held straight with stay braces, as shown. Either set a stud at the center, as shown, or use a straightedge instead, and apply the square, letting the horizontal arm intersect the bottom edge of the rafter. The distance from

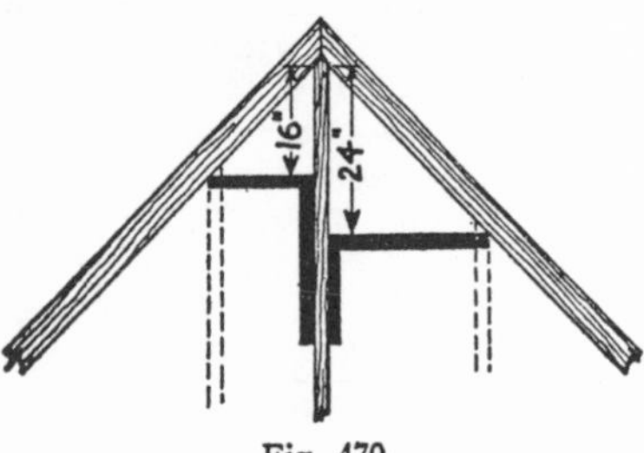

Fig. 470

the square to where the stud intersects the bottom edge of the rafter, as shown on the drawing, is the difference in the lengths of the studding.

Fig. 470 gives a detail, showing at a larger scale the two applications of the square—to the left for 16-inch spacing, and to the right for 2-foot spacing. The position of the studs to the right and left of the one in place, are shown by dotted lines. The stay braces shown in Fig. 469, should be left in place until the studding will hold the rafter straight.

A Roof Framing Problem.—Gable ends of a house and a garage are shown by Fig. 471. The front is to the right.

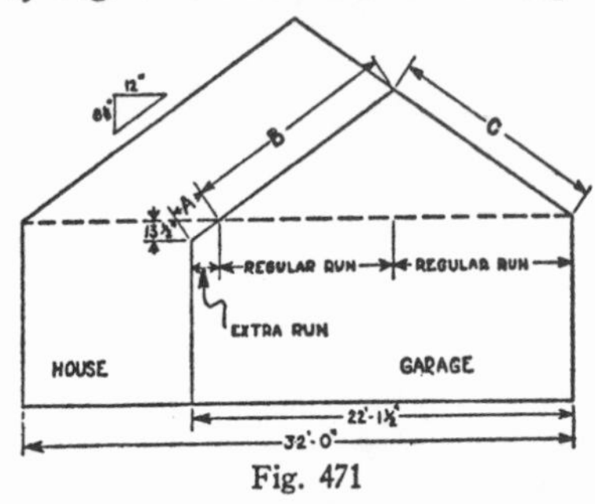

Fig. 471

The rise is 8¼ inches per foot run, as indicated on the drawing. The difficulty in framing the garage rafters is due to the 13½-inch drop of the rear rafter seat below the seats for the other rafters of the building. This drop is shown in figures on the drawing, directly below the dashed line. The first thing to do is to determine the extra length of the rear rafter, *A,* Fig. 471. To do this it is necessary to determine the extra run, because of the 13½-inch drop. Fig. 472 shows how these

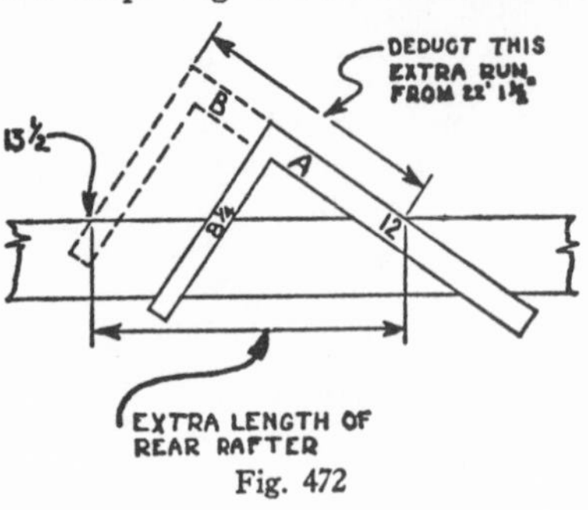

Fig. 472

distances are obtained. Apply the square to a straightedge, using 12 on the blade and 8¼ on the tongue, as shown by position *A*. Now slide the square up and to the left, until 13½ on the tongue intersects the edge of the straightedge, or from position *A* to position *B*, as indicated by dotted lines. The square in position *B*, gives the extra length of the rear rafter, and the extra run, as shown by the notes on the drawing. To get the regular runs, as shown by Fig. 471, deduct the extra run from the full span of the garage, 22 feet 1½ inches and divide by 2.

How to get the regular length of the rafters, *B* and *C*, as shown in Fig. 471, is illustrated by Fig. 473. Apply

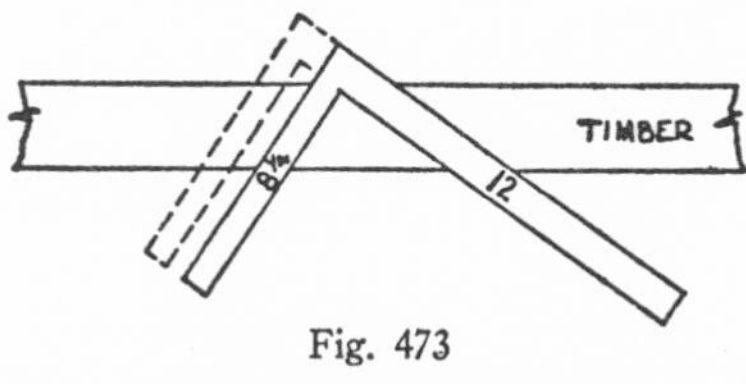

Fig. 473

the square to the rafter timber, using 12 on the blade and 8¼ on the tongue, and take as many steps as there are feet in the regular run. Before removing the square for the last step, slide the square to the left and up as much as the distance of the fraction of a foot in the run, as indicated by the dotted lines. Mark along the tongue for the plumb cut. The blade gives the foot cut.

Deducting for Ridge Board, Hips, and Valleys.—To determine the amount to deduct for ridge boards, hips, or valleys, make a full size drawing of the particular joint that you are working with, on the order shown in Fig. 474.

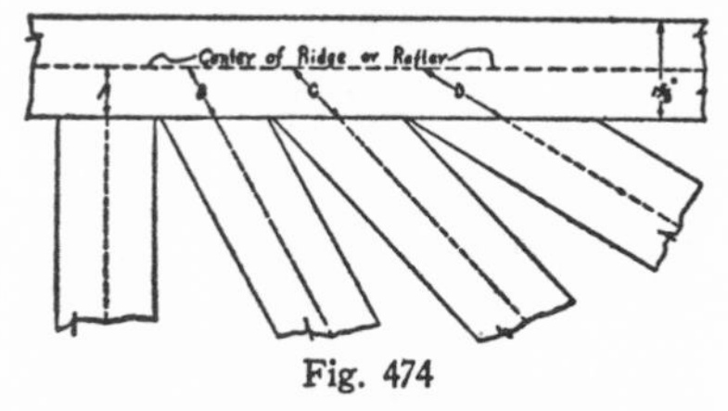
Fig. 474

In this drawing, as indicated, a 1⅝-inch ridge board (or rafter) is shown, to which four rafters at different angles are joined, as at *A, B, C*, and *D*. It doesn't make any difference whether these rafters join another rafter or a ridge board, always take half the distance through the timber at the angle the rafter joins the timber, and deduct that from the run of the last step. It will be noticed by referring to the drawing, that the distances *A, B, C*, and *D*, as indicated by the arrows, are not the same. It is important that the drawing be made full size, and in plan. All roof framing, theoretically, is done from the center of the rafter, and therefore the deductions are shown here in line with the centers, which are indicated by dotted lines.

Fig. 475 shows a square applied to a rafter timber, for, let's say, the last

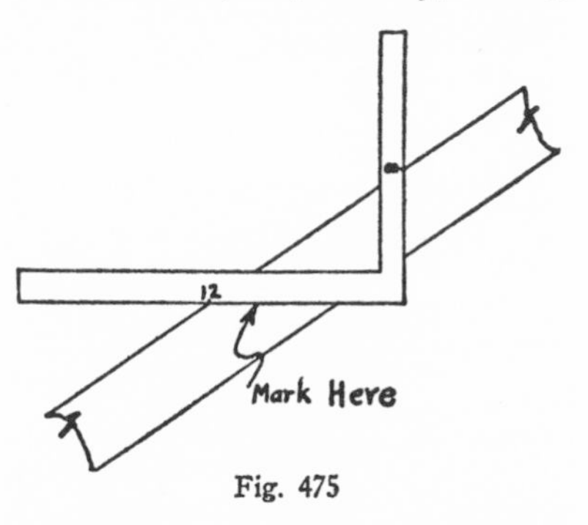

Fig. 475

step. But before the square is removed or the deducting is done, mark along the edge of the blade, as indicated on the drawing. Then, keeping the edge of the square on the pencil mark, pull the square back the distance that has to be deducted, and mark the plumb cut along the edge of the tongue.

Degrees Without a Protractor.—To measure angles in degrees without a protractor, make a drawing as follows: At a point called the vertex, strike a horizontal base line. Then set a compass at this point and strike a part-circle with a radius of 7³⁄₁₆ inches, as shown by Fig. 476. To get the angle, measure as many ⅛-inch spaces on the part-circle as there are degrees in the angle you want. In this case it is a 5-degree angle. To get a 25-degree angle, measure twenty-five ⅛-inch spaces on the part-circle, or 3⅛ inches. Twenty degrees added to the 5 degrees shown on the drawing, will make the 25 degrees.

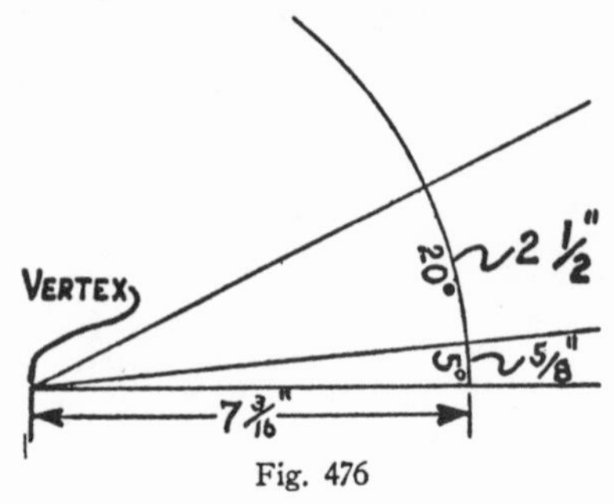

Fig. 476

To get more accurate results, multiply both 7 3/16 and ⅛ by 2, 3, 4, or more. Multiplying by 2, for example, would give a radius of 14⅜ inches, and for each degree a space of ¼-inch.

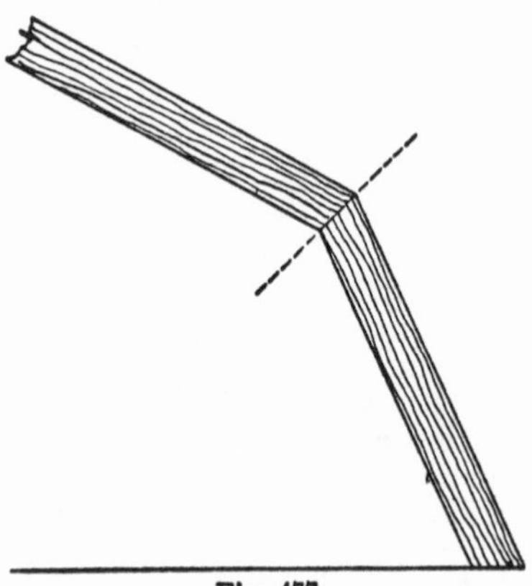

Fig. 477

Bisecting. — Fig. 477 shows a curb joint of a rafter for a gambrel roof. The problem is to obtain the cut that will be a perfect member cut.

In Fig. 478, lines *A-B,* and *B-C* diagram what is shown in Fig. 477. The first operation is to continue line *A-B* to *D,* as shown by dotted line. This done, take the steel square, using 12 on the blade as the starting point, and place it on the diagram in the two positions shown in Fig. 479, marking along the outside edge of the tongue in each application. Now strike the dotted line,

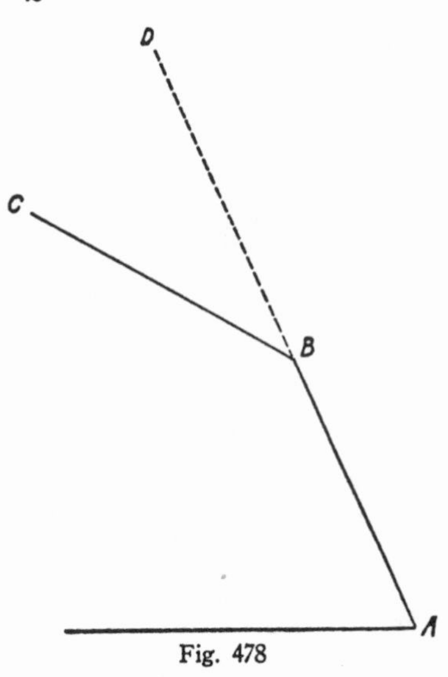

Fig. 478

a-b, in such a manner that it will intersect point *B* and cross the intersection of the two tongues. This will give the lines shown by Fig. 480.

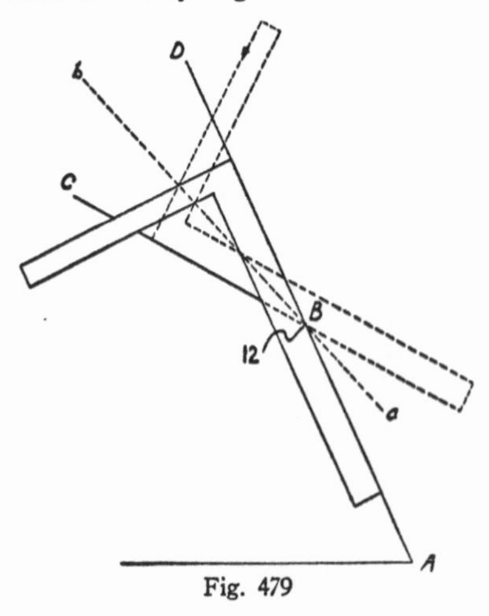

Fig. 479

Fig. 480 shows how to use the edge of a board for the line *A-B-D*. Having this line, draw line *B-C,* and bisect the angle *C-B-D* to obtain line *a-b*. To obtain the bevel, apply the square on line *a-b* as shown, making 12 intersect the angle at *B*. The shaded bevel shown in the angle between the tongue

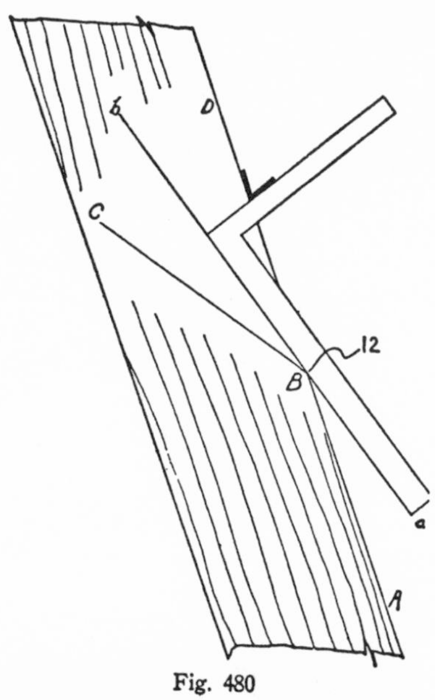

Fig. 480

of the square and the edge of the board is the bevel that will make a true member cut. In doing the marking, use 12 on the body of the square and the point where the tongue of the square intersects the edge of the board.

While Fig. 477 shows the curb joint of a rafter for a gambrel roof, the process is the same for obtaining the bevel for the cut of any kind of a miter joint.

Center and Radius.—Fig. 481 shows a part of a circle that runs from *A* to *B*. In case neither the center nor the radius is known, with a convenient radius, strike three smaller circles as shown on the drawing, setting the compass or the radius pole, whichever is used, at points *a, a, a.* Starting at *b* and *b*, draw two lines in such a manner that each line will cross the points where a side circle crosses the center circle. Where these two lines cross each other, as indicated at *c*, is the center of the large circle. The radius of

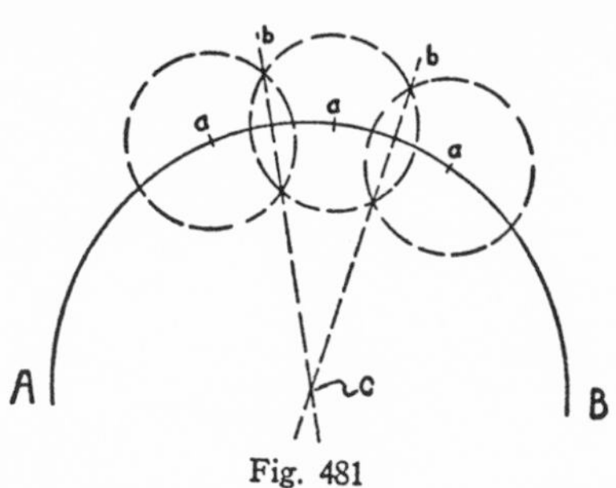

Fig. 481

the circle is the distance from point *c* to where either one of the two lines crosses the large circle.

Fig. 482 shows the same principle applied differently. At two convenient

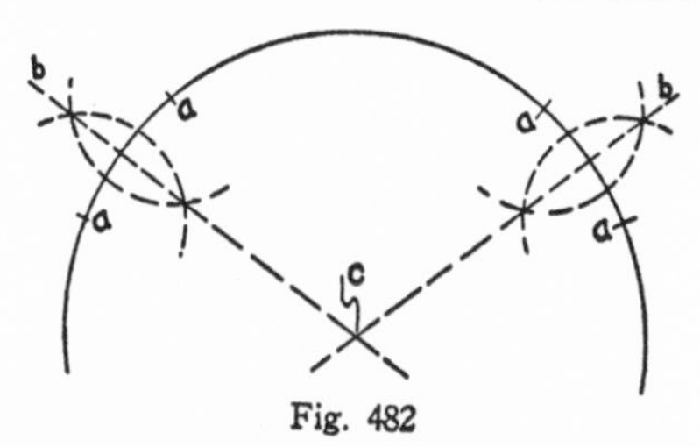

Fig. 482

points of the large circle, strike two part-circles, from points *a* and *a*, as shown. Draw two lines, starting at *b* and *b*, in such a way that they will cross the points where the part-circles intersect each other, and continue the lines until they cross. Where these two lines intersect is the center of the circle, as indicated at *c*. The distance from point *c* to where either one of these lines crosses the large circle, is the radius of the circle.

Chapter 47

RADIAL SQUARE

For many years this writer has felt the need of a tool measuring angles in degrees, without a protractor. That need is satisfied by the radial square. This square is a time-saver in marking the cuts for gable and regular hip and valley roofs. For irregular hip and valley roofs it serves as a short-cut bevel square. To mark any level or plumb cuts you simply set the tongue of the square to the degree that will give one or the other cut, and the square will mark both cuts.

Fig. 483 gives a face view of the radial square, set for marking the level and plumb cuts for a half pitch roof, which, as most mechanics know is 45 degrees. For a third pitch the tongue,

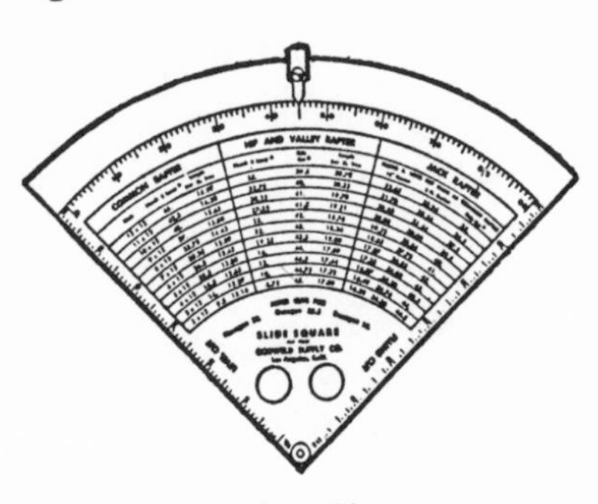

Fig. 483

or radial, is set at 33.75 degrees. A table on the face of the square gives the degrees to use for the different pitches.

Member Cuts. — To bisect angles, such as are shown by Fig. 484, measure, with the square, one or the other of the two angles in degrees, and divide by two — the quotient will give the point in degrees for marking the member cut, not only for the sharp (acute) angle, but also for the dull (obtuse) angle. The square to the right, with the face side up, is set for marking the member cut for the sharp angle. The member cut for the dull angle is marked, without changing the square, by swinging it into the position shown by dotted lines to the left.

Trisect Angles. — To trisect any angle, measure the angle in degrees

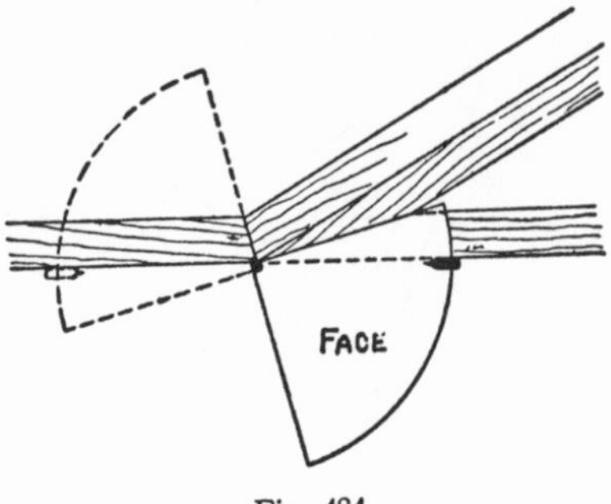

Fig. 484

and divide by three — the quotient giving the number of degrees in each of the three parts. This is shown by Fig. 485 where a 90 degree angle is shown trisected. The square was first set at 30 degrees. Then at 60 degrees, which divided the angle into three parts. To divide the same angle into six parts, set the square to one-sixth of 90, or 15 degrees, and mark the first line, as shown by the dotted line. The second line is made with the square set at 30 degrees. The third at 45 degrees, and so on until the angle is divided into six equal parts. In the

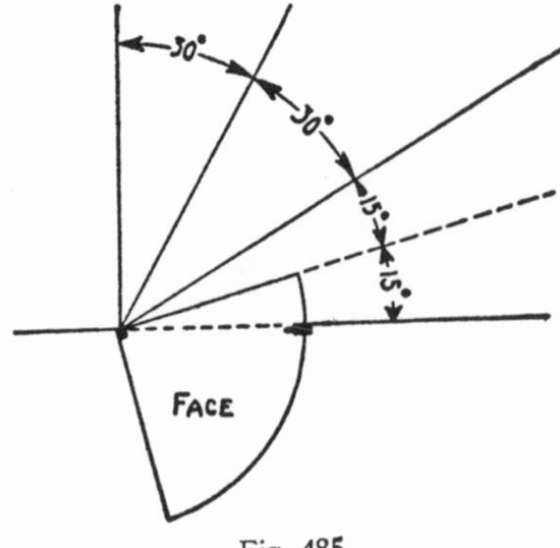

Fig. 485

same way any angle can be divided into as many parts as the case might require.

Cuts for Parallelograms. — To get the member cuts for the joints of a parallelogram, measure one of the angles in degrees, and set the square to half the number of degrees in the angle. The square set, as shown by Fig. 486 one edge will give the member cuts for the sharp angles, while the other edge will give the member cuts for the dull angles.

Every roof framer, stair builder, or finisher should have a radial square in his kit. In fact, every carpenter needs this tool to complete his set of tools.

Roof Framing by Degrees. — Roof framing by degrees is a simple method of laying out rafters with the radial square. See Fig. 483. This square is a combination of a square and a 90

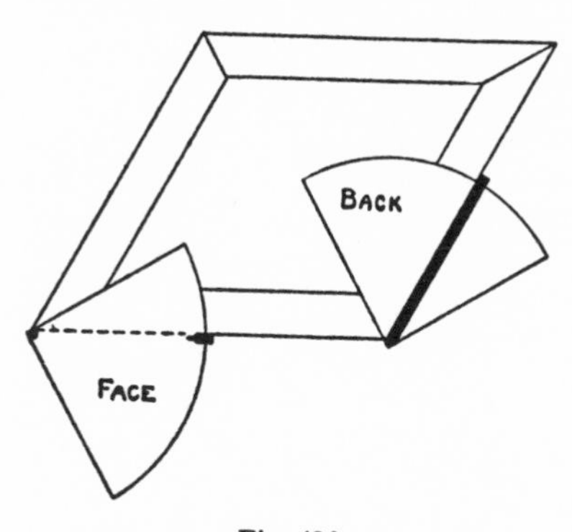

Fig. 486

degree protractor. On the face of it there are three roof framing tables. The first table covers common rafters. The second table deals with hip and valley rafters, while the third takes up jack rafters. To get the various cuts, you set the radial of the square to the degrees of the pitch you are using, shown on the table, and proceed to lay out the rafters.

To show how simple a matter roof framing becomes with this square, the easiest roof pitch to frame is used here, or the 12-9 pitch (12-inch run and 9-inch rise). With the radial square all of the other pitches are just as simple and easy to frame as the one used as a basis for the illustrations.

Common Rafter, Level and Plumb Cuts.—Fig. 487 shows the applications of the square for marking the level and plumb cuts of a common rafter. The radial of the square is set at 37

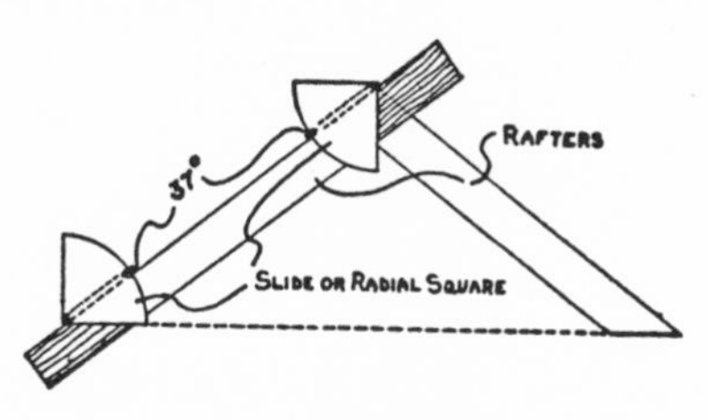

Fig. 487

degrees, as shown in the table. To the left the square is shown in position for marking the level cut. Without changing the radial, the square is applied to the timber for marking the plumb cut. This is shown at the top of the drawing. The parts to be cut off are shaded. Fig. 488 shows the square still set at 37 degrees, where it is applied for marking the cut of a tie or collar beam. Fig. 489 shows it in position for marking the bevel for gable studding, which are shown shaded. Fig. 490, upper left, shows the square applied for marking the plumb cut of the seat of a rafter with a tail. How the tail cut is marked is shown at the

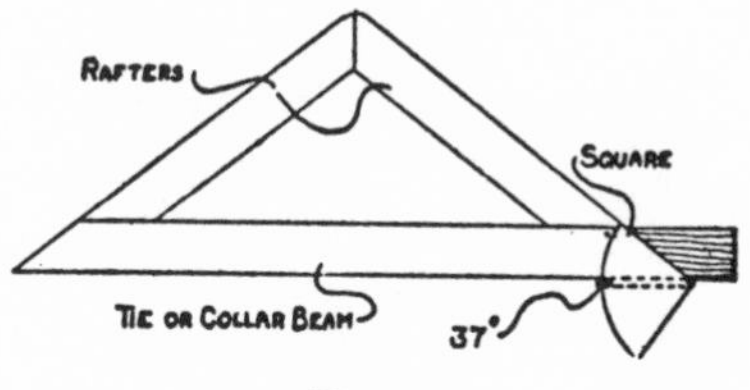

Fig. 488

bottom. The shaded part is to be cut off. The dotted-line square in the drawing to the right, shows the application of the square for marking the level cut of the seat. The part that is to be cut out for the seat is shown shaded.

All of the cuts shown in the last four illustrations are made with the radial set at 37 degrees. But if the radial must be changed for marking other bevels, you simply loosen the set-screw and adjust the radial to the degrees that will give the desired bevel.

Tables Explained.—The first table shown on the face of the radial square covers common rafters. The first col-

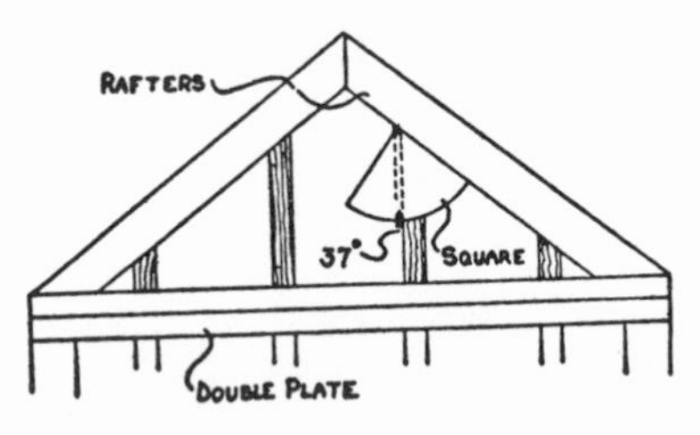

Fig. 489

umn of the table gives the pitch according to the steel square. The second column gives the degrees the radial of the square is set to for marking the level and plumb cuts, while the third column gives the length of the rafter per foot run. That is all you need to know to frame a common rafter.

The second table covers hip and valley rafters. The first column of this table gives the degrees to which the radial is set for marking the level and

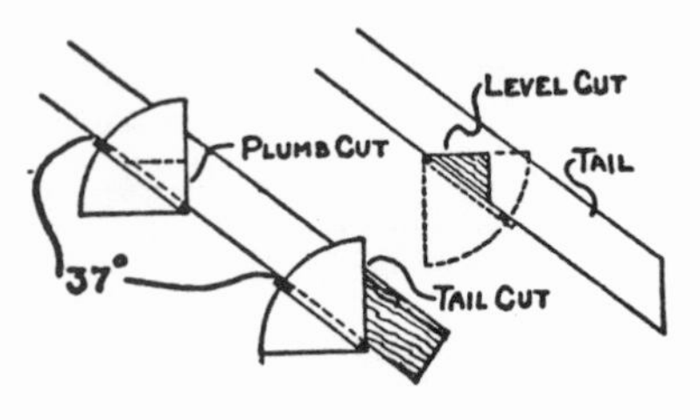

Fig. 490

plumb cuts. The second column gives the degrees for marking the edge bevel, or side cut, as it is also called. The last column gives the length of hips or valleys per foot run of the common rafter.

Cuts for Hips and Valleys.—Fig. 491 shows, left, the square applied to a timber for marking the level cut of hips or valleys. To the right it is shown applied for marking the plumb

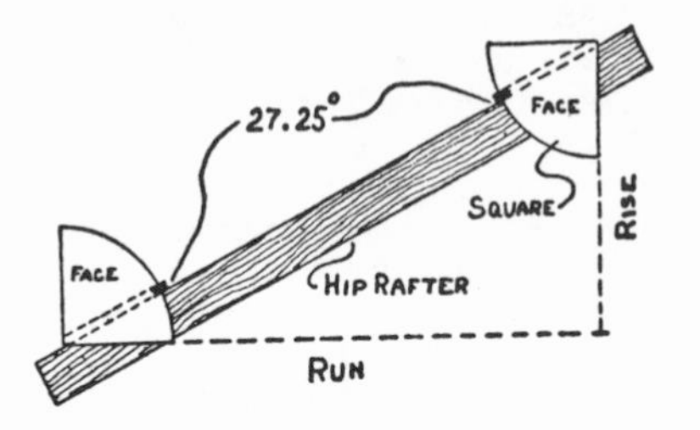

Fig. 491

cut. In both of these applications, it will be noticed, the radial of the square is set at 27.25 degrees. This means that both cuts can be marked without changing the radial. Fig. 492 shows a detail, in a little larger scale, of the tail of a hip rafter with the square applied to the left for marking the plumb cut of the tail, and to the upper right, it is shown in position for marking the plumb cut of the seat. The level cut of the seat is marked with the square in the position shown by dotted lines. These cuts can also be

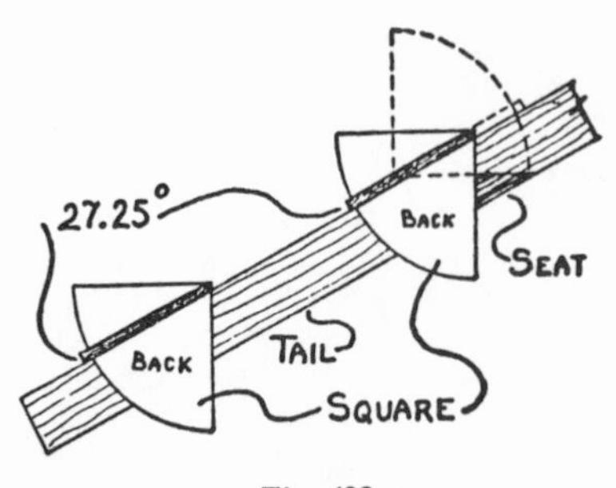

Fig. 492

marked by applying the square to the bottom edge of the timber. This will mean that the square would have to be swung around one-half turn. As in the other figure, the radial is set at 27.25 degrees.

Marking Edge Bevels.—Fig. 493, bottom drawing, shows a side view of a hip or valley rafter with the level and plumb cuts marked. This rafter

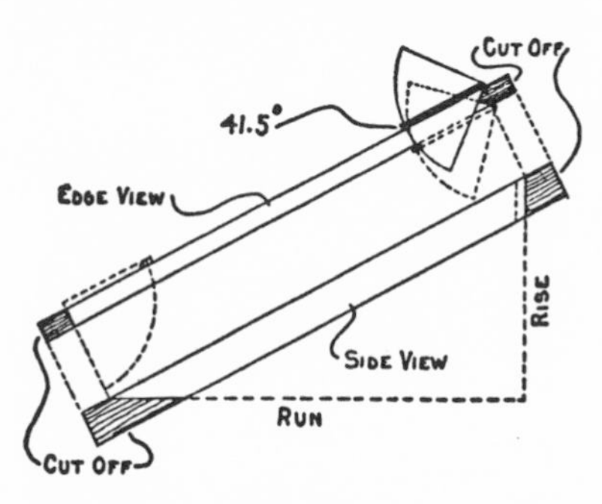

Fig. 493

has no tail. The upper drawing shows two applications of the square, right, for marking the edge bevels of hips or valleys. The shaded part cuts out for hips, while for valleys the cut is just the reverse, but the applications of the square are the same. Think this through. The radial is set at 41.5 degrees. To the left, by dotted lines, the square is shown in position for marking square across the timber. Here the radial is set flush with the edge of the blade.

Fig. 494, bottom drawing, shows a side view of the tail of a hip rafter marked for cutting. The shaded parts

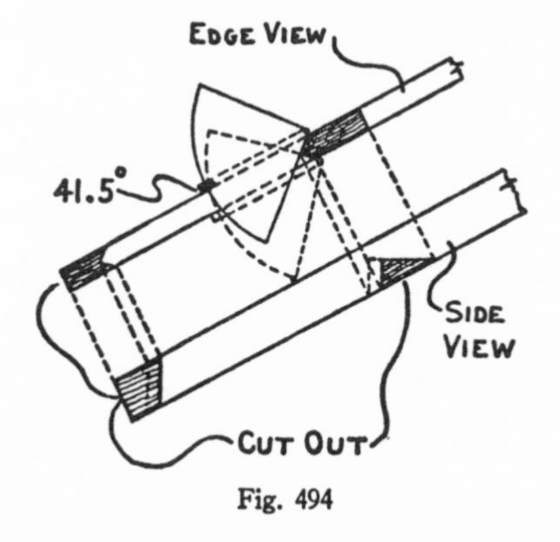

Fig. 494

are to be cut out, as shown. The upper drawing shows the bottom edge of the tail, and two applications of the square for marking the edge bevels of the seat cut. The seat is shown shaded. The radial in this case is again set at 41.5 degrees. The bevels for the valley, as any roof framer knows, are in reverse of the bevels of the hip, but the bevel is the same.

Cuts for Jack Rafters.—The third table on the face of the square covers jack rafters. The first column gives the difference in length of jacks spaced 16 inches on center. The second column gives the difference in length of jacks spaced 24 inches on center. The third column gives, in degrees, the setting of the radial for the side cut of jacks.

Fig.495, bottom drawing, shows a side view of a common rafter. The level and plumb cuts for jacks are always the same as for the common rafters. The upper drawing shows an

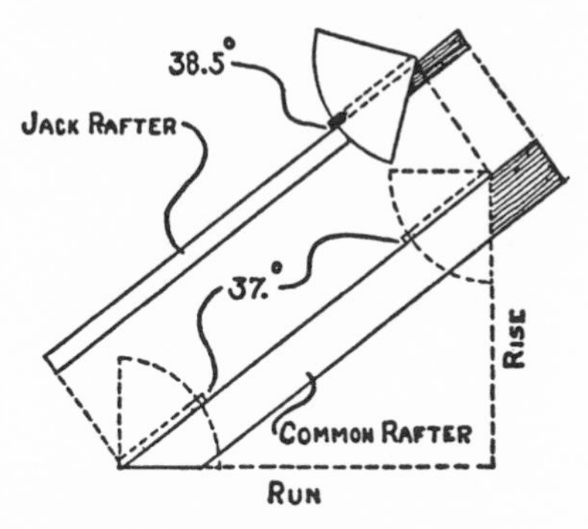

Fig. 495

edge view of a jack that is as long as the common rafter. At the top, right, the radial square is shown applied to this rafter for marking the edge bevel. The shaded parts of the two rafters are to be cut off.

Fig. 496, in a smaller scale, shows at the bottom a side view of a common rafter with a run of 48 inches, and a rise of 36 inches. At *A,* upper drawing, is shown a full-length jack with the bevel cut on the upper end, and marked for the second set of jacks for a spacing of 24 inches on center. It will be noticed by the figures given that the difference in the lengths of the jacks is 30 inches. At *B* is shown

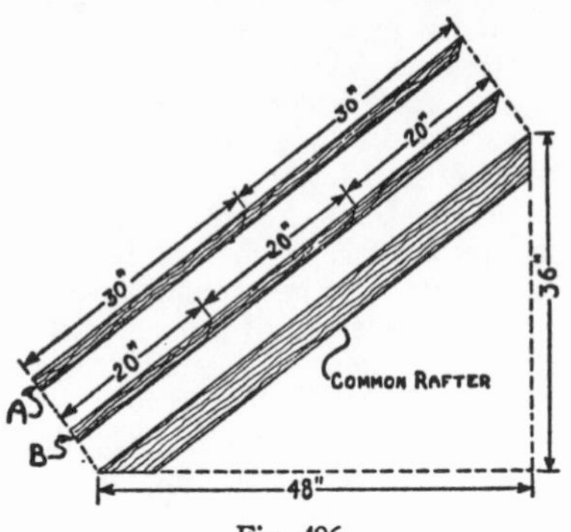

Fig. 496

an edge vew of a jack rafter marked for 16-inch spacing. Here the difference in the lengths is 20 inches. These two jacks show how roof framers mark the pattern, which is used for marking the different lengths of jacks.

Cuts for Polygons.—Fig. 497 shows at the center, left how the radial square

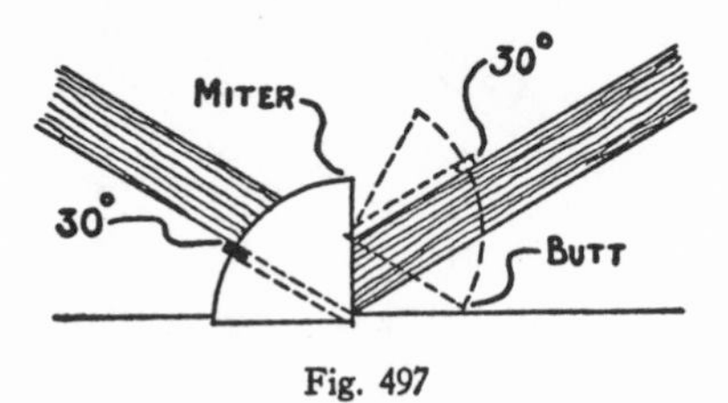

Fig. 497

is applied for marking the miter cut for a hexagon. To the right, by dotted lines, it is shown applied for marking the butt cut. In both of these cuts the radial is set at 30 degrees.

Fig. 498, center left, shows the application of the square for marking the miter joint for an octagon. To the right, by dotted lines, is shown the

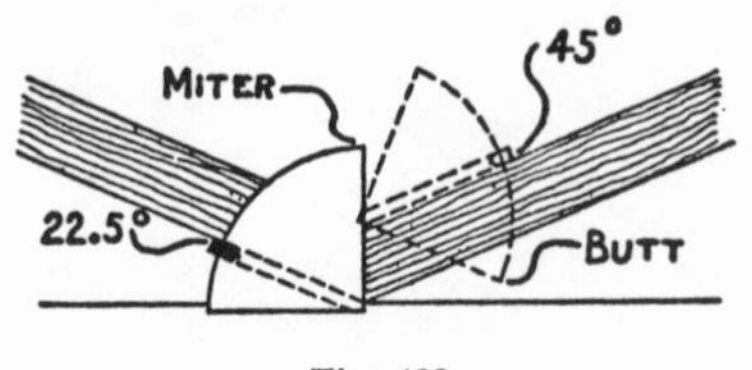

Fig. 498

application for marking the butt cut. Here the radial is set at 22.5 degrees for the miter cut, while for the butt cut it is set at 45 degrees.

The hexagon and the octagon are the most important polygons to carpenters. To get the degree to which the slide square is set for the miter cuts of any other polygon, divide 180 by the number of sides the polygon has. The answer gives, in degrees, the points for setting the radial.

Note:

The Radial Square is manufactured by the

Corweld Supply Company
8253 Crenshaw Drive
Inglewood 4, Calif.

INDEX

W